Dynamics of Immune Activation in Viral Diseases

Pallaval Veera Bramhachari
Editor

Dynamics of Immune Activation in Viral Diseases

Springer

Editor
Pallaval Veera Bramhachari
Department of Biotechnology
Krishna University
Machilipatnam, Andhra Pradesh, India

ISBN 978-981-15-1047-2 ISBN 978-981-15-1045-8 (eBook)
https://doi.org/10.1007/978-981-15-1045-8

This Springer imprint is published by the registered company Springer Nature Singapore Pte Ltd.
The registered company address is: 152 Beach Road, #21-01/04 Gateway East, Singapore 189721, Singapore

Preface

Viral infectious diseases signify an essential portion of global public health concern with millions of deaths annually. Viruses produce disease directly by killing the host cells they infect or by liberating toxins that can cause tissue damage and functional derangements in neighboring or distant cells and tissues. In the field of viral immunology, the researchers continue to make noteworthy and exhilarating contributions to our understanding of the fundamental biology of the immune system. Yet the practical translational applications of this fascinating and enthralling area of science are little disappointing with regard to the recurrent viral outbreaks.

The development of an infectious disease in an individual involves complex interactions between the virus and the host. These interactions are a dynamic interplay of host mechanisms aimed at eliminating infections and viral strategies designed to permit survival in the face of beneficial activation of innate and adaptive immune responses. Different types of infectious viral agents stimulate distinct types of immune responses and have evolved unique mechanisms for evading immunity. Noteworthy, in many instances, virus-induced proteins produced by emerging viruses can trickily antagonize specific immune recognition mechanisms or circumvent cell intrinsic restriction factors, thus influencing the host antiviral immune responses and complicating the viral replication kinetics. However, there is little understanding of within-host immunological processes underlying the reservoir host virus interactions, and this question is hardly addressed in several emerging viral diseases. Moreover, the viral genome itself plays an imperative role in the rapid evasion of adaptive immune responses, with the generation of diverse viral quasi-species that is an inherent consequence of error-prone RNA replication mechanisms. Moreover, elite immune controllers do not eliminate the virus from their system, because of the eventual emergence of virus variants that manage to escape the determined efforts of the immune system.

From serious pandemics and highly contagious infections to common influenza episodes, clinical prognosis often relies on early detection of the infectious agent. The recently developed viral diagnostic methods are reshaping the field of clinical microbiology and could contribute to reducing the prevalence of serious infectious diseases. Fortunately, the aspects of modernization that help drive pathogen emergence can also propel scientific innovation and current significant scientific advances have warranted our skill set to address the challenge of emerging viruses. From advanced genomic sequencing to novel methods in structural biology, we now have an increasingly classy toolkit with which to facilitate the detection and possible control of emerging viral diseases. Still, current outbreaks of viral diseases serve as powerful reminders of our ongoing vulnerability to emerging viral pathogens. Nonetheless, a number of additional viral diseases are considered for inclusion in the priority list every year. These events underscore the need for concerted efforts to develop and implement new interventions while continuing to invest in proven public health measures.

The field of viral immunology seeks to understand the mechanisms of virus–host interaction with a view of applying this knowledge to the design of effective vaccines to control viral infections. Whereas several viral infections are still in need of successful prophylactic vaccines, in many instances we also require therapeutic vaccines that could boost inadequate immunity. This book primarily emphasizes several areas of the field that hold substantial promise for translation, but where further work is critically required to find solutions. We emphasize that our fundamental understanding of virus–host relationships is moving in leaps and bounds, but we lag behind in applying this knowledge to the successful control of many viral infections. Nevertheless, elucidating the nature of immune responses in individual natural hosts may inform our understanding of how virus–host equilibria are established without substantially impacting host health. Furthermore, this may provide insights into the mechanisms of disease pathogenesis and immunity in humans.

We strongly believe that this book would provide enough insights into the current understanding of adaptive and innate immune response in viral diseases. This book is an attempt to compile the novel information available on recent advancements on various aspects of differential regulation of each immune cell during viral pathogenesis and immune evasion. This book aims to revitalize the interaction between fields of virology and immunology in order to advance our understanding of dynamics of viral immune pathogenesis, as well as innate and adaptive immune responses elicited by the host. The book also elucidates a comprehensive yet representative description of a large number of challenges associated with immune sensors, namely TLRs, DNA and RNA sensors, and other immune cells during viral infections, as it is indispensable to possess updated information on emerging viral diseases. This book could be an essential reading for the novice and experts in the field of viral immunology, immune interventions, and viral immunodiagnostics, including latest developments in vaccine research. With these objectives in mind, the content of this textbook has been arranged in a logical progression from fundamental to more advanced concepts. Finally, this book also outlines the most advanced immune techniques used in diagnostics of viral diseases and also primarily focuses on advancements of vaccine development research for emerging viral diseases.

We hope that this book stimulates your creativity and wish you success in your experiments. This book is a stunning reflection of the seriousness with which the several scientific minds are dedicated to the welfare of the scientific community. I am extremely thankful to the contributors for paying continuous attention to my request and showing faith in my capabilities. I shall always remain highly obliged to all of them forever. These words cannot justify the worthiness of their efforts.

We successfully compiled our creative and thoughtful research work due to genuine concern and painstaking effort of many more well-wishers whose names are not mentioned, but they are still in our heart. So, the reward is surely worth for their efforts. I want to dedicate this book to my mother, S. Jayaprada (late). Myself and the contributing authors hope from the bottom of our hearts that this book will be a good guidebook and compass for research studies in diagnostic virology and immunotechnology.

Machilipatnam, India Pallaval Veera Bramhachari

Acknowledgments

My sincere thanks are extended to all the academicians and scientists who have contributed the galaxy of topics in the form of chapters and happily agreed to share their work on *Dynamics of Immune Activation in Infectious Viral Diseases* in this volume.

This book is a stunning reflection of the seriousness with which the several scientific minds are dedicated from the immunology and virology scientific community. I am extremely thankful to the contributors for paying continuous attention to my requests and showing obsolete faith in my competencies and capabilities. I shall always remain highly obliged and indebted to all of them forever. These words cannot justify the worthiness of their untiring efforts. We appreciate the excellent work of the authors and coauthors who were invited to contribute diversified chapters in this book. The credit for making this book a reality goes to them. Me as an editor and the review team for the chapters especially appreciate sharing expertise with the contributors. Each chapter is informative and written as a stand-alone chapter, so the reader can begin anywhere in the book depending upon his or her interests and needs.

At the same time, I also express my deepest gratitude to my family members, especially my wife (Ramadevi Ramaswamy) and my kids (Ruthvik and Jayati), for their kind support which has prompted me to complete the assignment on time. I am also thankful to the Department of Biotechnology, Krishna University, for the support. I am equally thankful to the Springer Nature Publishing group for their full cooperation during the peer review and production of the volume.

I am thankful to my beloved teachers and mentors for their constant support and motivations at all stages of progress.

About the Book

Dynamics of Immune Activation in Viral diseases is an authoritative reference book in virology, which provides the current understanding of adaptive and innate immune response in viral diseases. This book also illustrates the differential regulation of each immune cell during viral pathogenesis and immune evasion. This book aims to revitalize the interaction between fields of virology and immunology in order to advance our understanding of dynamics of viral immune pathogenesis, as well as innate and adaptive immune responses elicited by the host. Recent advancements in immune intervention and viral immunodiagnostics, including latest developments in vaccine research, are discussed. It also covers a new area of immune biology where innate part of immune system helps adaptive part through cross talk with adaptive immune system. This book primarily emphasizes on the recent challenges of immune sensors, namely TLRs, DNA and RNA sensors, and other immune cells during viral infections, as it is indispensable to possess updated information on emerging viral diseases. Apart from the current understanding of immune response in human viral diseases, this book outlines the most advanced immune techniques used in diagnostics of viral diseases and also primarily focuses on advancements of vaccine development research for emerging viral diseases.

Contents

Editor and Contributors

About the Editor

Pallaval Veera Bramhachari is currently a Senior Faculty member of the Department of Biotechnology at Krishna University, Machilipatnam, India. He has earlier served as Postdoctoral Research Scientist at the Bacterial Pathogenesis Laboratory, Queensland Institute of Medical Research (QIMR-Berghofoer Research Center), Brisbane, Australia, and DBT Postdoctoral Fellow at the Department of Microbiology and Cell Biology, Indian Institute of Science (IISc), Bangalore. He completed his Ph.D. in Microbiology from Goa University, India. His major research interests are investigations on the applied research in microbiology and cell biology including biochemical mechanism of bacterial-biofilm formations, structure-function relationship of bacterial rhamnolipids, basic and translational cancer research. He has published more than 125 research articles including 5 international books in prestigious peer-reviewed international Scopus-indexed journals and presented 47 abstracts at various national and international conferences. Dr. Bramhachari served in the Editorial Board member and as Reviewer for a number of national and international peer-reviewed journals. He is a member of many international scientific societies and organizations, most importantly Indian Science Congress and Society of Biological Chemists, India. He successfully completed two major research projects.

Dr. Bramhachari was awarded a Travel scholarship from QIMR Australia in 2007 for attending the 4th Indo-Australian Biotechnology Conference at Brisbane, Australia, and also awarded with young scientist travel fellowship (2007) from the DST, Govt. of India for attending the XVII Lancefield International Symposium at Porto Heli, Greece, 2008. Dr. Bramhachari was conferred with various prestigious awards, notably Science Education Research Board (SERB), Government of India-Young Scientist award (2011) with a Research grant, Nominated as Associate Fellow of Andhra Pradesh Academy of Sciences (APAS) 2016, MASTER TRAINER: Andhra Pradesh, English Communications Skills Project, by the British Council & Andhra Pradesh State Council of Higher Education (APSCHE) in 2017, Andhra Pradesh State Best Scientist award by APCOST 2017, and Dr. V. Ramalingaswamy Memorial award in Biomedical Sciences 2019 by Science city of Andhra Pradesh and Andhra Pradesh Academy of Sciences. He also obtained two Indian patents in 2017. He has more than 13 years of teaching and research experience at the university level.

Contributors

Chanda Parulekar Berde Gogate Jogalekar College, Ratnagiri, Maharashtra, India

Sankar Bhattacharyya Vaccine and Infectious Disease Research Centre, Translational Health Science and Technology Institute, Faridabad, Haryana, India

Prudhvilal Bhukya Hepatitis Division, National Institute of Virology, Pune, Maharashtra, India

Samuel Ignatious Bolleddu Molecular Pharmacology and Physiology, University of South Florida, Tampa, FL, USA

Pallaval Veera Bramhachari Department of Biotechnology, Krishna University, Machilipatnam, Andhra Pradesh, India

Surekha Challa Department of Biochemistry and Bioinformatics, Institute of Science, Gandhi Institute of Technology and Management (GITAM) (Deemed to be University), Rushikonda, Visakhapatnam, Andhra Pradesh, India

Bojjibabu Chidipi Molecular Pharmacology and Physiology, University of South Florida, Tampa, FL, USA

Gururaj Rao Deshpande National Institute of Virology, Pune, Maharashtra, India

Suji George National Institute of Virology, Pune, Maharashtra, India

Samuel Ignatius Molecular Pharmacology and Physiology, University of South Florida, Tampa, FL, USA

Achanta Jagadeesh Tumor Microenvironment Global Core Research Center, College of Pharmacy, Seoul National University, Seoul, South Korea

Ruta Kulkarni Department of Communicable Diseases, Interactive Research School for Health Affairs (IRSHA), Bharati Vidyapeeth (Deemed to be University), Pune, Maharashtra, India

Madhavi Maddala Department of Zoology, Nizam College, Osmania University, Hyderabad, Telangana, India

M. Madhavi Department of Zoology, Nizam College, Osmania University, Hyderabad, Telangana, India

Devanabanda Mallaiah Department of Biotechnology and Bioinformatics, Yogi Vemana University, Kadapa, Andhra Pradesh, India

Alavala Mattareddy School of Life Sciences, Adikavi Nannaya University, Rajahmundry, Andhra Pradesh, India

Ganugula Mohana Sheela Department of Biotechnology, Krishna University, Machilipatnam, India

Preethika Nair NMIMS School of Science, Mumbai, India

Nageswara Rao Reddy Neelapu Department of Biochemistry and Bioinformatics, Institute of Science, Gandhi Institute of Technology and Management (GITAM) (Deemed to be University), Rushikonda, Visakhapatnam, Andhra Pradesh, India

Sudhakar Pola Department of Biotechnology, Andhra University, Visakhapatnam, Andhra Pradesh, India

A. M. V. N. Prathyusha Department of Biotechnology, Krishna University, Machilipatnam, Andhra Pradesh, India

Neelapu Nageswara Rao Reddy Department of Biochemistry and Bioinformatics, GITAM Institute of Science, Gandhi Institute of Technology and Management (GITAM) (Deemed-to-be-University), Visakhapatnam, Andhra Pradesh, India

Parul Sahu National Institute of Immunology, New Delhi, India

Pavani Sanapala Department of Biotechnology, Andhra University, Visakhapatnam, Andhra Pradesh, India

Gajanan N. Sapkal National Institute of Virology, Pune, Maharashtra, India

Siddhesh U. Sapre National Institute of Virology, Pune, India

K. Satish Kumar Department of Zoology, Adikavi Nannaya University, Rajamundry, Andhra Pradesh, India

Prakriti Sinha National Institute of Immunology, New Delhi, India

Divya Tiraki Department of Communicable Diseases, Interactive Research School for Health Affairs (IRSHA), Bharati Vidyapeeth Deemed University, Pune, Maharashtra, India

Harika Sai Vemuri Department of Biochemistry and Bioinformatics, Institute of Science, Gandhi Institute of Technology and Management (GITAM) (Deemed to be University), Rushikonda, Visakhapatnam, Andhra Pradesh, India

Abbreviations

ABCF1	ATP-binding cassette subfamily F member 1
ACLF	Acute-on-chronic liver failure
ADCC	Antibody-dependent cytotoxic cells
ADCP	Antibody-dependent cellular phagocytosis
ADE	Antibody-dependent enhancement
AIC	Anti-inflammatory cytokines
AIDS	Acquired immune deficiency syndrome
AKT	Serine/threonine kinases
ALRs	AIM2-like receptors
AP	Alternate pathway
AP-1	activation protein-1
APCs	Antigen-presenting cells
APOBEC3	Apolipoprotein B Editing Complex
ARDS	Acute respiratory distress syndrome
ATF-2	Activating transcription factor 2
BBB	Blood–brain barrier
BCR	B-cell receptors
BIV	Bovine immunodeficiency virus
CARDs	Caspase activation and recruitment domains
CCHF	Crimean–Congo hemorrhagic fever
CCHFV	Crimean–Congo hemorrhagic fever virus
CCL	Chemokine
CD	Clusters of differentiation
CDC	Complement-dependent cytotoxicity
CDNs	Cyclic di-nucleotides
cGAMP	Cyclic guanosine monophosphate-adenosine monophosphate
cGAS	Cyclic GMP-AMP synthase
cGAS	Cyclic GMP-AMP synthase
CH25H	Cholesterol-25-hydroxylase
CHIKV	Chikungunya virus
CIFN	Consensus IFN
CLIA	Chemiluminescent immunoassay
CLRs	C-type lectin receptors
CML	Complement-mediated lysis

CMV	Cytomegalovirus
CNS	Central nervous system
CP	Classical pathway
CpG	Unmethylated deoxycytidylate-phosphate-deoxyguanylate
CPPs	Cell-penetrating peptides
CS	Complement system
CSF	Colony-stimulating factor
CTL	Cytotoxic T lymphocytes
DAA	Directly acting antivirals
DAI	Z-DNA binding protein
DAMPS	Damage-associated molecular patterns
DCs	Dendritic cells
DDX	DExD/H-box
DDX41	DEAD box polypeptide 41
DENV	Dengue virus
DHF	Dengue hemorrhagic fever
DIV	Dog immunodeficiency virus
DNAse	Deoxyribonuclease
DSS	Dengue shock syndrome
EBOV	Ebola virus
EBV	Epstein–Barr virus
ECL	Electrochemiluminescence
EGF	Epidermal growth factor
EPO	Leptin and erythropoietin
Fab	Fragment antigen binding
Fc	Fragment crystalline
FEB	Field Effect Biosensing
FGF	Fibroblast growth factor
FIV	Feline immunodeficiency virus
FO	Follicular
FPIA	Fluorescence polarization immunoassay
GC	Germinal center
G-CSF	Granulocyte CSF
GEFs	Guanine nucleotide exchange factors
GH	Growth hormone
GM-CSF	Granulocyte macrophage colony-stimulating factor
HA	Hemagglutinin
HBcAg	Hepatitis B core antigen
HBsAg	Hepatitis B surface antigen
HCC	Hepatocellular carcinoma
HCV	Hepatitis C
HeV	Hendra virus
HFD	Hemagglutinin fusion domain
HIN	H-inversion
HIV	Human immunodeficiency virus

HLA	Human leukocyte antigen
HMGB	High mobility group B protein
HPV	Human papilloma virus
HSC	Hematopoietic stem cells
IF	Immunofluorescence
IFI16	IFN-γ-inducible protein 16
IFIT	IFN-induced protein with tetratricopeptide repeats ()
IFIT	Interferon-inducible transmembrane protein
IFITM families	IFN-induced transmembrane protein family
IFN	Interferon
IFN-γ	Interferon gamma
IKK	IκB kinase
IL	Interleukin
IRF	interferon regulatory factor
ISGs	Interferon-stimulated genes
ISGs	Interferon-stimulated genes
ISREs	IFN-stimulated response elements
JEV	Japanese encephalitis virus
JNK	Jun N-terminal kinase
KCs	Kupffer cells
LASV	Lassa virus
LBP	LPS-binding protein
LFIA	Lateral flow immunoassay
LGP2	laboratory of genetics and physiology 2
LILR-B1	Leukocyte immunoglobulin-like receptor B1
LP	Lectin pathway
LPS	Lipopolysaccharide
LRRFIP1	Leucine-rich repeat flightless-interacting protein 1
LRRs	Leucine-rich repeats
mAbs	Monoclonal antibodies
MAC	Membrane attack complex
MagLISA	Magnetic nano(e)nzyme-linked immunosorbent assay
MALDI	Matrix-assisted laser desorption ionization
MAP	Mitogen-activated protein
MASPs	MBL-associated proteins
MAVS	Mitochondrial antiviral signaling protein, *a.k.a.* IPS-1/VISA/Cardif
MDA-5	Melanoma differentiation-associated antigen 5
MDA-5	Melanoma differentiation-associated antigen 5
mDC	Myeloid dendritic cells
MEIA	Micro-particle immune assay
MHC	Major histocompatibility complex
MIA	Multiplex microsphere immunoassay
mNGS	Metagenomic next-generation sequencing assay
MNPs	Magnetic nanoparticles

MNV	Murine norovirus
Mϕ	Macrophages
MRE11A	Meiotic recombination 11 homolog A
MS	Mass spectrometry
MZ	Marginal zone
NA	Neuraminidase
NAbs	Neutralizing antibodies
NEMO	NF-κB essential modifier
NF-AT	Nuclear factor of activated T cell
NF-κB	Nuclear factor κ-B
NGS	Next-generation sequencing
NIRS	Near-infrared spectroscopy
NiV	Nipah virus
NLR	Neutrophil to lymphocyte ratio
NLRB	Baculovirus inhibitor of apoptosis repeats
NLRC	Caspase activation and recruitment domain CARD
NLRP	Pyrin domain PYD
NLRs	NOD-like receptors
NTCP	Sodium taurocholate co-transporting polypeptide
OAS	Original antigenic sin
OAS	RNase L/oligoadenylate synthase
PAMPS	Pathogen-associated molecular patterns
PBMCs	Peripheral blood mononuclear cells
PCRs	Polymerase chain reactions
pDC	Plasmacytoid dendritic cells
PDGF	Platelet-derived growth factor
PI3K	Phosphatidyl-inositol 3 kinase
PIC	Pro-inflammatory cytokine
PK	Protein kinase
PKC	Protein kinase C
PKR	Protein kinase R
PML	Promyelocytic leukemia protein
PRL	Prolactin
PRR	Pattern recognition receptors
PRRs	Pattern recognition receptors
py	Pyrin
PYHIN	Pyrin and H inversion domain
qPCR	Quantitative PCR
RCs	Replication complexes
RdRp	RNA-dependent RNA polymerase
RFFIT	Rapid fluorescent focus inhibition test
RIDT	Rapid influenza diagnostic test
RIG-I	Retinoic acid-inducible gene I
RIG-I	Retinoic acid-inducible gene I product
RLRs	RIG-I-like receptors

RNAse	Ribonuclease
RT-LAMP	Reverse transcription loop-mediated isothermal amplification
RT-RPA	Reverse transcription recombinase polymerase amplification assay
SARM	Sterile-alpha and Armadillo motif-containing protein
SARS-CoV	Severe Acute Respiratory Syndrome Coronavirus
SCF	Stem cell factor
SGPs	Small G-protein
SIV	Simian immunodeficiency virus
SNPs	Single nucleotide polymorphisms
SOCS3	Suppressor of cytokine signaling 3
SPR	Surface plasma resonance spectroscopy
ssRNA	Single-stranded RNA
STAT1	Signal transducer and activator of transcription 1
STING	Stimulator of interferon genes
Syk	Spleen tyrosine kinase
TBK1	TANK binding kinase 1
TEM	Transmission electron microscopy
T_{FH} cells	Typically follicular T helper cells
TGF	Transforming growth factor
TIR	Toll/IL-1 receptor
TLRs	Toll-like receptors
TNF	Tumor necrosis factor
TNP	T-cell epitope of nucleoprotein
TPR	Tetra tricopeptide
TRAF	TNF receptor-associated factor
TRAIL	TNF-related apoptosis-inducing ligand
Tregs	T regulatory cells
TREX1	Three-primer repair endonuclease 1
TRIF	TIR domain-containing adaptor protein-inducing IFN-β
TYK2	Tyrosine kinase2
VEEV	Venezuelan equine encephalitis virus
VEGF	Vascular endothelial growth factor
VHF	Viral hemorrhage fever
VHSV	Viral hemorrhagic septicemia virus
VSV	Vesicular stomatitis virus
WNV	West Nile virus
YFV	Yellow fever virus
ZIKV	Zika virus

Part I

Role of Innate Immunity in Combating Viral Diseases

Significance and Dynamics of Immune Responses During Viral Pathogenesis

1

Pallaval Veera Bramhachari

Abstract

Immune system is a homeostatic system which is active against numerous invading pathogens. Host immune system uses multiple immune responses (innate, adaptive, and complement system) to eliminate virus/viral particles. Additionally immune system is also well resourced with sensors that detect invading pathogens and direct responses for clearing numerous copies of virus recruits. Modern technologies help prevent pathogen emergence as well as thrust scientific improvements in understanding the viral immune responses; moreover, recent scientific advances in diagnostic virology have undeniably transformed the ability to address challenges of numerous emerging intricate viruses. The journey of diagnostic virology has started from serology, nucleic acid sequence-based amplification techniques, and genomic sequencing techniques to most advanced innovative methods (e.g., structural biology spectroscopy, NGS, microfluidics, metagenomics, CRISPR/Cas system, nanotechnology and structural biology), the world progressed way beyond with several classy diagnostic methods to help tackle the diagnosis and control of emerging viral diseases. However, the technical competencies alone are inadequate if not sustained by health promotion strategies to raise awareness of the significance of early detection and diagnosis, outbreak, and spread of virus. Yet, current outbreaks of viral diseases across the world dole out authoritative reminders to emerging viral pathogens.

Keywords

Immune system · Viruses · Diagnostic methods · Emerging viral diseases

---

P. V. Bramhachari (✉)
Department of Biotechnology, Krishna University, Machilipatnam, Andhra Pradesh, India

P. V. Bramhachari (ed.), *Dynamics of immune activation in viral diseases*,
https://doi.org/10.1007/978-981-15-1045-8_1

1.1 Introduction

Infectious diseases are reported to account for ~20% of global mortality, of which one-third of deaths are caused by viral diseases. The viral infections pose major public health risks and warrant further research and development, including surveillance and diagnostics. All the lethal viral diseases present an absurdity in mounting the pathogenesis there by killing their hosts, which is noticeably of no advantage to the virus. Viruses are infectious agents consisting of nucleic acids coated in a simple protein casing, infect, replicate in host cells, and cause acute, chronic infections. Since viral infection is an intricate and highly vibrant process, noticeably affected by the physical and chemical environment, studies into infectious viral biology ought to preferably occur in advanced research settings. Viruses replicate by hijacking the host cell's machinery and making host cells a huge virus factory and erupt as virus after… virus after… virus after… virus.

The immune system is also well resourced with sensors that detect invading pathogens and direct responses for clearing numerous copies of virus recruits. On the entry of virus, immune system of the host typically elicits both nonspecific innate and "specific" adaptive immune responses against foreign pathogens. Activation of varied immune responses, time, and the extent of response rely on how virus interacts and spreads within host cells. Host immune system uses multiple immune responses (innate, adaptive, and complement system) to eliminate virus/viral particles. The immune system employs most efficient mechanisms depending on the distinctiveness of infectious agents. Diverse actions occur during viral infections, equally toward free viral particles in addition to infected cells.

Race between virus and immune response establishes: whether the intruder will eliminate or establish a persistent infection. The host cell can also be damaged unswervingly by virus or by viral immune response. However, balance between good and bad antiviral immune response relies on the amount of viral load, chronicity of infection, and magnitude of tissues infected (Zinkernagel 1996). Therefore, a balance exists between immune activation vis-à-vis immune suppression during the setting of chronic infection. Viruses evolved numerous strategies to escape immune system to institute a chronic infection. Few viruses endure in definite cell types and hide from immune system, while some encode specific genes that target infected cells or immune system. A few viruses limit their replication, thus restricting accessibility of antigen to alert the immune system. However, a few viruses are error prone in their replication, escalating the possibility of making escape viral mutants. Nonetheless, these immune evasion strategies of virus are unprejudiced by regulation within the immune system (Finlay and McFadden 2006).

1.2 Innate Immune Responses by Virus

During early stages of infection "innate" response limits virus multiplication, which implicates synthesis of diffusible proteins called interferons, cytokines, and chemokines and stimulation of "natural killer" and dendritic lymphocytes (French and Yokoyama 2003). In order to limit viral replication, first line of defense system has to sense the pathogens/pathogen-associated molecules (Paladino et al. 2006). This provides temporary protection against the viral onslaught. This fundamental process is accomplished by sensing virus-associated nucleic acids in infected cells using a variety of pattern recognition receptors (PRRs). PRRs recognize distinct pathogens conserved structures known as pathogen-associated molecular patterns (PAMPs). There are several classes of PRRs, namely TLRs (toll-like receptors), RLRs (retinoid acid-inducible gene-I (RIG-I)-like receptors), NLRs (nucleotide oligomerization domain (NOD)-like receptors), and many DNA and RNA sensors (Mogensen 2009). Upon interaction with cognate viral ligands, TLRs, NLRs, and RLRs stimulate production of interferons, pro-inflammatory cytokines (NF-κB-dependent response), and chemokines that limit virus replication and dissemination (Shayakhmetov et al. 2010). Infected cells produce different classes of interferons such as INF-α, INF-β, and INF-γ. Interferons mediate the most effective innate immune response during viral infection by warning nearby cells about viral presence (Haller et al. 2006). This signals neighboring cells to increase MHC class I molecules on their surfaces and to facilitate surveying T cells to identify and eliminate viral infection. Interferons also limit viral replication by the activation of NK cells which is associated with the alterations in expression of SLA by infected cells. NK cells' cytotoxic mechanism against viral infected cells is very effective and does not depend on antigen (NK cells lack TCR) (Paul and Lal 2017). However, complement activation of an alternative pathway has the effect of activating destruction of viral particle. Innate defensive mechanism generates a severe evolutionary pressure on viruses to evolve efficient mechanisms to evade and also strategies permitting for subversion of host antiviral immune systems (Finlay and McFadden 2006).

1.3 Adaptive Immune Responses by Virus

Another important second line of antiviral response is adaptive immune system that includes $CD8^+$ and $CD4^+$ T cells and neutralizing antibodies which act against viral particles and infected cells in combination. Days to weeks are essential to mount an adaptive immune response tailored for specific virus. Adaptive immune response possesses two components: humoral response (utilities virus-specific antibodies by B lymphocytes) and cell-mediated response (specific cytotoxic T lymphocytes (CTL) that kill infected cells) (Kim et al. 1999). Both the components of adaptive defense system ensue production of long-lived "memory cells" that allow much more quick response for consequent infection with similar virus (Campos and Godson 2003). Antibodies are most vital mechanisms against viral particles, while cytotoxic mechanisms are noteworthy against infected cells.

1.3.1 Humoral Immunity

Virus or virus-infected cells stimulate B lymphocytes which synthesize IgG, IgM, and IgA Abs (specific for viral antigens). Antibody neutralization is a significant mechanism by which virus occurs in huge fluid spaces (e.g., serum) or on mucous surfaces (e.g., the gastrointestinal and respiratory tracts). Antibody can neutralize virus by (1) preventing host cell—virus interactions or (2) distinguishing viral antigens on infected cells that results in antibody-dependent cytotoxic cells (ADCC) or complement-mediated lysis (CML) (Parkin and Cohen 2001; Chaplin 2010).

1.3.2 Cell-Mediated Immunity

Viruses engage diverse strategies to restrain the presentation of virus-derived peptides. One of the important strategy entails the modulation of proteasome activity that produces peptides from full-length proteins; if there is adequate binding affinity, peptides bind to MHC class I molecules. However, a few viruses directly interact with and inhibit proteasome machinery likely for generation of peptides. Specialized immune cells, the CTL cells, recognize virus-infected cells with the help of specialized proteins (TCRs) on their surface and release cytotoxic factors to destroy infected cell and, therefore, avert survival of invading virus. The cytotoxic cells are specially armed with preformed mediators, namely perforins (forms pores in cell membranes) and granzymes (stored in and released from granules), permit entry of other factors into virus-infected cell to facilitate their destruction (Rosendahl Huber et al. 2014). Furthermore, CTL cells synthesize and release additional proteins, called cytokines, including interferon-ɣ and tumor necrosis factor-α, and transfer a signal from T cell to infected cell, or other neighboring cells, to further enhance killing mechanisms. Furthermore, cytokine signals, NK cells, act as evolutionary bridge linking innate and adaptive immunity (Belardelli and Ferrantini 2002; Sun and Lanier 2009).

It is well established that there is a mechanism for inhibiting immune response during chronic infection. In fact this mechanism is executed for two reasons: (1) To prevent immunopathology. CD8+ T-cell effector function can cause high levels of tissue damage through killing of infected cells and release of inflammatory cytokines. In reality, cytotoxicity and secretion of cytokines such as tumor necrosis factor (TNF) are often decreased if not lost in CD8+ T cells in a phenomenon known as T-cell exhaustion (Ou et al. 2008). (2) To prevent excessive proliferation of virus-specific T cells. During acute infection, virus-specific T cells can increase tenfold each day. This level of proliferation is dangerous and is therefore greatly reduced upon continued exposure to antigen (Thimme et al. 2012).

1.4 Significance

Viruses are intracellular pathogens that invade and infect host cells. Immune system is a homeostatic system which is active against numerous invading pathogens inside the body to clear infections. Propitiously, the aspects of recent modernization technologies that helped prevent pathogen emergence can also impel scientific improvements; moreover, recent scientific advances in diagnostic virology undeniably transformed the ability to address the challenges of numerous emerging intricate viruses across the globe. Starting from serology, nucleic acid sequence-based amplification techniques, genomic sequencing techniques to most advanced innovative methods (structural biology spectroscopy, NGS, microfluidics, metagenomics, CRISPR/Cas system, nanotechnology, and structural biology), the world has progressed way beyond with more classy diagnostic methods to help tackle the detection and control of emerging viral diseases. However, the technical competencies alone are inadequate if not sustained by health promotion strategies to raise awareness of the significance of early detection, outbreak, and spread of virus. Still, current outbreaks of viral diseases across the world dole out powerful reminders of our ongoing susceptibility to emerging viral pathogens. Reflecting the diversity of viruses and viral diseases, these are notoriously difficult drug targets since they modify and adapt themselves quickly to build up resistance and emerge as new serotypes. Nonetheless, better perceptiveness of host–pathogen interactions, viral protein, and nucleic acid functions led to additional rational drug designs, resulting in important therapeutic advances against viral diseases. Furthermore, novel platforms for vaccine design, namely nanoparticles and virus-like particles, have embarked avenues for development of new vaccine targets. Treatment strategies for viral infections comprise hindering binding of virus to host cells by inhabiting on host cell receptor with an additional molecule, or use of vaccine developed for a particular virus or an analogous target of diverse viruses. Serious actions should accentuate the call for strenuous efforts to develop and execute novel interventions in viral disease diagnosis and vaccine development keeping public health perspective in view.

Dynamics of Immune Activation in Viral Diseases is an authoritative reference book in virology, which provides the current understanding of adaptive and innate immune response in viral diseases. This book also illustrates about differential regulation of each immune cells during viral pathogenesis and immune evasion. This book aims to revitalize the interaction between fields of virology and immunology in order to advance our understanding of dynamics of viral immune pathogenesis, as well as innate and adaptive immune responses elicited by host. Recent advancements in immune intervention and viral immunodiagnostics, including latest developments in vaccine research, will be discussed. It also covers new areas of immune biology where innate part of immune system helps adaptive part through cross talk with adaptive immune system. This book primarily emphasizes on the recent challenges of immune sensors, namely TLRs, DNA and RNA sensors, and other immune cells during viral infections, as it is indispensable to possess updated information on emerging viral diseases. Apart from the current understanding of immune response in human viral diseases, this book also outlines the most advanced immune techniques used in diagnostics of viral diseases and also primarily focuses on advancements of vaccine development research for emerging viral diseases.

Acknowledgement DR. PVBC is grateful to Krishna University, Machilipatnam, for the support extended.

Conflict of Interest The author declares that he has no competing interests.

References

Belardelli F, Ferrantini M (2002) Cytokines as a link between innate and adaptive antitumor immunity. Trends Immunol 23(4):201–208
Campos M, Godson DL (2003) The effectiveness and limitations of immune memory: understanding protective immune responses. Int J Parasitol 33(5–6):655–661
Chaplin DD (2010) Overview of the immune response. J Allergy Clin Immunol 125(2):S3–S23
Finlay BB, McFadden G (2006) Anti-immunology: evasion of the host immune system by bacterial and viral pathogens. Cell 124(4):767–782
French AR, Yokoyama WM (2003) Natural killer cells and viral infections. Curr Opin Immunol 15(1):45–51
Haller O, Kochs G, Weber F (2006) The interferon response circuit: induction and suppression by pathogenic viruses. Virology 344(1):119–130
Kim JJ, Nottingham LK, Tsai A, Lee DJ, Maguire HC, Oh J, Dentchev T, Manson KH, Wyand MS, Agadjanyan MG, Ugen KE, Weiner DB (1999) Antigen-specific humoral and cellular immune responses can be modulated in rhesus macaques through the use of IFN-γ, IL-12, or IL-18 gene adjuvants. J Med Primatol 28(4–5):214–223
Mogensen TH (2009) Pathogen recognition and inflammatory signaling in innate immune defenses. Clin Microbiol Rev 22(2):240–273
Ou R, Zhang M, Huang L, Moskophidis D (2008) Control of virus-specific CD8+ T-cell exhaustion and immune-mediated pathology by E3 ubiquitin ligase Cbl-b during chronic viral infection. J Virol 82(7):3353–3368
Paladino P, Cummings DT, Noyce RS, Mossman KL (2006) The IFN-independent response to virus particle entry provides a first line of antiviral defense that is independent of TLRs and retinoic acid-inducible gene I. J Immunol 177(11):8008–8016
Parkin J, Cohen B (2001) An overview of the immune system. Lancet 357(9270):1777–1789
Paul S, Lal G (2017) The molecular mechanism of natural killer cells function and its importance in cancer immunotherapy. Front Immunol 8:1124
Rosendahl Huber S, van Beek J, de Jonge J, Luytjes W, van Baarle D (2014) T cell responses to viral infections–opportunities for peptide vaccination. Front Immunol 5:171
Shayakhmetov DM, Di Paolo NC, Mossman KL (2010) Recognition of virus infection and innate host responses to viral gene therapy vectors. Mol Ther 18(8):1422–1429
Sun JC, Lanier LL (2009) Natural killer cells remember: an evolutionary bridge between innate and adaptive immunity? Eur J Immunol 39(8):2059–2064
Thimme R, Binder M, Bartenschlager R (2012) Failure of innate and adaptive immune responses in controlling hepatitis C virus infection. FEMS Microbiol Rev 36(3):663–683
Zinkernagel RM (1996) Immunology taught by viruses. Science 271(5246):173–178

Antibody-Dependent Enhancement of Viral Infections

2

Ruta Kulkarni

Abstract

Antiviral antibodies constitute an important component of the host immune response against viral infections and serve to neutralize and reduce infectivity of the virus. However, these antibodies, intended to protect the host, may sometimes prove beneficial to the virus, by facilitating viral entry and replication in the target cell. This phenomenon, known as antibody-dependent enhancement (ADE) of infection, is a result of interaction of virus–antibody immune complexes with Fcγ and/or complement receptors on certain types of host cells and promotes viral entry into the host cells. The internalized immune complexes then modulate host immune response so as to enhance viral replication and aggravate disease severity. The possibility of induction of ADE remains a concern in the development and implementation of viral vaccines and immunotherapeutics.

Keywords

Viral infection · Neutralization · Antibody-dependent enhancement · Immune response · Vaccine · Immunotherapeutic

2.1 Introduction

Antibody-dependent enhancement (ADE) of infection represents a paradoxical phenomenon in host–pathogen biology, in which, antibody, an important pillar of the host defense against invading pathogen, actually allows entry of the pathogen into host territory. This traitorous behavior of the antibody further serves to weaken the host defense system and thus generates an environment conducive for enhanced

R. Kulkarni (✉)
Department of Communicable Diseases, Interactive Research School for Health Affairs (IRSHA), Bharati Vidyapeeth (Deemed to be University), Pune, Maharashtra, India

P. V. Bramhachari (ed.), *Dynamics of immune activation in viral diseases*,
https://doi.org/10.1007/978-981-15-1045-8_2

growth of the pathogen and consequently exacerbates disease in the host. This phenomenon has far-reaching implications for disease control and prevention, as therapeutic antibodies deployed to protect the host may aid the pathogen instead. Similarly, antibodies induced by vaccination may actually increase the risk and/or severity of disease in subsequent host–pathogen encounters.

This chapter aims to provide insights into the current knowledge on ADE of viral infections, with a focus on its molecular mechanisms and contribution to viral pathogenesis and disease, as well as implications for disease control strategies.

2.2 Antibody-Mediated Control of Viral Infections

Antibodies are an important component of the host adaptive immune response against infecting virus. Virus neutralization is considered to be the key mechanism of antibody-mediated protection. The antiviral activity of antibody is further augmented by Fc-mediated effector mechanisms such as complement-mediated lysis of viral particles, phagocytosis, and antibody-dependent cellular cytotoxicity (ADCC) (Braciale et al. 2013).

Neutralization occurs when antibody molecule(s) bind to the virus particle at its surface epitope and block viral attachment (to cellular receptor) and entry into host cell, and/or post-entry processes such as fusion and uncoating, thus resulting in loss of virion infectivity (Fig. 2.1). Two models have been proposed to describe the kinetics of neutralization. According to the single-hit model, binding of a single antibody molecule to a critical site on the virion is sufficient for neutralization. The more accepted multi-hit model postulates that neutralization can be achieved only when the virion is bound by multiple antibodies, exceeding the stoichiometric threshold of neutralization (Klasse and Sattentau 2002). The potency of neutralization is determined by the affinity of antibody binding and accessibility of the neutralization epitope(s) on the viral surface (Dowd et al. 2011).

2.3 The Antibody-Dependent Enhancement Phenomenon

The phenomenon of antibody-dependent enhancement (ADE) was discovered in flaviviruses during the 1960s–1970s. While observational studies on dengue virus (DENV) infections revealed a significant association between severe illness and antibody responses in secondary infection (Halstead et al. 1967), in vitro experiments demonstrated increased infectivity of Murray Valley encephalitis virus (Hawkes 1964; Hawkes and Lafferty 1967) and DENV (Halstead et al. 1973a) in the presence of antibody. In vivo studies in rhesus monkeys showing association of antibody response with higher dengue viremia offered further supportive evidence for this phenomenon (Halstead et al. 1973b; Halstead 1979; Marchette et al. 1973).

ADE ensues when antibodies binding to the virus particle fail to efficiently neutralize the virus (Halstead 2003; Taylor et al. 2015). This may occur due to the non-neutralizing nature of the antibody (binding to viral epitopes other than those

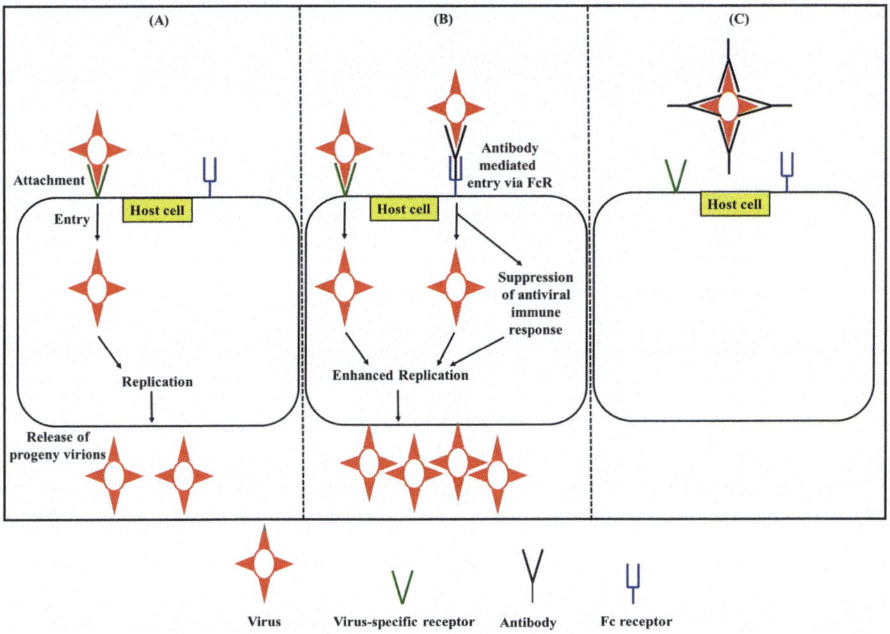

Fig. 2.1 Schematic representation of effect of antibody on virus–host cell interaction. (**a**) In the absence of virus-specific antibody, virus enters the host cell via interaction with cell surface (virus-specific) receptor, followed by replication and release of progeny virions. (**b**) In the presence of non-neutralizing or sub-neutralizing concentrations of antibody, an additional pathway of entry into host cell is available to the virus. Virus–antibody immune complexes are internalized via antibody interaction with cell surface Fc receptor (FcR), resulting in antibody-dependent enhancement (ADE) of viral infection and replication. Internalized immune complexes bring about suppression of cellular innate antiviral immune response further boosting viral replication. (**c**) In the presence of neutralizing concentrations of antibody, virus attachment and entry into host cell is blocked

involved in cell attachment and entry) or the presence of sub-neutralizing concentrations of antibodies (binding to viral epitopes below the threshold for neutralization). Either way, these antibodies aid entry of the virus into target cells, resulting in enhancement of viral infection, known as antibody-dependent enhancement (Fig. 2.1.). The mechanism of ADE involves internalization of virus–antibody immune complexes into cells via interaction of the antibody Fc region with the cellular Fc receptors (FcRs). Consequently, FcR-bearing myeloid cells such as monocytes, macrophages, dendritic cells, and certain granulocytes are permissive to ADE of infection, through phagocytic uptake of the immune complexes. Such ADE is principally mediated by IgG antibodies, however, IgM along with complement, and IgA antibodies have also been shown to be capable of ADE (Hawkes and Lafferty 1967; Cardosa et al. 1983; Janoff et al. 1995; Kozlowski et al. 1995).

Initially, the role of antibody in enhancing viral infection was believed to be limited to facilitation of virus entry into host cells, resulting in increase in the number of infected cells and, consequently, higher virus production, in a process termed

Table 2.1 Summary of key features of antibody-dependent enhancement (ADE) of viral infections

Virus (family)	Mechanism of ADE	Clinical significance of ADE	Implications for vaccines
Dengue virus (family *Flaviviridae*)	1. Antibodies against viral EDI, EDII, prM proteins involved in ADE 2. Extrinsic ADE: FcR-mediated virus internalization of virus–antibody immune complexes into monocytes, macrophages, dendritic cells 3. Intrinsic ADE: Modulation of host antiviral response by LILR-B1 co-ligation, downregulation of TLR-dependent, RIG-I/MDA5 signaling pathways, induction of IL-10 production 4. Antibody-mediated enhancement of viral fusion suggested 5. ADE associated with massive release of inflammatory cytokines leading to alteration of vascular permeability and plasma leakage which are the hallmarks of severe dengue	1. Association of ADE with severe dengue (DHF/DSS) during secondary heterotypic infections has been demonstrated 2. Severe disease during primary infections among infants born to dengue-immune mothers has been attributed to ADE	1. Increased risk of severe dengue among seronegative Dengvaxia (CYD-TDV) vaccine recipients led to WHO recommending adoption of a pre-vaccination screening strategy, and vaccine administration only to dengue seropositives 2. Identification of stronger correlates of protection, development of more robust assays for estimation of neutralizing/enhancing ability, long-term follow-up of vaccine recipients in clinical trials suggested to better predict vaccine-related ADE 3. Subunit EDIII, NS1 vaccines, chimeric vaccines with replaced DENV pr gene, T-cell response-inducing vaccines suggested to reduce ADE, but in vivo protection not yet demonstrated
Human immunodeficiency virus (family *Retroviridae*)	Antibodies against viral gp41 and gp120 enhance viral entry into T cells, monocytes/macrophages, and granulocytic cells, by the following mechanisms: 1. Complement-mediated, C-ADE: Interaction of antibody-opsonized virus with complement component C3d,g, cellular complement receptor CR2 and virus-specific receptor CD4, co-receptor CXCR4 2. FcR-mediated, FcR-ADE: By CD4-dependent or independent process	Positive correlation of enhancing antibody level with plasma viral load and negative correlation with CD4 cell count suggest association of ADE with accelerated immunosuppression and disease progression	ADE indicated by observation of higher infection rate/risk of infection among vaccine recipients in AIDSVAX, RV144 clinical trials, but not yet confirmed

Influenza virus (family *Orthomyxoviridae*)	1. Antibodies against viral hemagglutinin (HA) and neuraminidase (NA) mediate virus uptake via FcRs into macrophages, possibly leading to increased antigen presentation and T-cell activation 2. Enhancement of viral fusion by anti-HA2 antibodies suggested	Increased risk of medically attended illness among individuals with prior influenza-like illness during 1918 pandemic, and among seasonal influenza vaccine recipients during 2009 H1N1 pandemic, is suggestive of ADE	ADE suggested to be responsible for vaccine-associated enhanced respiratory disease in immunized pigs and ferrets, but mechanism not yet clearly understood
Respiratory syncytial virus (family *Paramyxoviridae*)	Antibodies against viral glycoproteins G and F mediate virus uptake via FcRs into monocytes, macrophages, dendritic cells leading to immune response modulation	Clinical relevance remains unclear in light of contradictory reports regarding association of maternal antibody-induced ADE with severe disease in infants	Enhanced disease in formalin-inactivated RSV vaccine recipients initially attributed to ADE, but other mechanisms also recently suggested
Ebola virus (family *Filoviridae*)	Antibodies against viral glycoprotein (GP) promote virus internalization by FcR-mediated or complement component C1q/C1q receptor-mediated process into monocytes/macrophages, endothelial, epithelial cells, and hepatocytes	Clinical relevance not yet understood	Vaccine-related ADE not yet demonstrated, but remains a concern. Avoiding induction of known infectivity-enhancing antibodies, while retaining T-cell epitopes in vaccines proposed for ADE mitigation
SARS—Coronavirus (family *Coronaviridae*)	Antibodies against viral spike (S) glycoprotein mediate virus uptake via FcRs into immune cells such as B cells, monocytes, macrophages	Clinical relevance still debated	Impact on vaccine safety not yet understood
Chikungunya virus (family *Togaviridae*)	FcR-mediated internalization into B cells, monocytes suggested	Clinical relevance not yet understood	Impact on vaccine safety not yet understood

EDI envelope domain I, *EDII* envelope domain II, *prM* precursor membrane, *FcR* Fc receptor, *LILR-B1* leukocyte immunoglobulin-like receptor B1, *TLR* Toll-like receptor, *RIG-I* retinoic acid-inducible gene I, *MDA5* melanoma differentiation-associated antigen 5, *DHF* dengue hemorrhagic fever, *DSS* dengue shock syndrome, *EDIII* envelope domain III as "extrinsic ADE." However, studies on Ross River Virus suggested that internalized immune complexes further serve to enhance viral replication by suppression of the cellular innate antiviral immune responses (Lidbury and Mahalingam 2000; Suhrbier and La Linn 2003). This mechanism is termed as "intrinsic ADE" and has been subsequently observed in flaviviruses as well (Chareonsirisuthigul et al. 2007; Ubol et al. 2010). Thus, ADE is a complex phenomenon comprising of extrinsic and intrinsic components, which together contribute to augmentation of viral infection and replication. The consequence of this increased virus production is the massive release of inflammatory and vasoactive mediators by host cells, which ultimately leads to exacerbation of viral pathogenesis and disease severity.

Prior sensitization of the humoral immune response is a prerequisite for ADE (Halstead 2003; Taylor et al. 2015). This phenomenon is thus widely observed during secondary infection with a heterotypic virus of the same genus, wherein preexisting antibodies against the primary (sensitizing) infection bind to the (secondary) virus, but fail to neutralize it. The pathogenicity and outcome of secondary infection is influenced by the time interval between primary/secondary infections, with increasing time being associated with more severe disease, and may be explained by the waning of broadly neutralizing antibodies over time. Passively acquired antibodies are also capable of inducing ADE, as indicated by the enhancement of viral disease by preexisting maternal antibody in infants born to dengue-immune mothers (Kliks et al. 1988; Chau et al. 2008, 2009).

The ADE phenomenon has implications for the use of antiviral immunoglobulins as therapy against viral infection, due to the associated risk of enhancement of disease (Taylor et al. 2015). ADE also poses a major challenge for implementation of vaccination programs, as vaccine-induced antibodies may enhance subsequent viral infection, thus placing vaccine recipients at increased risk of severe disease. Indeed, concerns regarding the safety of the world's first licensed dengue vaccine "Dengvaxia" have forced authorities to reconsider the mass vaccination strategy and issue specific recommendations for safe implementation of the vaccine in dengue control programs (Wilder-Smith et al. 2019).

ADE has been exploited by a variety of viruses belonging to different virus families. The mechanism and clinical significance of ADE of selected viruses will be reviewed in this chapter and are also summarized in Table 2.1. The ADE phenomenon has been most extensively studied in dengue virus and will be considered in detail here.

2.4 Dengue Virus

Dengue virus (DENV) is a member of family *Flaviviridae*, genus *Flavivirus*, and is transmitted to humans through the bite of *Aedes* mosquitoes, mainly *Aedes aegypti*. Infection with any of the four serotypes of this virus, DENV-1 to DENV-4, may be asymptomatic, or lead to dengue fever with symptoms ranging from mild to high-degree fever with headache, myalgia, arthralgia, rash, and retro-orbital pain. In some patients, the illness may progress to life-threatening severe dengue [earlier classified as dengue hemorrhagic fever (DHF)/dengue shock syndrome (DSS)], characterized by increased vascular permeability, plasma leakage, extensive pleural effusion, severe hemorrhages, respiratory distress, and organ failure (WHO 2009). Dengue disease is endemic in more than 125 countries and is considered a major public health problem worldwide, with an estimated 96 million cases and 20,000 deaths reported annually (Bhatt et al. 2013). In the absence of effective antivirals, dengue treatment mainly relies on symptomatic interventions. Though a licensed vaccine is now available, implementation of mass immunization programs remains complicated due to ADE, and is discussed in later sections.

The DENV genome is a single-stranded, positive-sense RNA molecule (~11 kb in size) and contains a single open reading frame coding for a large polyprotein, which is subsequently processed into three structural proteins, capsid (C), membrane (M), and envelope (E), and at least seven nonstructural proteins, NS1, NS2A, NS2B, NS3, NS4A, NS4B, and NS5 (Lindenbach et al. 2013). The E glycoprotein is the major surface-exposed region of the virus and displays the viral antigenic determinants. It functions to bind cellular receptors and fuse with host cell membranes during virus penetration and also directs viral assembly and budding. The E protein monomer can be divided into three structural/functional domains—EDI, the central region; EDII, the site of fusion; and EDIII, the site of receptor binding (Modis et al. 2004). The M protein is initially expressed as a membrane precursor (prM), which is present in the intracellular immature virus particle. Virion assembly and maturation involves cleavage of the precursor (pr) peptide, resulting in release of the mature virus particle containing the M protein into the extracellular environment (Li et al. 2008; Junjhon et al. 2010).

2.4.1 Epidemiological and Experimental Evidence for DENV-ADE

The possibility of immune enhancement of DENV infection was first suggested by observation of strong association of severe disease (DHF/DSS) with secondary infection among dengue patients in Bangkok, Thailand, during 1962–1964 (Halstead et al. 1967; Guzman et al. 2013). These initial findings were supported by prospective sero-epidemiological studies showing a higher rate of DHF/DSS during secondary infections (Sangkawibha et al. 1984; Graham et al. 1999; Halstead 2008). The association of secondary heterotypic DENV (different serotype) infection with ADE and severe dengue was further strengthened by reports of DHF/DSS cases in Cuba during the 1981/1997 DENV-2 outbreaks and 2001–2002 DENV 3 outbreak in a population immune to DENV-1 (1981/1997) and DENV-1/2 (2001–02), respectively (Guzman et al. 1990, 2000; Alvarez et al. 2006). Further, it has been demonstrated that such secondary DHF/DSS cases have higher viremia and levels of pro-inflammatory cytokines, suggesting that greater infected cell mass and subsequent increase in cytokine release contribute to disease severity in these patients (Vaughn et al. 2000; Wang et al. 2006; Rothman 2011). However, interestingly, tertiary/quaternary DENV infections have been rarely associated with severe disease, presumably due to sufficient cross-protective immunity acquired after two (primary/secondary) different DENV infections (Gibbons et al. 2007).

The most compelling evidence for ADE is provided by observation of severe dengue during primary infection in infants born to dengue-immune mothers (Kliks et al. 1988; Chau et al. 2008, 2009). During the first 3–4 months after birth, passively acquired maternal antibodies have been shown to protect infants from symptomatic dengue. Thereafter, the maternal antibodies begin to decline and reach sub-neutralizing levels, at which the antibodies are capable of enhancing DENV infection. Such enhancing antibodies persist till ~12 months of age, placing the infants at increased risk of severe dengue. Indeed, this enhancing activity of the infant sera has been demonstrated in vitro.

DENV-ADE is supported extensively through in vitro experiments. Enhancement of DENV infection by anti-DENV monoclonal antibodies or polyclonal human immune sera has been observed in primary human myeloid cells as well as cell lines such as FcγR-expressing BHK-21 cells, monocytic U937, THP-1, K562 cells, and macrophage-like P388D1 cells (Halstead et al. 1973a; Morens et al. 1987; Cardosa 1987; Kliks 1990; Goncalvez et al. 2007; Moi et al. 2013; Guzman et al. 2013; Acosta and Bartenschlager 2016). Moreover, in vitro ADE by human sera has been associated with severe dengue in patients. The ADE phenomenon has also been demonstrated in animal studies, with higher viremia being recorded during experimental secondary infection in rhesus monkeys as well as in passively immunized monkeys (Halstead et al. 1973b; Halstead 1979; Marchette et al. 1973; Goncalvez et al. 2007). Further, DENV-specific antibodies have been shown to enhance DENV infection and disease severity in AG129 mouse model (Zellweger et al. 2010). In another study using AG129 mice, maternally acquired anti-DENV antibodies were found to enhance viremia and vascular leakage, leading to earlier death (Ng et al. 2014).

The ultimate verification for DENV-ADE in humans has been provided only recently in a long-term pediatric cohort study in dengue-endemic Nicaragua (Katzelnick et al. 2017). This study observed highest risk for severe dengue among individuals with suboptimal anti-DENV antibody titers, and protection from symptomatic dengue at high antibody titers, thus proving that the level of preexisting anti-DENV antibodies is directly associated with the severity of secondary dengue disease in humans.

2.4.2 Molecular Mechanism of DENV-ADE

2.4.2.1 Antibodies Enhancing DENV Infection

Studies using monoclonal antibodies obtained from primary/secondary dengue patients have identified DENV surface proteins E and prM as major targets of host immune response (Serafin and Aaskov 2001; Crill and Chang 2004; Beltramello et al. 2010; Matsui et al. 2010; Wahala and Silva 2011; Fibriansah et al. 2015; Gan et al. 2017). These studies have suggested that highly neutralizing antibodies are predominantly raised against the complex quaternary epitopes of the viral envelope and the EDIII (lateral ridge) region involved in cellular attachment. These antibodies are mostly serotype-specific; however, antibodies raised against the quaternary "envelope dimer epitope" (EDE) have been found to potently neutralize all four DENV serotypes (Dejnirattisai et al. 2015). Such protective antibodies represent only a small subset of the antibody repertoire in dengue patients, with the immune response being dominated by cross-reactive, non-neutralizing antibodies targeting EDI, EDII (fusion loop) and prM, which have the potential to enhance viral infection (Smith et al. 2014). Moreover, all neutralizing antibodies are also capable of enhancing infection when present at sub-neutralizing concentration.

The role of antibodies against prM protein is particularly important because of its ability to enhance infection of immature virions, which are otherwise noninfectious. The prM protein, present in the intracellular immature virus particle, covers the fusion domain of the E protein, thus preventing its fusion with the host cell membrane. Cleavage of the precursor (pr) peptide from the prM protein during the process of virus maturation renders the released mature virus particle "infectious" to host cells for further rounds of replication (Li et al. 2008; Junjhon et al. 2010). Inefficient processing of the prM protein results in the production of immature or partially immature virus particles, which are normally noninfectious due to their inability to fuse with host cell. However, studies using human monoclonal antibodies obtained from persons with secondary dengue have revealed that these immature virus particles are capable of infecting host cells in the presence of the non-neutralizing anti-prM antibodies, and are thus infectious during secondary infections with the possibility of contributing to severe disease (Dejnirattisai et al. 2010; Schmidt 2010). Similarly, antibodies specific to the fusion loop (FL) in EDII have also been shown to enhance the infectivity of immature DENV particles (Rodenhuis-Zybert et al. 2011).

2.4.2.2 Fcγ Receptors and Their Role in ADE

The Fcγ receptors (FcγRs) present on the surface of cells of the myeloid lineage function to internalize antibody-opsonized virus via receptor-mediated endocytosis or phagocytosis and thus aid in clearance of the virus from blood circulation. However, these FcγRs can also contribute to ADE of DENV infection, by offering the virus an alternative pathway of entry into its target cells, the FcγR-bearing monocytes, macrophages, and dendritic cells (Taylor et al. 2015; Gan et al. 2017). There are three classes of human FcγRs, namely, FcγRI, FcγRII, and FcγRIII. FcγRI is a high-affinity activating receptor that can bind to monomeric IgG molecules (Vogelpoel et al. 2015). FcγRII is a class of low-affinity receptors that require high avidity binding by IgG immune complexes. Of these, FcγRIIA and FcγRIIC are activating receptors, while FcγRIIB is an inhibitory receptor (Guilliams et al. 2014). The activating receptors are involved in intracellular signal transduction via the immunoreceptor tyrosine-based activation motif (ITAM), either accessory (for FcγRI) or present within the receptor cytoplasmic domain (for FcγRIIA), while the inhibitory receptor contains immunoreceptor tyrosine-based inhibitory motif (ITIM) in its cytoplasmic domain (Boonnak et al. 2013). Activating receptors efficiently internalize antibody-opsonized virus, and can be involved in neutralization or enhancement, depending on the antibody concentration and receptor properties. FcγRI being a high-affinity receptor requires significantly less antibody for complete neutralization and thus plays a predominant role in DENV neutralization (Chawla et al. 2013). In contrast, FcγRIIA has been shown to be an effective mediator of ADE, with its ITAM-containing cytoplasmic tail playing a crucial role (Boonnak et al. 2013).

The monocytes, which are the major site of DENV infection and replication, express FcγRI and FcγRIIA at high levels and FcγRIIB at intermediate levels. FcγRIII is expressed on only ~10% of the monocyte population and has little impact on DENV neutralization or enhancement (van de Winkel and Anderson 1991). Macrophages express FcγRI, FcγRII, and FcγRIII. Dendritic cells express FcγRIIA and FcγRIIB, but FcγRIIB decreases with maturity (Guilliams et al. 2014).

2.4.2.3 Antibody-Mediated Viral Entry into Target Cell

Cells of the mononuclear phagocyte lineage, like monocytes, macrophages, and dendritic cells, are the main targets for DENV infection in humans. Attachment of DENV E protein (EDIII) to putative cell surface receptors such as heparan sulfate, C-type lectins, heat shock protein 70/90, and phosphatidylserine receptors constitutes the first step of viral infection (Castilla et al. 2015; Cruz-Oliveira et al. 2015). This is followed by penetration of the host cell by receptor-dependent, clathrin- and dynamin-mediated endocytic pathway. Acidic pH of the endocytic vesicle then triggers fusion of the viral envelope with the endosomal membrane, thus facilitating virion uncoating and release of the viral genome into the cellular cytoplasm.

During ADE, the antibody-opsonized DENV binds to the FcγRs present on the surface of the target cells and is internalized by FcγR-mediated endocytosis/phagocytosis. The entry pathway is similar to that of free virus; however, it shows slight alterations, depending on the cell type as well as the FcγR engaged. In vitro studies in monocytic cell lines U937 and K562 have shown that FcγRII-mediated entry occurs through clathrin-coated vesicles, while FcγRI-mediated entry is clathrin-independent. Also, as FcγRII translocates into lipid rafts upon immune complex binding, entry via this receptor is affected by the membrane cholesterol content (Carro et al. 2018). In macrophage-like P388D1 cells, antibody-opsonized virus is shown to be internalized by phagocytic uptake facilitated by actin-mediated membrane protrusions which actively search and capture the antibody-bound virus particles, possible by chemotaxis (Ayala-Nunez et al. 2016).

In addition to FcγR, other primary cell receptors may also be required for cell penetration. In fact, one theory suggests that FcγR plays an auxiliary role in concentrating the immune complexes to the cell surface and viral entry is then mediated by the primary cellular receptors (Chotiwan et al. 2014).

2.4.2.4 Suppression of Innate Antiviral Immune Response in ADE

The suppression of the host antiviral immune response to facilitate higher virus production during ADE of DENV infection was first suggested in 2007 through studies in monocytic THP-1 cells, which revealed upregulation of anti-inflammatory cytokine production (IL-6, IL-10) but downregulation of anti-DENV free radicals (nitric oxide) and pro-inflammatory cytokine production (IL-12, IFN-γ, TNF-α) (Chareonsirisuthigul et al. 2007). Subsequent studies in THP-1 cells offered further insights into the mechanism of this "intrinsic" DENV-ADE (Ubol et al. 2010; Modhiran et al. 2010; Tsai et al. 2014; Chan et al. 2014). Overall, these studies have suggested Type I interferon and pro-inflammatory cytokine production as the main targets of immune suppression during DENV-ADE, with four mechanisms being proposed for the same: (1) leukocyte immunoglobulin-like receptor B1 (LILR-B1) co-ligation and consequent inhibition of FcγR-signaling and Type I interferon-stimulated gene (ISG) expression, (2) downregulation of toll-like receptor (TLR)-dependent antiviral signaling pathway, (3) downregulation of retinoic acid-inducible gene I (RIG-I)/melanoma differentiation-associated antigen 5 (MDA-5) signaling pathway resulting in suppression of Type I interferon (IFN) production, and (4) induction of IL-10 production resulting in suppressor of cytokine signaling 3 (SOCS3)-mediated suppression of pro-inflammatory cytokines important for antiviral response.

Binding of immune complexes to cellular activating FcγRs is known to trigger ITAM-spleen tyrosine kinase (Syk)-signal transducer and activator of transcription 1 (STAT1) signaling pathway which leads to ISG induction and inhibition of viral replication. Inhibition of this FcγR-dependent ISG transcription is thus crucial for successful DENV replication and is proposed to be brought about by co-ligation of LILR-B1 by the FcγR-bound antibody-opsonized DENV (Chan et al. 2014; Nimmerjahn and Lux 2014). LILR-B1 belongs to a family of ITIM-containing inhibitory receptors and is expressed on immune effector cells including monocytes, macrophages, and dendritic cells. At sub-neutralizing antibody concentrations, the DENV–antibody immune complex binds to two cellular receptors: FcγR (through antibody) and LILR-B1 (through DENV E protein). Binding to LILR-B1 recruits Src homology phosphatase-1 (SHP-1) which dephosphorylates and inactivates Syk and prevents ISG expression. SHP-1 signaling also prevents rapid acidification and lysosomal degradation of DENV (Ong et al. 2017). On the contrary, at high antibody concentrations, DENV E protein epitopes are completely blocked by the antibody; consequently, LILR-B1 binding is not possible, resulting in neutralization of the FcγR-internalized virus.

The TLR-dependent pathway is a key mediator of the innate antiviral immune response and stimulates production of Type I IFN and pro-inflammatory cytokines via nuclear factor-κβ (NF-κβ) family of transcription factors. Studies using monocytic cell lines and patient peripheral blood mononuclear cells (PBMCs) have shown that this pathway is suppressed during DENV-ADE (Modhiran et al. 2010). Engagement of FcR by DENV–antibody immune complexes or FcR-mediated entry of immune complexes into target cells results in downregulation of expression of TLR-3,-4,-7 and TLR signaling molecules. This is accompanied by activation of Sterile-alpha and Armadillo motif-containing protein (SARM) and TRAF family member-associated NF-κβ activator (TANK), which are negative regulators of the TLR adaptor molecule, TIR-domain-containing adapter-inducing interferon-β (TRIF), and effector molecule, TNF receptor-associated factor 6 (TRAF6), respectively. SARM/TANK activation thus results in suppression of TLR-dependent IFN stimulating pathway.

The RIG-I/MDA-5 antiviral signaling pathway is also downregulated during DENV-ADE, as demonstrated in vitro and observed among patients of severe secondary DENV infection (Ubol et al. 2010). Entry of immune complexes into target cells activates the negative regulators, dihydroxyacetone kinase (DAK) and autophagy-related 5–autophagy-related 12 (Atg5-Atg12), which then disrupt the RIG-I/MDA-5 signaling cascade. This results in suppression of Type I IFN production and IFN-mediated antiviral responses.

IL-10 is known to induce T-helper cell 2 (Th2) cytokine response and inhibit production of pro-inflammatory cytokines such as IFN-γ, IL-12, TNF-α as well as nitric oxide radicals, which are potent inhibitors of DENV replication. IL-10 induction during DENV-ADE thus exerts an immunosuppressive effect and favors viral replication. Indeed, high IL-10 levels have been associated with severe disease in secondary dengue patients (Nguyen et al. 2004; Ubol et al. 2010). The mechanism of IL-10-mediated immune suppression has been demonstrated through in vitro studies in monocytic cell lines and patient PBMCs. These studies have shown that IL-10 induction during ADE occurs through enhanced Syk-regulated phosphoinositide 3-kinase (PI3K)/protein kinase B (PKB)/glycogen synthase kinase (GSK)-3β/cyclic adenosine monophosphate response element-binding (CREB) signaling (Tsai et al. 2014). IL-10 then stimulates expression of SOCS3, which subsequently inactivates Janus kinase (Jak)-STAT signaling, resulting in suppression of pro-inflammatory cytokine production. In addition, IL-10 also suppresses inducible nitric oxide synthase (iNOS) gene expression and thus nitric oxide production, by inhibiting STAT-1 and Interferon Regulatory Factor 1 (IRF-1) activity (Chareonsirisuthigul et al. 2007; Ubol et al. 2010).

Majority of the studies elucidating the mechanism of intrinsic ADE have been carried out in monocytic cell lines. Experiments using primary monocyte-derived human macrophages have demonstrated lower IFN-β and higher IL-6 and SOCS3 levels during DENV-ADE, consistent with the observations in monocytic cells (Rolph et al. 2011). However, IL-10 induction and RIG-I/MDA-5 downregulation were not observed. Thus, IL-6 has been proposed as critical regulator of DENV-ADE in macrophages, responsible for SOCS-3-mediated suppression of pro-inflammatory cytokines, indicating that mechanisms of intrinsic ADE may be cell type-dependent.

Contradictory to the well-accepted mechanism of ADE-mediated immune suppression, a recent study observed that entry of DENV immune complexes in primary human macrophages did not trigger immunosuppressive signaling, but resulted in enhanced fusion of viral envelope with the host cell membrane (Flipse et al. 2016). Further investigations into the different mechanisms involved in ADE of DENV infection are thus warranted.

2.4.2.5 Enhancement of Disease Severity: ADE and Cytokine Storm

The extrinsic and intrinsic mechanisms of ADE together contribute to enhancement of viral replication and consequently higher viremia levels, which correlate with the incidence of severe dengue in patients (Vaughn et al. 2000). ADE is also associated with triggering of the "cytokine storm" which involves massive release of inflammatory cytokines and other chemical mediators at high levels (Pang et al. 2007). ADE of DENV infection and replication results in increased presentation of viral antigens on the surface of infected cells, leading to activation and proliferation of T cells sensitized during a prior infection. This results in release of cytokines such as IFN-γ, TNF-α, IL-2, IL-6, IL1-β, IL-8 and immunoregulators such as IL-12p70, which directly act upon the vascular endothelial cells resulting in plasma leakage, the hallmark of severe dengue. Increased levels of such cytokines have been associated with plasma leakage in severe dengue patients (Chaturvedi et al. 2000), while effect of the cytokine response on alteration of vascular permeability has been demonstrated in vitro and in mice models, with disruption of the apical junction complexes in endothelial cells being the suggested mechanism (Dewi et al. 2004; Appanna et al. 2012; Puerta-Guardo et al. 2013).

2.4.3 Implications for Dengue Vaccines

Generation of a long-term protective immune response against all four serotypes of DENV is a prerequisite for a dengue vaccine and assumes greater importance due to the associated risk of vaccine-induced ADE of DENV infection and disease. During the six-decade long quest for an effective dengue vaccine, several candidates such as live-attenuated, inactivated, recombinant subunit, virus-like particle (VLP)-based, and DNA vaccines have been explored, with Sanofi Pasteur's "Dengvaxia" or CYD-TDV, a chimeric yellow fever 17D-tetravalent DENV vaccine, National Institutes of Health's TV003/TV005, and Takeda Pharmaceutical's TAK-003, both live attenuated tetravalent vaccines, leading the development pipeline (McArthur et al. 2013). Of these, CYD-TDV has been approved for human use and is licensed in 20 countries, while TV003/TV005 and TAK-003 are currently undergoing Phase III trials.

CYD-TDV, the world's first licensed dengue vaccine, was approved by the World Health Organization (WHO) in May 2016, following Phase III trials in dengue-endemic regions (Capeding et al. 2014; Villar et al. 2015; Hadinegoro et al. 2015). These trials had demonstrated high vaccine efficacy among 9–16-year-olds during the first 2 years after vaccine (first of three doses) administration, but indicated increased risk of hospitalized dengue in the 2–5 year age group during the third year of follow-up. These results suggested the possibility of the vaccine sensitizing dengue seronegatives to subsequent enhanced DENV infection and disease. However, considering the population-level benefits in reducing disease burden, the vaccine was approved by WHO Strategic Advisory Group of Experts (SAGE) on immunization, with recommendations for use in dengue-endemic regions with ≥70% seroprevalence and administration to individuals in the age range of 9–45 years. This approval was followed by additional studies to investigate the long-term benefit–risk ratio of dengue vaccination in seronegative population. These studies revealed that vaccine performance was indeed affected by serostatus, with vaccine efficacy being significantly higher in seropositives than seronegatives. Furthermore, an increased risk of hospitalized, severe dengue was observed among seronegative vaccine recipients as compared to seronegative controls (Sridhar et al. 2018). In light of this new evidence regarding vaccine-related ADE, WHO-SAGE issued updated recommendations on the use of CYD-TDV vaccine in April 2018 (Wilder-Smith et al. 2019). For countries considering vaccination as part of their dengue control program, WHO now recommends a "pre-vaccination screening strategy," in which all potential vaccine recipients are tested for anti-DENV IgG, preferably by ELISA, and only dengue-seropositive persons are vaccinated. Further, for trials of new vaccines, WHO advises stratification of participants according to pre-vaccination serostatus during data analysis, as well as long-term follow-up of all vaccine recipients.

The occurrence of breakthrough infections in the CYD-TDV vaccinees despite having high titers of neutralizing antibodies has raised concerns regarding the value of the currently used in vitro neutralization tests as correlates of protection. The neutralizing potential of antibodies is commonly evaluated using the plaque reduction neutralization test (PRNT) in LLC-MK2, Vero, or BHK-21 cell lines. However, neutralization of DENV infection in these FcγR-negative cells may not reflect the ability of the antibodies to prevent infection of myeloid cells, which are primary targets of DENV, and capable of internalizing antibody-opsonized DENV through FcγR-mediated phagocytosis. The use of FcγR-bearing cell lines or primary myeloid cells for in vitro neutralization has thus been proposed for measuring the neutralizing versus enhancing ability of anti-DENV antibodies (Mahalingam et al. 2013; Gan et al. 2017). Furthermore, identification of stronger immune correlates for disease risk and protection is urgently needed (Katzelnick and Harris 2017).

While the results of Phase III trials of TV003/TV005 and TAK-003 vaccines are awaited, the concern of vaccine-related ADE has promoted development of vaccines specifically designed to minimize ADE. Monoclonal antibody-based studies have helped identify neutralizing and enhancing viral epitopes, making design of such tailored vaccines possible. DENV EDIII induces predominantly neutralizing antibodies, thus making recombinant EDIII an attractive subunit vaccine candidate (Gil et al. 2017; Frei et al. 2018). DENV NS1 has also been considered as a vaccine candidate as it avoids ADE (Hertz et al. 2017). Chimeric DENV constructs with pr gene replaced with Japanese Encephalitis virus pr and DNA vaccines with substitutions or removal of enhancing epitopes have demonstrated reduction in ADE (Crill et al. 2012; Tang et al. 2015; Wang et al. 2015). Another strategy for evasion of ADE is the development of vaccine candidates inducing protective CD4+ and CD8+ T-cell responses. DENV C protein is an immunodominant T-cell epitope, and C-based vaccines have shown induction of protective T-cell-mediated immunity in monkeys (Gil et al. 2014). Such next-generation vaccines do face the disadvantage of not retaining the intact virion structure and the quaternary epitopes, which are important for potent neutralization. However, these vaccines have shown encouraging results in preclinical studies, and whether they are capable of conferring in vivo protection remains to be seen.

2.4.4 Implications for Dengue Therapeutics

Humanized monoclonal antibodies (mAbs) are attractive therapeutic options against DENV infection. Therapeutic mAbs must neutralize DENV, without increasing the risk of ADE, in order to offer protection against dengue. Several mAbs targeting the DENV E and/or prM proteins are being developed as therapeutic candidates, with reduction of ADE as the major focus of development (Chan et al. 2013; Sun et al. 2018). Use of serotype-specific mAbs or high dose administration of cross-reactive mAbs has demonstrated reduced risk of ADE in vitro and in animal models. Further, antibodies with modifications in the Fc region such as deletion of nine amino acids (at positions 231–239) in the N terminus, mutation of asparagine to glutamine at position 297 (N297Q), or mutation of leucine to alanine at positions 234 and 235 (LALA mutants) have also shown reduced risk of ADE and exhibited prophylactic and therapeutic efficacy in animal models (Balsitis et al. 2010; Shi et al. 2016; Xu et al. 2017; Injampa et al. 2017; Lu et al. 2018). Fc region modifications that extend the antibody half-life also help minimize ADE by maintaining the mAbs at high levels.

2.4.5 ADE in Other Flaviviruses

ADE has also been reported in other viruses of family *Flaviviridae* including Murray Valley Encephalitis virus, Japanese encephalitis virus (JEV), West Nile virus (WNV), yellow fever virus (YFV), and Zika virus (ZIKV) of genus *Flavivirus,* as well as hepatitis C virus (HCV), of genus *Hepacivirus* (Hawkes and Lafferty 1967; Cardosa et al. 1983; Gould and Buckley 1989; Meyer et al. 2008; Bardina et al. 2017). Further, infection with one *Flavivirus* has also been demonstrated to enhance subsequent infection with another virus of the same genus, as observed for DENV-ZIKV in in vitro studies, mice models, and rhesus macaques (Bardina et al. 2017; George et al. 2017). Such DENV-ZIKV cross-enhancement has also been observed in primary human PBMCs (Li et al. 2018); however, in the absence of clinical or epidemiological evidence, the clinical relevance of these findings is unknown. Interestingly, a beneficial effect of cross-enhancement among flaviviruses has been reported recently in a clinical trial, with preexisting JEV vaccine-induced antibodies enhancing immunogenicity of subsequently administered YFV vaccine (Chan et al. 2016).

2.5 Human Immunodeficiency Virus

Human immunodeficiency virus (HIV), a member of family *Retroviridae*, genus *Lentivirus*, is primarily a sexually transmitted virus, but nonsexual transmission through intravenous inoculation or mother-to-child exposure can also occur. HIV infects the CD4+ T cells leading to their progressive depletion and thus hampers the host adaptive immune response. The final stage of HIV infection is acquired immunodeficiency syndrome (AIDS), characterized by immune system failure and occurrence of life-threatening opportunistic infections. As of 2016, ~37 million people worldwide have been estimated to be living with HIV (Ghosn et al. 2018). While a protective vaccine or a complete cure of HIV infection remains elusive, implementation of combination antiretroviral therapy (cART) has been successful in prolonging the duration between HIV infection and AIDS and reducing AIDS-related deaths.

2.5.1 ADE of HIV Infection: Experimental Evidence and Molecular Mechanism

ADE of HIV infection was first reported in the 1980s, shortly after the identification and isolation of HIV, and two different mechanisms, namely, complement-mediated ADE (C-ADE) (Robinson Jr. et al. 1987, 1988) and FcR-mediated ADE (FcR-ADE) (Takeda et al. 1988; Homsy et al. 1989), were described. Through either of these mechanisms, virus-specific antibodies present in sera of HIV-infected individuals were shown to enhance viral entry and in some cases replication within T cells, monocytes/macrophages, and granulocytic cells (Beck et al. 2008). Further, both these mechanisms have also been shown to enhance HIV infection of human syncytiotrophoblast cells, thus suggesting contribution of ADE to materno-fetal transmission (Tóth et al. 1994). Both these forms of ADE are mediated by antibodies targeting the viral surface glycoproteins gp41 and gp120 (Trischmann et al. 1995; Takada and Kawaoka 2003). Such enhancing antibodies have been detected in 72% of HIV patients (Subbramanian et al. 2002).

C-ADE of HIV infection has been well characterized through in vitro studies in T-cell lines such as MT-2 and SupT1/R5, showing enhancement of cell line-adapted HIV strains or primary virus isolates by human monoclonal antibodies or sera from HIV-infected individuals, and has been associated with increased synthesis of viral RNA and protein and enhanced release of infectious virions (Robinson Jr. et al. 1989, 1990a, b; Robinson Jr. 2006; Willey et al. 2011). Such C-ADE is primarily mediated by antibodies targeting the N-terminal immunodominant domain of viral gp41. Antibody binding to gp41 initiates the complement cascade leading to deposition of complement component C3d,g on the virion. Interaction of this opsonized virus with the cell surface complement receptor type 2 (CR2) facilitates virus attachment and internalization. Engagement of the HIV receptor CD4 and co-receptor CXCR4 is also required for this mode of virus uptake. Increased attachment of the virus to target cell, rather than CR2-mediated signaling, has been suggested to be responsible for enhancement of infection through C-ADE. An alternative route of complement component C1q binding followed by interaction with C1q receptor has also been suggested for C-ADE (Prohaszka et al. 1997).

FcR-ADE has been demonstrated in vitro mainly in monocyte/macrophage cell lines and primary cultures, with all three classes of FcγR, namely, FcγRI, FcγRII, and FcγRIII, reported to support ADE (Homsy et al. 1989; Takeda et al. 1988, 1990; Laurence et al. 1990; Connor et al. 1991; Trischmann et al. 1995). ADE by FcγRI and FcγRII requires viral interaction with CD4, indicating that the FcRs facilitate virus entry by potentiating attachment to CD4 receptors. Alternatively, FcR-mediated endocytosis of virus–antibody complexes followed by intracellular fusion with endosomal membrane occurring as a result of binding of virus with CD4 receptors on the endosomal membranes has also been suggested. FcγRIII-mediated ADE has been shown to be CD4-independent. FcR-ADE assumes particular importance in the viral life cycle as it mediates infection of macrophages, which are important for maintaining a viral reservoir during persistent infections.

Another mechanism of complement-independent, FcR-independent ADE, mediated by antibodies specific to viral gp120, has been described (Sullivan et al. 1998; Guillon et al. 2002). Antibody binding has been shown to modulate interaction of gp120 with HIV co-receptor CCR5 or induce conformational changes in gp120, leading to its activation and subsequent promotion of membrane fusion. Further, FcαR-mediated ADE has also been reported, with human serum IgA showing low-level enhancement of HIV replication in vitro (Janoff et al. 1995; Kozlowski et al. 1995).

An important feature of HIV is the rapid viral evolution during chronic infection. While neutralizing antibodies are known to exert a selection pressure facilitating emergence of neutralization escape mutants, enhancing antibodies have been suggested to favor emergence of ADE-susceptible variants. Indeed, at later stages of infection, patient sera have demonstrated increased proportion of enhancing antibodies and potential for ADE, while virus strains isolated at such later stage have exhibited increased susceptibility to ADE in vitro (Davis et al. 2001; Subbramanian et al. 2002; Willey et al. 2011).

2.5.2 Clinical Significance of HIV-ADE

The HIV disease outcome is suggested to be influenced by the balance of neutralizing and enhancing antibodies in HIV patients, with studies showing lower neutralizing and higher enhancing antibody levels among AIDS patients (Homsy et al. 1990; Toth et al. 1991; Fust et al. 1994; Szabo et al. 1999; Subbramanian et al. 2002). These studies have observed strong negative correlation between the extent of ADE and CD4 cell count and positive correlation between ADE and plasma viral load in HIV-positive individuals. As CD4 cell count and HIV plasma viral load are considered the most potent predictors of HIV disease progression, such observations indicate contribution of ADE to accelerated immunosuppression and disease progression. Further support for this association was provided by reports of decline in enhancing antibody levels and improvement in disease prognosis in patients receiving highly active antiretroviral therapy (HAART) (Banhegyi et al. 2003). Contradictory observations of absence of any clinical correlation between ADE and disease have also been reported (Montefiori et al. 1991); however, it has been suggested to be the result of conduction of experiments in conditions which do not sufficiently resemble in vivo conditions (Takada and Kawaoka 2003; Beck et al. 2008).

2.5.3 Implications for HIV Vaccines

The impact of ADE on HIV vaccine safety is not clearly understood. Studies in Rhesus monkeys have demonstrated the association of active and passive immunization with enhanced disease progression on subsequent infection (Staprans et al. 2004; Sholukh et al. 2014). In humans, ADE has been indicated in two vaccine clinical trials. In the AIDSVAX trial, the rate of HIV infection was observed to be higher among vaccinees with low (non-protective) antibody responses (Gilbert et al. 2005). In the RV144 trial, positive correlation of vaccine-elicited plasma IgA responses with risk for HIV acquisition was documented (Shmelkov et al. 2014). Further, a study investigating the effect of passive immunization of HIV-positive pregnant women on mother-to-infant transmission suggested the possibility of antibody-enhanced in utero transmission (Onyango-Makumbi et al. 2011). These findings are suggestive of vaccine-induced ADE, but cannot confirm the same. Assays capable of estimating neutralizing and enhancing potential of immune serum, such as those involving use of complement and complement receptor carrying target cells for C-ADE, have been proposed for a better understanding of the clinical relevance of vaccine-related ADE. Such assays would also provide stronger immune correlates during preclinical and clinical studies. Further, design of vaccine immunogens that will elicit protective and not enhancing antibodies, by avoiding inclusion of known enhancing epitopes, has been put forth as a mitigation strategy for HIV-ADE (Beck et al. 2008).

2.6 Influenza Virus

Influenza is a highly communicable acute respiratory disease caused by infection with influenza virus of family *Orthomyxoviridae*. The disease is a serious health threat of epidemic and pandemic proportions, with severe cases associated with pneumonia and lung tissue damage. Influenza A viruses are the most common in human infections and are further classified into subtypes on the basis of the two surface proteins, hemagglutinin (H) and neuraminidase (N), with 18 H and 11 N subtypes identified (Shao et al. 2017). Influenza A virus undergoes rapid evolution due to mechanisms such as antigenic drift, shift, and genetic reassortment, resulting in periodic emergence of novel viral strains. Annual vaccination against the circulating viral strain is thus implemented for protection against the disease.

Epithelial cells of the respiratory tract are the primary targets of influenza virus infection; however, in vitro ADE of this virus has been reported in primary macrophages and macrophage-like cell lines (P388D1), under the presence of monoclonal antibodies, mice, rabbit, or human immune sera (Ochiai et al. 1988, 1990, 1992; Tamura et al. 1991, 1994; Gotoff et al. 1994). Engagement of FcγRs by antibodies specific to viral HA and NA has been shown to mediate virus uptake, with subtype cross-reactive non-neutralizing antibodies being particularly involved in infection enhancement of heterotypic viruses (different HA/NA type). While such internalization usually results in abortive infection of macrophages, productive replication in these cells has been shown to be possible in the presence of appropriate protease for HA cleavage. Increased presentation of viral antigens by the infected macrophages leading to augmented virus-specific T-cell activation is also suggested to contribute to viral pathogenesis. Enhancement of virus fusion by anti-HA2 subunit antibodies is another mechanism of influenza-ADE and has been shown to be responsible for the occurrence of vaccine-associated enhanced respiratory disease (VAERD) in pigs immunized with inactivated H1N2 vaccine and challenged with pandemic 2009 H1N1 virus (Gauger et al. 2011; Khurana et al. 2013). Influenza-ADE is also suggested by epidemiologic observations in humans. Retrospective analysis of medical records from the 1918 influenza pandemic revealed association of prior influenza-like illness with increased risk of medically attended illness during the pandemic period (Shanks et al. 2016). More recently, prior receipt of seasonal influenza vaccine was found to increase the risk of medically attended pandemic H1N1 illness during 2009 in Canada (Skowronski et al. 2010). Furthermore, presence of high titer non-protective serum antibodies was associated with immune complex-mediated severe H1N1 illness during this pandemic (Monsalvo et al. 2011; To et al. 2012).

Observation of VAERD in pigs and ferrets immunized with inactivated or HA subunit vaccines raises concern regarding the safety of vaccines, particularly universal influenza vaccines (Gauger et al. 2011; Khurana et al. 2013; Rajao et al. 2014; Skowronski et al. 2014). However, vaccine-mediated protection and absence of VAERD is also reported in few studies in pigs, suggesting that vaccine-induced ADE occurs only in certain conditions, which are not yet clearly understood (Reeth et al. 2004; Kyriakis et al. 2010; Ricklin et al. 2016). Regarding therapeutic candidates, Phase 2 studies of anti-HA stalk mAbs have indicated possibility of enhanced viral shedding in some treated human subjects. Screening for high potency protective antibodies, defining optimal dose range for effective neutralization, using cocktails of mAbs with different antigen specificities, has been suggested for mitigation of ADE (Chan-Hui and Swiderek 2016).

2.7 Respiratory Syncytial Virus

Respiratory syncytial virus (RSV), a member of family *Paramyxoviridae*, genus *Pneumovirus*, is the leading cause of lower respiratory tract infections among infants and young children worldwide, with severe disease requiring hospitalization occurring most frequently between 6 weeks and 6 months of life (Glezen et al. 1986; Hall et al. 2009). RSV disease is also a major health problem in adults and the elderly. Severe bronchiolitis is one of the major outcomes of RSV infection.

Observation of enhanced disease among children receiving the formalin-inactivated RSV (FI-RSV) vaccine in the 1960s (Kim et al. 1969), as well as the frequent occurrence of severe disease among infants <6 months of age, when maternal antibodies are present, prompted investigations on ADE of RSV infections. Subsequently, enhancement of RSV infection by RSV-specific monoclonal antibodies and human immune sera was demonstrated in vitro in monocytic (U937, THP-1) and macrophage-like (P388D1) cell lines (Gimenez et al. 1989, 1996; Krilov et al.

1989; Osiowy et al. 1994). The phenomenon has also been reported in Bonnet monkeys developing enhanced disease following FI-RSV immunization (Ponnuraj et al. 2001, 2003). FcγR-mediated internalization of immune complexes is the suggested mechanism, with antibodies against the viral surface glycoproteins, the attachment glycoprotein (G) and fusion glycoprotein (F), playing an important role. Although epithelial cells of the respiratory tract are primary targets of RSV infection, it has been suggested that alveolar macrophages are infected through ADE, resulting in activation of Th2 response and increased expression of TNF-α, IL-6, thus causing enhanced disease (Gimenez et al., 1996). Further, ADE-mediated infection of lung dendritic cells (DCs) has been demonstrated to negatively modulate DC cell function, resulting in impaired T-cell activation, and has been suggested to contribute to RSV pathogenesis (Gomez et al. 2016). However, the clinical effects of RSV-ADE still remain unclear with different studies reporting contradictory observations regarding the association of maternal antibody-induced ADE and disease severity in infants (Chanock et al. 1970; van Erp et al. 2017).

Enhanced RSV disease in FI-RSV vaccine recipients was initially attributed to vaccine-induced ADE; however, recent studies have suggested a role of other mechanisms involving T cells (Huisman et al. 2009). Whether ADE would affect the efficacy of other RSV vaccines in development remains to be seen. As far as immunotherapeutics are concerned, the F protein-specific humanized monoclonal antibody, Palivizumab, is the only clinically approved intervention. This mAb has demonstrated ADE at sub-neutralizing concentrations in vitro, yet has been found to have a protective effect in animal models and humans (TI-RS Group 1998; Mejıas et al. 2004; van Mechelen et al. 2016).

2.8 Ebola Virus

Ebola virus is a member of family *Filoviridae.* Of the four Ebola virus species infecting humans, *Zaire ebolavirus* (EBOV) is the most virulent and is the causative agent of a severe hemorrhagic fever disease, Ebola virus disease (EVD), with a fatality rate of ~90% (Feldmann and Geisbert 2011). The disease is zoonotic, with initial cases occurring due to contact with infected fruit bats (reservoir) or their contaminated material, followed by human-to-human transmission. The virus is mainly endemic in Africa; however, due to its increased incidence and fast spread, as observed during the 2014–2016 African outbreak, the virus is considered a pandemic threat.

ADE of Ebola virus infection has been demonstrated in vitro in a variety of cell lines and is shown to be mediated by antibodies specific to the viral envelope glycoprotein (GP) (Takada et al. 2001, 2003, 2007). Such enhancing antibodies have been identified in the sera of EVD patients (Takada et al. 2003; Furuyama et al. 2016). Two different mechanisms have been proposed for ADE of EBOV infection: (a) FcγR-mediated ADE and (b) C1q-mediated ADE. In FcγR-mediated ADE, antibody–virus complexes bind cell surface FcγRIIA triggering phosphorylation of Src family protein tyrosine kinases (PTKs) and activation of Src signaling pathways leading to virus uptake through phagocytosis and/or micropinocytosis (Furuyama et al. 2016). C1q-mediated ADE involves interaction of virus-bound antibodies with complement component C1q and subsequent attachment to cell surface C1q receptors, leading to virus internalization by C1q receptor-mediated endocytosis or via virus-specific receptor (Takada et al. 2003, 2007). C1q receptors are present on monocytes/macrophages, which are primary targets of EBOV, as well as most other mammalian cells, including endothelial cells, epithelial cells, and hepatocytes. C1q-ADE mediated dissemination of EBOV infection to these "secondary target cells" has been suggested to exacerbate disease.

Although clinical relevance of EBOV-ADE is not yet understood, repeated in vitro demonstrations raise concerns regarding safety of vaccines and immunotherapeutics. Several candidate vaccines under development have demonstrated protective efficacy in preclinical studies and immunogenicity in clinical trials (Dhama et al. 2018). The protective effect of these vaccines seems to depend on cytotoxic T-cell responses. Avoiding induction of known infectivity-enhancing antibodies while retaining T-cell epitopes should thus be considered during vaccine design. For immunotherapy, the use of well-characterized mAbs instead of polyclonal convalescent human serum has been suggested to reduce ADE.

2.9 Severe Acute Respiratory Syndrome: Coronavirus

Severe acute respiratory syndrome (SARS) is a respiratory disease caused by SARS-coronavirus (SARS-CoV) of family *Coronaviridae*, which caused outbreaks mainly in China, during 2002–2003. Although public health measures successfully contained further spread and human outbreaks, future reemergence of SARS-CoV or other related viruses remains a credible threat due to the zoonotic origin of this virus.

ADE of SARS-CoV infection was first demonstrated in vitro in human B cells using sera from immunized animals as well as convalescent patients (Kam et al. 2007). Further studies showed that antibodies specific to the viral surface spike glycoprotein (S) are capable of enhancing viral infection of immune cells, particularly monocytes and macrophages, thus allowing the virus to broaden its cellular tropism (Jaume et al. 2011; Wang et al. 2014; Yip et al. 2014). Internalization of the immune complexes by cellular FcγRs, mainly FcγRIIA, is the suggested mechanism, with intracellular signaling by the FcγR cytosolic domain playing an important role. This FcγR-mediated internalization follows a pH- and cysteine protease-independent pathway and differs from the endosomal/lysosomal pathway utilized by angiotensin I-converting enzyme 2 (ACE2), the accepted receptor for SARS-CoV. Such antibody-mediated infection of macrophages, however, was not found to support productive virus replication or modify expression of cytokines/chemokines (Yip et al. 2016). Although demonstrated recently in rhesus monkeys (Wang et al. 2016), the occurrence of SARS-ADE and its association with disease severity in humans are still debated, with different clinical studies reporting both protective and disease-enhancing effects of anti-SARS-CoV antibodies (Lee et al. 2006; Zhang et al. 2006).

2.10 Chikungunya Virus

Chikungunya virus (CHIKV) is an arbovirus belonging to family *Togaviridae*, genus *Alphavirus*. The virus is recognized as a significant global health threat since the last decade. It is primarily transmitted through the bite of infected *Aedes* mosquitoes, *Aedes aegypti* and *Aedes albopictus*. CHIKV infection is associated with a febrile illness accompanied by myalgia, arthralgia, and cutaneous rash (Kam et al. 2009). Debilitating persistent polyarthralgia is the characteristic feature of Chikungunya disease.

ADE of CHIKV infection was suggested by the association of low titers of anti-CHIKV IgG in immunized mice with the early onset of disease upon subsequent CHIKV challenge (Hallengard et al. 2014). Another recent study observed enhanced attachment of CHIKV to primary human monocytes and B cells and increased viral replication in the murine macrophage cell line, RAW264.7, in the presence of anti-CHIKV antibodies (Lum et al. 2018). Further, this study also detected higher viral RNA load and enhancement of disease severity in mice infected with CHIKV in the presence of sub-neutralizing concentrations of antibody, thus providing the first evidence for ADE of CHIKV infection and disease. The enhancement was found to be FcγR-mediated, with FcγRII playing a predominant role. However, ADE of CHIKV infection has not been reported to date in humans, and thus, the clinical significance of these findings remains unknown and warrants further investigation.

2.11 Concluding Remarks

Virus-specific antibodies function to neutralize virus infection and are an important component of the host antiviral immune response. However, under sub-neutralizing or non-neutralizing conditions, the antibodies are capable of enhancing viral infection by providing an alternative route of virus entry into host cells. This phenomenon of antibody-dependent enhancement facilitates viral infection of a larger number of primary target cells and also allows viral entry into otherwise non-susceptible secondary target cells, thus broadening the virus tropism. Moreover, ADE also modulates the host intracellular signaling pathways, thereby turning down the antiviral immune response. Together these mechanisms serve to augment viral replication and pathogenesis, with the balance between neutralizing and enhancing antibodies influencing the outcome of viral infection and disease.

ADE has been demonstrated in vitro and in animal models for a number of viruses including DENV, HIV, influenza, RSV, Ebola, SARS-CoV, and CHIKV. While the occurrence of DENV-ADE in humans and its association with severe disease have been clearly demonstrated, the clinical relevance of ADE in other viruses is not completely understood. Nevertheless, ADE remains a concern in development and use of vaccines and immunotherapeutics against these viruses. Recent advances in identification of neutralizing and enhancing viral epitopes are expected to guide future design of non-enhancing, protective viral vaccines and therapeutic mAbs. Comprehensive evaluation of humoral and cell-mediated immune response to vaccines in preclinical and clinical studies, with the help of robust assays that can detect ADE, is suggested to be crucial for the assessment of vaccine safety and efficacy. Such evaluations would require a better understanding of the immune correlates of protection. Lastly, investigation of long-term effects of vaccine implementation is advisable while considering approval for future vaccines.

Acknowledgments The author is grateful to Dr. A.C. Mishra, Director, IRSHA, and Dr. V.A. Arankalle, Head, Department of Communicable Diseases, IRSHA, for their constant support.

Conflict of Interest The authors declare that they have no competing interests.

References

Acosta EG, Bartenschlager R (2016) Paradoxical role of antibodies in dengue virus infections: considerations for prophylactic vaccine development. Expert Rev Vaccines 15(4):467–482

Alvarez M, Rodriguez-Roche R, Bernardo L, Vazquez S, Morier L, Gonzalez D, Castro O, Kouri G, Halstead SB, Guzman MG (2006) Dengue hemorrhagic fever caused by sequential dengue 1–3 virus infections over a long time interval: Havana epidemic, 2001–2002. Am J Trop Med Hyg 75:1113–1117

Appanna R, Wang SM, Ponnampalavanar SA, Lum LC, Sekaran SD (2012) Cytokine factors present in dengue patient sera induces alterations of junctional proteins in human endothelial cells. Am J Trop Med Hyg 87:936–942

Ayala-Nunez NV, Hoornweg TE, van de Pol DP, Sjollema KA, Flipse J, van der Schaar HM, Smit JM (2016) How antibodies alter the cell entry pathway of dengue virus particles in macrophages. Sci Rep 6:28768

Balsitis SJ, Williams KL, Lachica R, Flores D, Kyle JL, Mehlhop E, Johnson S, Diamond MS, Beatty PB, Harris E (2010) Lethal antibody enhancement of dengue disease in mice is prevented by Fc modification. PLoS Pathog 6(2):e1000790

Banhegyi D, Bacsi A, Toth FD, Prohaszka Z, Horvath A, Beck Z, Konya J, Fust G (2003) Significant decrease of the enhancement/neutralization index in HIV patients during highly active antiretroviral therapy (HAART). Immunol Lett 89(1):25–30

Bardina SV, Bunduc P, Tripathi S, Duehr J, Frere JJ, Brown JA, Nachbagauer R, Foster GA, Krysztof D, Tortorella D et al (2017) Enhancement of Zika virus pathogenesis by preexisting antiflavivirus immunity. Science 356(6334):175–180

Beck Z, Prohászka Z, Füst G (2008) Traitors of the immune system-enhancing antibodies in HIV infection: their possible implication in HIV vaccine development. Vaccine 26(24):3078–3085

Beltramello M, Williams KL, Simmons CP, Macagno A, Simonelli L, Quyen NT, Sukupolvi-Petty S, Navarro-Sanchez E, Young PR, de Silva AM et al (2010) The human immune response to dengue virus is dominated by highly cross-reactive antibodies endowed with neutralizing and enhancing activity. Cell Host Microbe 8(3):271–283

Bhatt S, Gething PW, Brady OJ, Messina JP, Farlow AW, Moyes CL, Drake JM, Brownstein JS, Hoen AG, Sankoh O (2013) The global distribution and burden of dengue. Nature 496(7446):504–507

Boonnak K, Slike BM, Donofrio GC, Marovich MA (2013) Human FcγRII cytoplasmic domains differentially influence antibody-mediated dengue virus infection. J Immunol 190(11):5659–5665
Braciale TJ, Hahn YS, Burton DR (2013) Adaptive immune response to viral infections. In: Knipe DM, Howley PM (eds) Fields Virology, 6th edn. Lippincott Williams and Wilkins, Philadelphia, pp 214–253
Capeding MR, Tran NH, Hadinegoro SR, Ismail HI, Chotpitayasunondh T, Chua MN, Luong CQ, Rusmil K, Wirawan DN, Nallusamy R et al (2014) Clinical efficacy and safety of a novel tetravalent dengue vaccine in healthy children in Asia: a phase 3, randomised, observer-masked, placebo-controlled trial. Lancet 384:1358–1365
Cardosa MJ (1987) Dengue virus isolation by antibody-dependent enhancement of infectivity in macrophages. Lancet 329(8526):193–194
Cardosa MJ, Porterfield JS, Gordon S (1983) Complement receptor mediates enhanced flavivirus replication in macrophages. J Exp Med 158:258–263
Carro AC, Piccini LE, Damonte EB (2018) Blockade of dengue virus entry into myeloid cells by endocytic inhibitors in the presence or absence of antibodies. PLoS Negl Trop Dis 12(8):e0006685
Castilla V, Piccini LE, Damonte EB (2015) Dengue virus entry and trafficking: perspectives as antiviral target for prevention and therapy. Future Virol 10:625–645
Chan KR, Ong EZ, Ooi EE (2013) Therapeutic antibodies as a treatment option for dengue fever. Expert Rev Anti-Infect Ther 11(11):1147–1157
Chan KR, Ong EZ, Tan HC, Zhang SL, Zhang Q, Tang KF, Kaliaperumal N, Lim AP, Hibberd ML, Chan SH et al (2014) Leukocyte immunoglobulin-like receptor B1 is critical for antibody-dependent dengue. Proc Natl Acad Sci U S A 111(7):2722–2727
Chan KR, Wang X, Saron WAA, Gan ES, Tan HC, Mok DZ, Zhang SL, Lee YH, Liang C, Wijaya L et al (2016) Cross-reactive antibodies enhance live attenuated virus infection for increased immunogenicity. Nat Microbiol 1:16164
Chan-Hui PY, Swiderek KM (2016) Immunological considerations for developing antibody therapeutics for Influenza A. Hum Vaccin Immunother 12(2):474–477
Chanock RM, Kapikian AZ, Mills J, Kim HW, Parrott RH (1970) Influence of immunological factors in respiratory syncytial virus disease. Arch Environ Health 21(3):347–355
Chareonsirisuthigul T, Kalayanarooj S, Ubol S (2007) Dengue virus (DENV) antibody-dependent enhancement of infection upregulates the production of anti-inflammatory cytokines, but suppresses anti-DENV free radical and pro-inflammatory cytokine production, in THP-1 cells. J Gen Virol 88:365–375
Chaturvedi UC, Agarwal R, Elbishbishi EA, Mustafa AS (2000) Cytokine cascade in dengue hemorrhagic fever: implications for pathogenesis. FEMS Immunol Med Microbiol 28:183–188
Chau TN, Quyen NT, Thuy TT, Tuan NM, Hoang DM, Dung NT, Lien le B, Quy NT, Hieu NT, Hieu LT et al (2008) Dengue in Vietnamese infants—results of infection enhancement assays correlate with age-related disease epidemiology, and cellular immune responses correlate with disease severity. J Infect Dis 198:516–524
Chau TN, Hieu NT, Anders KL, Wolbers M, Lien le B, Hieu LT, Hien TT, Hung NT, Farrar J, Whitehead S et al (2009) Dengue virus infections and maternal antibody decay in a prospective birth cohort study of Vietnamese infants. J Infect Dis 200:1893–1900
Chawla T, Chan KR, Zhang SL, Tan HC, Lim APC, Hanson BJ, Ooi EE (2013) Dengue virus neutralization in cells expressing fc gamma receptors. PLoS One 8(5):e65231
Chotiwan N, Roehrig JT, Schlesinger JJ, Blair CD (2014) Molecular determinants of dengue virus 2 envelope protein important for virus entry in FcγRIIA-mediated antibody-dependent enhancement of infection. Virology 456–457:238–246
Connor RI, Dinces NB, Howell AL, Romet-Lemonne J-L, Pasquali J-L, Fanger MW (1991) Fc receptors for IgG (FcγRs) on human monocytes and macrophages are not infectivity receptors for human immunodeficiency virus type 1 (HIV-1): studies using bispecific antibodies to target HIV-1 to various myeloid cell surface molecules, including the FcyR. Proc Natl Acad Sci U S A 88:9593–9597

Crill WD, Chang G-JJ (2004) Localization and characterization of flavivirus envelope glycoprotein cross-reactive epitopes. J Virol 78(24):13975–13986

Crill WD, Hughes HR, Trainor NB, Davis BS, Whitnet MT, Chang G-JJ (2012) Sculpting humoral immunity through dengue vaccination to enhance protective immunity. Front Immunol 3:334

Cruz-Oliveira C, Freire JM, Conceicão TM, Higa LM, Castanho MA, Da Poian AT (2015) Receptors and routes of dengue virus entry into the host cells. FEMS Microbiol Rev 39:155–170

Davis D, Trischmann H, Stephens DM, Lachmann PJ (2001) Antibodies raised to short synthetic peptides with sequences derived from HIV-1 SF2 gp120 can both neutralize and enhance HIV-1 SF13: a later variant isolated from the same host. J Med Virol 64(3):207–216

Dejnirattisai W, Jumnainsong A, Onsirisakul N, Fitton P, Vasanawathana S, Limpitikul W, Puttikhunt C, Edwards C, Duangchinda T, Supasa S et al (2010) Cross-reacting antibodies enhance dengue virus infection in humans. Science 328:745–748

Dejnirattisai W, Wongwiwat W, Supasa S, Zhang X, Dai X, Rouvinski A, Jumnainsong A, Edwards C, Quyen NTH, Duangchinda T et al (2015) A new class of highly potent, broadly neutralizing antibodies isolated from viremic patients infected with dengue virus. Nat Immunol 16:170–177

Dewi BE, Takasaki T, Kurane I (2004) In vitro assessment of human endothelial cell permeability: effects of inflammatory cytokines and dengue virus infection. J Virol Methods 121:171–180

Dhama K, Karthik K, Khandia R, Chakraborty S, Munjal A, Latheef SK, Kumar D, Ramakrishnan MA, Malik YS, Singh R et al (2018) Advances in designing and developing vaccines, drugs, and therapies to counter Ebola virus. Front Immunol 9:1803

Dowd KA, Jost CA, Durbin AP, Whitehead SS, Pierson TC (2011) A dynamic landscape for antibody binding modulates antibody-mediated neutralization of West Nile virus. PLoS Pathog 7:e1002111

Feldmann H, Geisbert TW (2011) Ebola haemorrhagic fever. Lancet 377(9768):849–862

Fibriansah G, Tan JL, Smith SA, de Alwis R, Ng TS, Kostyuchenko VA, Jadi RS, Kukkaro P, de Silva AM, Crowe JE et al (2015) A highly potent human antibody neutralizes dengue virus serotype 3 by binding across three surface proteins. Nat Commun 6:6341

Flipse J, Diosa-Toro MA, Hoornweg TE, van de Pol DP, Urcuqui-Inchima S, Smit JM (2016) Antibody-dependent enhancement of dengue virus infection in primary human macrophages; balancing higher fusion against antiviral responses. Sci Rep 6:29201

Frei JC, Wirchnianski AS, Govero J, Vergnolle O, Dowd KA, Pierson TC, Kielian M, Girvin ME, Diamond MS, Lai JR (2018) Engineered dengue virus domain III proteins elicit cross-neutralizing antibody responses in mice. J Virol 92(18):e01023–e01018

Furuyama W, Marzi A, Carmody AB, Maruyama J, Kuroda M, Miyamoto H, Nanbo A, Manzoor R, Yoshida R, Igarashi M et al (2016) Fcγ-receptor IIa-mediated Src signaling pathway is essential for the antibody-dependent enhancement of Ebola virus infection. PLoS Pathog 12(12):e1006139

Fust G, Toth FD, Kiss J, Ujhelyi E, Nagy I, Banhegyi D (1994) Neutralizing and enhancing antibodies measured in complement-restored serum samples from HIV-1-infected individuals correlate with immunosuppression and disease. AIDS 8(5):603–609

Gan ES, Ting DH, Chan KR (2017) The mechanistic role of antibodies to dengue virus in protection and disease pathogenesis. Expert Rev Anti-Infect Ther 15(2):111–119

Gauger PC, Vincent AL, Loving CL, Lager KM, Janke BH, Kehrli ME Jr, Roth JA (2011) Enhanced pneumonia and disease in pigs vaccinated with an inactivated human-like (delta-cluster) H1N2 vaccine and challenged with pandemic 2009 H1N1 influenza virus. Vaccine 29(15):2712–2719

George J, Valiant WG, Mattapallil MJ, Walker M, Huang Y-JS, Vanlandingham DL, Misamore J, Greenhouse J, Weiss DE, Verthelyi D et al (2017) Prior exposure to Zika virus significantly enhances peak dengue-2 viremia in rhesus macaques. Sci Rep 7:10498

Ghosn J, Taiwo B, Seedat S, Autran B, Katlama C (2018) HIV. Lancet 392(10148):685–697

Gibbons RV, Kalanarooj S, Jarman RG, Nisalak A, Vaughn DW, Endy TP, Mammen MP Jr, Srikiatkhachorn A (2007) Analysis of repeat hospital admissions for dengue to estimate the frequency of third or fourth dengue infections resulting in admissions and dengue hemorrhagic fever, and serotype sequences. Am J Trop Med Hyg 77:910–913

Gil L, Izquierdo A, Lazo L, Valdes I, Ambala P, Ochola L, Marcos E, Suzarte E, Kariuki T, Guzmán G et al (2014) Capsid protein: evidences about the partial protective role of neutralizing antibody independent immunity against dengue in monkeys. Virology 456–457:70–76

Gil L, Lazo L, Valdés I, Suzarte E, Yen P, Ramírez R, Álvarez M, Dung LT, Cobas K, Marcos E et al (2017) The tetravalent formulation of domain III-capsid proteins recalls memory B- and T-cell responses induced in monkeys by an experimental dengue virus infection. Clin Transl Immunology 6(6):e148

Gilbert PB, Peterson ML, Follmann D, Hudgens MG, Francis DP, Gurwith M, Heyward WL, Jobes DV, Popovic V, Self SG et al (2005) Correlation between immunologic responses to a recombinant glycoprotein 120 vaccine and incidence of HIV-1 infection in a phase 3 HIV-1 preventive vaccine trial. J Infect Dis 191:666–677

Gimenez HB, Keir HM, Cash P (1989) In vitro enhancement of respiratory syncytial virus infection of U937 cells by human sera. J GenVirol 70:89–96

Gimenez HB, Chisholm S, Dornan J, Cash P (1996) Neutralizing and enhancing activities of human respiratory syncytial virus-specific antibodies. Clin Diagn Lab Immunol 3:280–286

Glezen WP, Taber LH, Frank AL, Kasel JA (1986) Risk of primary infection and reinfection with respiratory syncytial virus. Am J Dis Child 140:543–546

Gomez RS, Ramirez BA, Cespedes PF, Cautivo KM, Riquelme SA, Prado CE, Gonzalez PA, Kalergis AM (2016) Contribution of Fcgamma receptors to human respiratory syncytial virus pathogenesis and the impairment of T-cell activation by dendritic cells. Immunology 147:55–72

Goncalvez AP, Engle RE, St Claire M, Purcell RH, Lai CJ (2007) Monoclonal antibody-mediated enhancement of dengue virus infection in vitro and in vivo and strategies for prevention. Proc Natl Acad Sci U S A 104:9422–9427

Gotoff R, Tamura M, Janus J, Thompson J, Wright P, Ennis FA (1994) Primary influenza A virus infection induces cross-reactive antibodies that enhance uptake of virus into Fc receptor/bearing cells. J Infect Dis 169:200–203

Gould EA, Buckley A (1989) Antibody-dependent enhancement of yellow fever and Japanese encephalitis virus neurovirulence. J Gen Virol 70(Pt 6):1605–1608

Graham RR, Juffrie M, Tan R, Hayes CG, Laksono I, Ma'roef C, Erlin, Sutaryo, Porter KR, Halstead SB (1999) A prospective seroepidemiologic study on dengue in children four to nine years of age in Yogyakarta, Indonesia I. studies in 1995–1996. Am J Trop Med Hyg 61:412–419

Guilliams M, Bruhns P, Saeys Y, Hammad H, Lambrecht BN (2014) The function of Fcγ receptors in dendritic cells and macrophages. Nat Rev Immunol 14:94–108

Guillon C, Schutten M, Boers PH, Gruters RA, Osterhaus AD (2002) Antibody-mediated enhancement of human immunodeficiency virus type 1 infectivity is determined by the structure of gp120 and depends on modulation of the gp120-CCR5 interaction. J Virol 76:2827–2834

Guzman MG, Kouri GP, Bravo J, Soler M, Vazquez S, Morier L (1990) Dengue hemorrhagic fever in Cuba, 1981: a retrospective seroepidemiologic study. Am J Trop Med Hyg 42:179–184

Guzman MG, Kouri G, Valdes L, Bravo J, Alvarez M, Vazquez S, Delgado I, Halstead SB (2000) Epidemiologic studies on dengue in Santiago de Cuba, 1997. Am J Epidemiol 152:793–799

Guzman MG, Alvarez M, Halstead SB (2013) Secondary infection as a risk factor for dengue hemorrhagic fever/dengue shock syndrome: an historical perspective and role of antibody-dependent enhancement of infection. Arch Virol 158(7):1445–1459

Hadinegoro SR, Arredondo-Garcia JL, Capeding MR, Deseda C, Chotpitayasunondh T, Dietze R, Muhammad Ismail HI, Reynales H, Limkittikul K, Rivera-Medina DM et al (2015) Efficacy and long-term safety of dengue vaccine in regions of endemic disease. N Engl J Med 373:1195–1206

Hall CB, Weinberg GA, Iwane MK, Blumkin AK, Edwards KM, Staat MA, Auinger P, Griffin MR, Poehling KA, Erdman D, Grijalva CG, Zhu Y, Szilagyi P (2009) The burden of respiratory syncytial virus infection in young children. N Engl J Med 360:588–598

Hallengard D, Lum FM, Kümmerer BM, Lulla A, Lulla V, García-Arriaza J, Fazakerley JK, Roques P, Le Grand R, Merits A et al (2014) Prime-boost immunization strategies against Chikungunya virus. J Virol 88(22):13333–13343

Halstead SB (1979) In vivo enhancement of dengue virus infection in rhesus monkeys by passively transferred antibody. J Infect Dis 140:527–533

Halstead SB (2003) Neutralization and antibody-dependent enhancement of dengue viruses. Adv Virus Res 60:421–467

Halstead SB (2008) Pathogenesis: risk factors prior to infection. In: Halstead SB (ed) Dengue. Imperial College Press, London, pp 219–256

Halstead SB, Nimmannitya S, Yamarat C, Russell PK (1967) Hemorrhagic fever in Thailand: recent knowledge regarding etiology. Jpn J Med Sci Biol 20:96–103

Halstead SB, Chow J, Marchette NJ (1973a) Immunologic enhancement of dengue virus replication. Nat New Biol 243:24–26

Halstead SB, Shotwell H, Casals J (1973b) Studies on the pathogenesis of dengue infection in monkeys. II. Clinical laboratory responses to heterologous infection. J Infect Dis 128:15–22

Hawkes RA (1964) Enhancement of the infectivity of arboviruses by specific antisera produced in domestic fowls. Aust J Exp Biol Med Sci 42:465–482

Hawkes RA, Lafferty KJ (1967) The enhancement of virus infectivity by antibody. Virology 33:250–261

Hertz T, Beatty PR, MacMillen Z, Killingbeck SS, Wang C, Harris E (2017) Antibody epitopes identified in critical regions of dengue virus nonstructural 1 protein in mouse vaccination and natural human infections. J Immunol 198(10):4025–4035

Homsy J, Meyer M, Tateno M, Clarkson S, Levy JA (1989) The Fc and not CD4 receptor mediates antibody enhancement of HIV infection in human cells. Science 244(4910):1357–1360

Homsy J, Meyer M, Levy JA (1990) Serum enhancement of human immunodeficiency virus (HIV) infection correlates with disease in HIV-infected individuals. J Virol 64:1437–1440

Huisman W, Martina BEE, Rimmelzwaan GF, Gruters RA, Osterhaus ADME (2009) Vaccine-induced enhancement of viral infections. Vaccine 27:505–512

Injampa S, Muenngern N, Pipattanaboon C, Benjathummarak S, Boonha K, Hananantachai H, Wongwit W, Ramasoota P, Pitaksajjakul P (2017) Generation and characterization of cross neutralizing human monoclonal antibody against 4 serotypes of dengue virus without enhancing activity. Peer J 5:e4021

Janoff EN, Wahl SM, Thomas K, Smith PD (1995) Modulation of human immunodeficiency virus type 1 infection of human monocytes by IgA. J Infect Dis 172:855–858

Jaume M, Yip MS, Cheung CY, Leung HL, Li PH, Kien F, Dutry I, Callendret B, Escriou N, Altmeyer R et al (2011) Anti-severe acute respiratory syndrome coronavirus spike antibodies trigger infection of human immune cells via a pH- and cysteine protease-independent FcγR pathway. J Virol 85:10582–10597

Junjhon J, Edwards TJ, Utaipat U, Bowman VD, Holdaway HA, Zhang W, Keelapang P, Puttikhunt C, Perera R, Chipman PR et al (2010) Influence of pr-M cleavage on the heterogeneity of extracellular dengue virus particles. J Virol 84(16):8353–8358

Kam YW, Kien F, Roberts A, Cheung YC, Lamirande EW, Vogel L, Chu SL, Tse J, Guarner J, Zaki SR et al (2007) Antibodies against trimeric S glycoprotein protect hamsters against SARS-CoV challenge despite their capacity to mediate FcgammaRII-dependent entry into B cells in vitro. Vaccine 25:729–740

Kam YW, Ong EK, Rénia L, Tong JC, Ng LF (2009) Immuno-biology of chikungunya and implications for disease intervention. Microbes Infect 11(14–15):1186–1196

Katzelnick LC, Harris E (2017) Immune correlates of protection for dengue: state of the art and research agenda. Vaccine 35:4659–4669

Katzelnick LC, Gresh L, Halloran ME, Mercado JC, Kuan G, Gordon A, Balmaseda A, Harris E (2017) Antibody-dependent enhancement of severe dengue disease in humans. Science 358(6365):929–932

Khurana S, Loving CL, Manischewitz J, King LR, Gauger PC, Henningson J, Vincent AL, Golding H (2013) Vaccine-induced anti-HA2 antibodies promote virus fusion and enhance influenza virus respiratory disease. Sci Transl Med 5(200):200ra114

Kim HW, Canchola JG, Brandt CD, Pyles G, Chanock RM, Jensen K, Parrott RH (1969) Respiratory syncytial virus disease in infants despite prior administration of antigenic inactivated vaccine. Am J Epidemiol 89:422–434
Klasse PJ, Sattentau QJ (2002) Occupancy and mechanism in antibody-mediated neutralization of animal viruses. J Gen Virol 83:2091–2108
Kliks S (1990) Antibody-enhanced infection of monocytes as the pathogenetic mechanism for severe dengue illness. AIDS Res Hum Retrovir 6(8):993–998
Kliks SC, Nimmanitya S, Nisalak A, Burke DS (1988) Evidence that maternal dengue antibodies are important in the development of dengue hemorrhagic fever in infants. Am J Trop Med Hyg 38:411–419
Kozlowski PA, Black KP, Shen L, Jackson S (1995) High prevalence of serum IgA HIV-1 infection-enhancing antibodies in HIV-infected persons. Masking by IgG. J Immunol 154:6163–6173
Krilov LR, Anderson LJ, Marcoux L, Bonagura VR, Wedgwood JF (1989) Antibody-mediated enhancement of respiratory syncytial virus infection in two monocyte/macrophage cell lines. J Infect Dis 160:777–782
Kyriakis CS, Gramer MR, Barbe F, Van Doorsselaere J, Van Reeth K (2010) Efficacy of commercial swine influenza vaccines against challenge with a recent European H1N1 field isolate. Vet Microbiol 144(1–2):67–74
Laurence J, Saunders A, Early E, Salmon JE (1990) Human immunodeficiency virus infection of monocytes: relationship to Fc-gamma receptors and antibody-dependent viral enhancement. Immunology 70:338–343
Lee N, Chan PK, Ip M, Wong E, Ho J, Ho C, Cockram CS, Hui DS (2006) Anti-SARS-CoV IgG response in relation to disease severity of severe acute respiratory syndrome. J Clin Virol 35:179–184
Li L, Lok SM, Yu I-M, Zhang Y, Kuhn RJ, Chen J, Rossmann MG (2008) The flavivirus precursor membrane-envelope protein complex: structure and maturation. Science 319(5871):1830–1834
Li M, Zhao L, Zhang C, Wang X, Hong W, Sun J, Liu R, Yu L, Wang J, Zhang F et al (2018) Dengue immune sera enhance Zika virus infection in human peripheral blood monocytes through Fc gamma receptors. PLoS One 13(7):e0200478
Lidbury BA, Mahalingam S (2000) Specific ablation of antiviral gene expression in macrophages by antibody-dependent enhancement of Ross River virus infection. J Virol 74:8376–8381
Lindenbach BD, Murray CL, Thiel H-J, Rice CM (2013) Flaviviridae. In: Knipe DM, Howley PM (eds) Fields virology, 6th edn. Lippincott Williams and Wilkins, Philadelphia, pp 712–746
Lu J, Wang R, Xia B, Yu Y, Zhou X, Yang Z, Huang P (2018) Potent neutralization ability of a human monoclonal antibody against serotype 1 dengue virus. Front Microbiol 9:1214
Lum FM, Couderc T, Chia BS, Ong RY, Her Z, Chow A, Leo YS, Kam YW, Rénia L, Lecuit M et al (2018) Antibody-mediated enhancement aggravates chikungunya virus infection and disease severity. Sci Rep 8(1):1860
Mahalingam S, Herring BL, Halstead SB (2013) Call to action for dengue vaccine failure. Emerg Infect Dis 19:1335–1337
Marchette NJ, Halstead SB, Falkler WA Jr, Stenhouse A, Nash D (1973) Studies on the pathogenesis of dengue infection in monkeys. III. Sequential distribution of virus in primary and heterologous infections. J Infect Dis 128:23–30
Matsui K, Gromowski GD, Li L, Barrett ADT (2010) Characterization of a dengue type-specific epitope on dengue 3 virus envelope protein domain III. J Gen Virol 91(Pt 9):2249–2253
McArthur MA, Sztein MB, Edelman R (2013) Dengue vaccines: recent developments, ongoing challenges and current candidates. Expert Rev Vaccines 12(8):933–953
Mejias A, Chavez-Bueno S, Rios AM, Saavedra-Lozano J, Fonseca Aten M, Hatfield J, Kapur P, Gómez AM, Jafri HS, Ramilo O (2004) Anti-respiratory syncytial virus (RSV) neutralizing antibody decreases lung inflammation, airway obstruction, and airway hyperresponsiveness in a murine RSV model. Antimicrob Agents Chemother 48:1811–1822
Meyer K, Ait-Goughoulte M, Keck Z-Y, Foung S, Ray R (2008) Antibody-dependent enhancement of hepatitis C virus infection. J Virol 82(5):2140–2149

Modhiran N, Kalayanarooj S, Ubol S (2010) Subversion of innate defenses by the interplay between DENV and pre-existing enhancing antibodies: TLRs signaling collapse. PLoS Negl Trop Dis 4(12):e924

Modis Y, Ogata S, Clements D, Harrison SC (2004) Structure of the dengue virus envelope protein after membrane fusion. Nature 427:313–319

Moi ML, Takasaki T, Saijo M, Kurane I (2013) Dengue virus infection enhancing activity of undiluted sera obtained from patients with secondary dengue virus infection. Trans R Soc Trop Med Hyg 107(1):51–58

Monsalvo AC, Batalle JP, Lopez MF, Krause JC, Klemenc J, Hernandez JZ, Maskin B, Bugna J, Rubinstein C, Aguilar L et al (2011) Severe pandemic 2009 H1N1 influenza disease due to pathogenic immune complexes. Nat Med 17:195–199

Montefiori DC, Lefkowitz LB, Keller RE, Holmberg V, Sandtstrom E, Phair JP, Multicenter AIDS Cohort Study (1991) Absence of a clinical correlation for complement-mediated, infection-enhancing antibodies in plasma and sera from HIV-infected persons. AIDS 5:513–517

Morens DM, Larsen LK, Halstead SB (1987) Study of the distribution of antibody-dependent enhancement determinants on dengue 2 isolates using dengue 2-derived monoclonal antibodies. J Med Virol 22:163–167

Ng JK, Zhang SL, Tan HC, Yan B, Martinez JM, Tan WY, Lam JH, Tan GK, Ooi EE, Alonso S (2014) First experimental in vivo model of enhanced dengue disease severity through maternally acquired heterotypic dengue antibodies. PLoS Pathog 10:e1004031

Nguyen TH, Lei HY, Nguyen TL, Lin YS, Huang KJ, Le BL, Lin CF, Yeh TM, Do QH, Vu TQ (2004) Dengue hemorrhagic fever in infants: a study of clinical and cytokine profiles. J Infect Dis 189:221–232

Nimmerjahn F, Lux A (2014) LILR-B1 blocks activating FcγR signaling to allow antibody dependent enhancement of dengue virus infection. Proc Natl Acad Sci U S A 111(7):2404–2405

Ochiai H, Kurokawa M, Hayashi K, Niwayama S (1988) Antibody mediated growth of influenza A NWS virus in macrophage-like cell line P388D1. J Virol 62:20–26

Ochiai H, Kurokawa M, Kuroki Y, Niwayama S (1990) Infection enhancement of influenza A H1 subtype viruses in macrophage-like P388D1 cells by cross-reactive antibodies. J Med Virol 30:258–265

Ochiai H, Kurokawa M, Matsui S, Yamamoto T, Kuroki Y, Kishimoto C, Shiraki K (1992) Infection enhancement of influenza A NWS virus in primary murine macrophages by anti-hemagglutinin monoclonal antibody. J Med Virol 36(3):217–221

Ong EZ, Zhang SL, Tan HC, Gan ES, Chan KR, Ooi EE (2017) Dengue virus compartmentalization during antibody-enhanced infection. Sci Rep 7:40923

Onyango-Makumbi C, Omer SB, Mubiru M, Moulton LH, Nakabiito C, Musoke P, Mmiro F, Zwerski S, Wigzell H, Falksveden L et al (2011) Safety and efficacy of HIV hyper immune globulin for prevention of mother-to-child HIV transmission in HIV-1-infected pregnant women and their infants in Kampala, Uganda (HIVIGLOB/NVP STUDY). J Acquir Immune Defic Syndr 58:399–407

Osiowy C, Horne D, Anderson R (1994) Antibody-dependent enhancement of respiratory syncytial virus infection by sera from young infants. Clin Diagn Lab Immunol 1:670–677

Pang T, Cardosa MJ, Guzman MG (2007) Of cascades and perfect storms: the immunopathogenesis of dengue haemorrhagic fever-dengue shock syndrome (DHF/DSS). Immunol Cell Biol 85(1):43–45

Ponnuraj EM, Hayward AR, Raj A, Wilson H, Simoes EAF (2001) Increased replication of respiratory syncytial virus (RSV) in pulmonary infiltrates is associated with enhanced histopathological disease in bonnet monkeys (Macaca radiata) pre-immunized with a formalin-inactivated RSV vaccine. J Gen Virol 82:2663–2674

Ponnuraj EM, Springer J, Hayward AR, Wilson H, Simoes EA (2003) Antibody dependent enhancement, a possible mechanism in augmented pulmonary disease of respiratory syncytial virus in the bonnet monkey model. J Infect Dis 187:1257–1263

Prohaszka Z, Nemes J, Hidvegi T, Toth FD, Kerekes K, Erdei A, Szabó J, Ujhelyi E, Thielens N, Dierich MP et al (1997) Two parallel routes of the complement-mediated antibody-dependent enhancement of HIV-1 infection. AIDS 11(8):949–958

Puerta-Guardo H, Raya-Sandino A, González-Mariscal L, Rosales VH, Ayala-Dávila J, Chávez-Mungía B, Martínez-Fong D, Medina F, Ludert JE, del Angel RM (2013) The cytokine response of U937-derived macrophages infected through antibody-dependent enhancement of dengue virus disrupts cell apical-junction complexes and increases vascular permeability. J Virol 87(13):7486–7501

Rajao DS, Loving CL, Gauger PC, Kitikoon P, Vincent AL (2014) Influenza A virus hemagglutinin protein subunit vaccine elicits vaccine-associated enhanced respiratory disease in pigs. Vaccine 32(40):5170–5176

Reeth KV, Brown I, Essen S, Pensaert M (2004) Genetic relationships, serological cross-reaction and cross-protection between H1N2 and other influenza A virus subtypes endemic in European pigs. Virus Res 103(1–2):115–124

Ricklin ME, Vielle NJ, Python S, Brechbühl D, Zumkehr B, Posthaus H, Zimmer G, Summerfield A (2016) Partial protection against porcine influenza A virus by a hemagglutinin-expressing virus replicon particle vaccine in the absence of neutralizing antibodies. Front Immunol 7:253

Robinson WE Jr (2006) Mechanism for complement-mediated, antibody dependent enhancement of human immunodeficiency virus type 1 infection in MT2 cells is enhanced entry through CD4, CD21, and CXCR4 chemokine receptors. Viral Immunol 19(3):434–447

Robinson WE Jr, Montefiori DC, Mitchell WM (1987) A human immunodeficiency virus type 1 (HIV-1) infection-enhancing factor in seropositive sera. Biochem Biophys Res Commun 149(2):693–699

Robinson WE Jr, Montefiori DC, Mitchell WM (1988) Antibody dependent enhancement of human immunodeficiency virus type 1 infection. Lancet 1(8589):790–794

Robinson WE Jr, Montefiori DC, Gillespie DH, Mitchell WM (1989) Complement mediated, antibody-dependent enhancement of HIV-1 infection in vitro is characterized by increased protein and RNA syntheses and infectious virus release. J Acquir Immune Defic Syndr 2:33–42

Robinson WE Jr, Kawamura T, Lake D, Masuho Y, Mitchell WM, Hersh EM (1990a) Antibodies to the primary immunodominant domain of human immunodeficiency virus type 1 (HIV-1) glycoprotein gp4l enhance HIV-1 infection in vitro. J Virol 64(11):5301–5305

Robinson WE Jr, Kawamura T, Gorny MK, Lake D, Xu JY, Matsumoto Y, Sugano T, Masuho Y, Mitchell WM, Hersh E et al (1990b) Human monoclonal antibodies to the human immunodeficiency virus type 1 (HIV-1) transmembrane glycoprotein gp41 enhance HIV-1 infection in vitro. Proc Natl Acad Sci U S A 87(8):3185–3189

Rodenhuis-Zybert IA, Moesker B, da Silva Voorham JM, van der Ende-Metselaar H, Diamond MS, Wilschut J, Smit JM (2011) A fusion-loop antibody enhances the infectious properties of immature flavivirus particles. J Virol 85:11800–11808

Rolph MS, Zaid A, Rulli NE, Mahalingam S (2011) Downregulation of interferon-β in antibody-dependent enhancement of dengue viral infections of human macrophages is dependent on interleukin-6. J Infect Dis 204(3):489–491

Rothman AL (2011) Immunity to dengue virus: a tale of original antigenic sin and tropical cytokine storms. Nat Rev Immunol 11:532–543

Sangkawibha N, Rojanasuphot S, Ahandrik S, Viriyapongse S, Jatanasen S, Salitul V, Phanthumachinda B, Halstead SB (1984) Risk factors in dengue shock syndrome: a prospective epidemiologic study in Rayong, Thailand. I. the 1980 outbreak. Am J Epidemiol 120:653–669

Schmidt AC (2010) Response to dengue fever - the good, the bad, and the ugly? N Engl J Med 363(5):484–487

Serafin IL, Aaskov JG (2001) Identification of epitopes on the envelope (E) protein of dengue 2 and dengue 3 viruses using monoclonal antibodies. Arch Virol 146(12):2469–2479

Shanks GD, Burroughs SA, Sohn JD, Waters NC, Smith VF, Waller M, Brundage JF (2016) Enhanced risk of illness during the 1918 influenza pandemic after previous influenza-like illnesses in three military populations. Epidemiol Infect 144(10):2043–2048

Shao W, Li X, Goraya ML, Wang S, Chen J-L (2017) Evolution of influenza A virus by mutation and re-assortment. Int J Mol Sci 18(8):E1650

Shi X, Deng Y, Wang H, Ji G, Tan W, Jiang T, Li X, Zhao H, Xia T, Meng Y et al (2016) A bispecific antibody effectively neutralizes all four serotypes of dengue virus by simultaneous blocking virus attachment and fusion. MAbs 8(3):574–584

Shmelkov E, Nadas A, Cardozo T (2014) Could vaccination with AIDSVAX immunogens have resulted in antibody dependent enhancement of HIV infection in human subjects? Hum Vaccin Immunother 10(10):3013–3016

Sholukh AM, Byrareddy SN, Shanmuganathan V, Hemashettar G, Lakhashe SK, Rasmussen RA, Watkins JD, Vyas HK, Thorat S, Brandstoetter T et al (2014) Passive immunization of macaques with polyclonal anti-SHIV IgG against a heterologous tier 2 SHIV: outcome depends on IgG dose. Retrovirology 11:8

Skowronski DM, De Serres G, Crowcroft NS, Janjua NZ, Boulianne N, Hottes TS, Rosella LC, Dickinson JA, Gilca R, Sethi P et al (2010) Association between the 2008–09 seasonal influenza vaccine and pandemic H1N1 illness during spring–summer 2009: four observational studies from Canada. PLoS Med 7(4):e1000258

Skowronski DM, Hamelin ME, De Serres G, Janjua NZ, Li G, Sabaiduc S, Bouhy X, Couture C, Leung A, Kobasa D (2014) Randomized controlled ferret study to assess the direct impact of 2008-09 trivalent inactivated influenza vaccine on A (H1N1)pdm09 disease risk. PLoS One 9:e86555

Smith SA, de Alwis AR, Kose N, Jadi RS, de Silva AM, Crowe JE (2014) Isolation of dengue virus-specific memory B cells with live virus antigen from human subjects following natural infection reveals the presence of diverse novel functional groups of antibody clones. J Virol 88(21):12233–12241

Sridhar S, Luedtke A, Langevin E, Zhu M, Bonaparte M, Machabert T, Savarino S, Zambrano B, Moureau A, Khromava A et al (2018) Effect of dengue serostatus on dengue vaccine safety and efficacy. N Engl J Med 379(4):327–340

Staprans SI, Barry AP, Silvestri G, Safrit JT, Kozyr N, Sumpter B, Nguyen H, McClure H, Montefiori D, Cohen JI et al (2004) Enhanced SIV replication and accelerated progression to AIDS in macaques primed to mount a CD4 T cell response to the SIV envelope protein. Proc Natl Acad Sci U S A 101(35):13026–13031

Subbramanian RA, Xu J, Toma E, Morisset R, Cohen EA, Menezes J, Ahmad A (2002) Comparison of human immunodeficiency virus (HIV)-specific infection-enhancing and -inhibiting antibodies in AIDS patients. J Clin Microbiol 40(6):2141–2146

Suhrbier A, La Linn M (2003) Suppression of antiviral responses by antibody-dependent enhancement of macrophage infection. Trends Immunol 24:165–168

Sullivan N, Sun Y, Binley J, Lee J, Barbas CF 3rd, Parren PW, Burton DR, Sodroski J (1998) Determinants of human immunodeficiency virus type 1 envelope glycoprotein activation by soluble CD4 and monoclonal antibodies. J Virol 72:6332–6338

Sun H, Chen Q, Lai H (2018) Development of antibody therapeutics against flaviviruses. Int J Mol Sci 19(1):E54

Szabo J, Prohaszka Z, Toth FD, Gyuris A, Segesdi J, Banhegyi D, Ujhelyi E, Minarovits J, Fust G (1999) Strong correlation between the complement-mediated antibody-dependent enhancement of HIV-1 infection and plasma viral load. AIDS 13:1841–1849

Takada A, Kawaoka Y (2003) Antibody-dependent enhancement of viral infection: molecular mechanisms and in vivo implications. Rev Med Virol 13(6):387–398

Takada A, Watanabe S, Okazaki K, Kida H, Kawaoka Y (2001) Infectivity-enhancing antibodies to Ebola virus glycoprotein. J Virol 75(5):2324–2330

Takada A, Feldmann H, Ksiazek TG, Kawaoka Y (2003) Antibody-dependent enhancement of Ebola virus infection. J Virol 77(13):7539–7544

Takada A, Ebihara H, Feldmann H, Geisbert TW, Kawaoka Y (2007) Epitopes required for antibody-dependent enhancement of Ebola virus infection. J Infect Dis 196(Suppl 2):S347–S356

Takeda A, Tuazon CU, Ennis FA (1988) Antibody-enhanced infection by HIV-1 via Fc receptor-mediated entry. Science 242(4878):580–583

Takeda A, Sweet RW, Ennis FA (1990) Two receptors are required for antibody-dependent enhancement of human immunodeficiency virus type 1 infection: CD4 and FcγR. J Virol 64(11):5605–5610

Tamura M, Webster RG, Ennis FA (1991) Antibodies to HA and NA augment uptake of influenza A viruses into cells via Fc receptor entry. Virology 182:211–219

Tamura M, Webster RG, Ennis FA (1994) Subtype cross-reactive, infection-enhancing antibody responses to influenza A viruses. J Virol 68(6):3499–3504

Tang C-T, Li P-C, Liu I-J, Liao M-Y, Chiu CY, Chao D-Y, Wu HC (2015) An epitope-substituted DNA vaccine improves safety and immunogenicity against dengue virus type 2. PLoS Negl Trop Dis 9(7):e0003903

Taylor A, Foo SS, Bruzzone R, Dinh LV, King NJ, Mahalingam S (2015) Fc receptors in antibody-dependent enhancement of viral infections. Immunol Rev 268(1):340–364

TI-RS Group (1998) Palivizumab, a humanized respiratory syncytial virus monoclonal antibody, reduces hospitalization from respiratory syncytial virus infection in high-risk infants. Pediatrics 102(3 Pt 1):531–537

To KK, Zhang AJ, Hung IF, Xu T, Ip WC, Wong RT, Ng JC, Chan JF, Chan KH, Yuen KY (2012) High titer and avidity of nonneutralizing antibodies against influenza vaccine antigen are associated with severe influenza. Clin Vaccine Immunol 19(7):1012–1018

Toth FD, Szabo B, Ujhelyi E, Paloczi K, Horvath A, Fust G, Kiss J, Bánhegyi D, Hollán SR (1991) Neutralizing and complement-dependent enhancing antibodies in different stages of HIV infection. AIDS 5(3):263–268

Tóth FD, Mosborg-Petersen P, Kiss J, Aboagye-Mathiesen G, Zdravkovic M, Hager H, Aranyosi J, Lampé L, Ebbesen P (1994) Antibody-dependent enhancement of HIV-1 infection in human term syncytiotrophoblast cells cultured in vitro. Clin Exp Immunol 96(3):389–394

Trischmann H, Davis D, Lachmann PJ (1995) Lymphocytotropic strains of HIV type 1 when complexed with enhancing antibodies can infect macrophages via Fc gamma RIII. AIDS Res Hum Retrovir 11(3):343–352

Tsai T-T, Chuang Y-J, Lin Y-S, Chang C-P, Wan S-W, Lin S-H, Chen C-L, Lin C-F (2014) Antibody-dependent enhancement infection facilitates dengue virus-regulated signaling of IL-10 production in monocytes. PLoS Negl Trop Dis 8(11):e3320

Ubol S, Phuklia W, Kalayanarooj S, Modhiran N (2010) Mechanisms of immune evasion induced by a complex of dengue virus and preexisting enhancing antibodies. J Infect Dis 201:923–935

van de Winkel JG, Anderson CL (1991) Biology of human immunoglobulin G Fc receptors. J Leukoc Biol 49(5):511–524

van Erp EA, van Kasteren PB, Guichelaar T, Ahout IML, de Haan CAM, Luytjes W, Ferwerda G, Wicht O (2017) In vitro enhancement of respiratory syncytial virus infection by maternal antibodies does not explain disease severity in infants. J Virol 91(21):e00851–e00817

van Mechelen L, Luytjes W, de Haan CA, Wicht O (2016) RSV neutralization by palivizumab, but not by monoclonal antibodies targeting other epitopes, is augmented by Fc gamma receptors. Antivir Res 132:1–5

Vaughn DW, Green S, Kalayanarooj S, Innis BL, Nimmannitya S, Suntayakorn S, Endy TP, Raengsakulrach B, Rothman AL, Ennis FA et al (2000) Dengue viraemia titer, antibody response pattern, and virus serotype correlate with disease severity. J Infect Dis 181(1):2–9

Villar L, Dayan GH, Arredondo-Garcia JL, Rivera DM, Cunha R, Deseda C, Reynales H, Costa MS, Morales-Ramírez JO, Carrasquilla G et al (2015) Efficacy of a tetravalent dengue vaccine in children in Latin America. N Engl J Med 372:113–123

Vogelpoel LT, Baeten DL, de Jong EC, den Dunnen J (2015) Control of cytokine production by human Fc gamma receptors: implications for pathogen defense and autoimmunity. Front Immunol 6:79

Wahala WM, Silva AM (2011) The human antibody response to dengue virus infection. Viruses 3(12):2374–2395

Wang WK, Chen HL, Yang CF, Hsieh SC, Juan CC, Chang SM, Yu CC, Lin LH, Huang JH, King CC (2006) Slower rates of clearance of viral load and virus-containing immune complexes in patients with dengue hemorrhagic fever. Clin Infect Dis 43:1023–1030

Wang SF, Tseng SP, Yen CH, Yang JY, Tsao CH, Shen CW, Chen KH, Liu FT, Liu WT, Chen YM et al (2014) Antibody-dependent SARS coronavirus infection is mediated by antibodies against spike proteins. Biochem Biophys Res Commun 451:208–214

Wang Y, Si L, Luo Y, Guo X, Zhou J, Fang D, Yan H, Zeng G, Jiang L (2015) Replacement of pr gene with Japanese encephalitis virus pr using reverse genetics reduces antibody-dependent enhancement of dengue virus 2 infection. Appl Microbiol Biotechnol 99(22):9685–9698

Wang Q, Zhang L, Kuwahara K, Li L, Liu Z, Li T, Zhu H, Liu J, Xu Y, Xie J et al (2016) Immunodominant SARS coronavirus epitopes in humans elicited both enhancing and neutralizing effects on infection in non-human primates. ACS Infect Dis 2(5):361–376

WHO (2009) Dengue: guidelines for diagnosis, treatment, prevention and control. WHO, Geneva. www.who.int/tdr/publications/documents/dengue-diagnosis.pdf

Wilder-Smith A, Hombach J, Ferguson N, Selgelid M, O'Brien K, Vannice K, Barrett A, Ferdinand E, Flasche S, Guzman M et al (2019) Deliberations of the strategic advisory group of experts on immunization on the use of CYD-TDV dengue vaccine. Lancet Infect Dis 19(1):e31–e38

Willey S, Aasa-Chapman MM, O'Farrell S, Pellegrino P, Williams I, Weiss RA, Neil SJ (2011) Extensive complement-dependent enhancement of HIV-1 by autologous non-neutralising antibodies at early stages of infection. Retrovirology 8:16

Xu M, Zuest R, Velumani S, Tukijan F, Toh YX, Appanna R, Tan EY, Cerny D, MacAry P, Wang C et al (2017) A potent neutralizing antibody with therapeutic potential against all four serotypes of dengue virus. NPJ Vaccin 2:2

Yip MS, Leung NH, Cheung CY, Li PH, Lee HH, Daëron M, Peiris JS, Bruzzone R, Jaume M (2014) Antibody-dependent infection of human macrophages by severe acute respiratory syndrome coronavirus. Virol J 11:82

Yip MS, Leung HL, Li PH, Cheung CY, Dutry I, Li D, Daëron M, Bruzzone R, Peiris JS, Jaume M (2016) Antibody-dependent enhancement of SARS coronavirus infection and its role in the pathogenesis of SARS. Hong Kong Med J 22(3 Suppl 4):25–31

Zellweger RM, Prestwood TR, Shresta S (2010) Enhanced infection of liver sinusoidal endothelial cells in a mouse model of antibody-induced severe dengue disease. Cell Host Microbe 7:128–139

Zhang L, Zhang F, Yu W, He T, Yu J, Yi CE, Ba L, Li W, Farzan M, Chen Z et al (2006) Antibody responses against SARS coronavirus are correlated with disease outcome of infected individuals. J Med Virol 78:1–8

Inflammation During Virus Infection: Swings and Roundabouts

3

Sankar Bhattacharyya

Abstract

Inflammation constitutes a concerted series of cellular and molecular responses that follow disturbance of systemic homeostasis, by either toxins or infectious organisms. Leukocytes modulate inflammation through production of secretory mediators, like cytokines and chemokines, which work in an autocrine and/or paracrine manner. These mediators can either promote or attenuate the inflammatory response and depending on differential temporal and spatial expression play a crucial role in the outcome of infection. Even though the objective is clearance of the pathogen with minimum damage to host, the pathogenesis of multiple human pathogenic viruses has been suggested to emanate from a dysregulation of the inflammatory response, sometimes with fatal consequences. This review discusses the nature and the outcome of inflammatory response, which is triggered in the human host subsequent to infection by single-sense plus-strand RNA viruses. In view of such harmful effects of a dysregulated inflammatory response, an exogenous regulation of these reactions by either interference or supplementation of critical regulators has been suggested. Currently multiple such factors are being tested for their beneficial and adverse effects. A successful use of such an approach in diseases of viral etiology can potentially protect the affected individual without directly affecting the virus life cycle. Further, such approaches whenever applicable would be useful in mitigating death and/or debility that is caused by the infection of those viruses which have proven particularly difficult to control by either prophylactic vaccines and/or therapeutic strategies using specific antiviral drugs.

S. Bhattacharyya (✉)
Vaccine and Infectious Disease Research Centre, Translational Health Science and Technology Institute, Faridabad, Haryana, India
e-mail: sankar@thsti.res.in

P. V. Bramhachari (ed.), *Dynamics of immune activation in viral diseases*,
https://doi.org/10.1007/978-981-15-1045-8_3

Keywords
Inflammation · RNA virus · Cytokine therapy

3.1 Introduction

The mammalian immune system has evolved arsenal and strategies to make a distinction between microbes that are either beneficial or benign or bad, an integral part of which is differential treatment of "self" and "non-self." Whereas recognition of "self" as "non-self" can cause autoimmunity, the converse results in microbial colonization. In fact the human gut does harbor multiple variety of microbes as natural part of the biological ecosystem (Scarpellini et al. 2015). The recognized non-self are counteracted by adaptive and innate effectors of the immune system, using dedicated cells and biochemicals, which attempt to restrict the growth and impede colonization by the pathogen. The innate response is nonspecific, while the secondary adaptive response is specific for the pathogen or closely related species. The cellular component includes innate immune cells like the monocytes/macrophages, neutrophils, and natural killer (NK) cells and adaptive immune cells like B- and T-lymphocytes, which coordinate for an effective response. Cytokines are a dedicated group of biochemicals involved in this coordination and include interferons (IFNs), interleukins (ILs), and chemokines that are responsible for synchronizing the initiation, regulation, and termination of an immune response. A group (~100) of small polypeptides (<20 kDa) produced predominantly although not exclusively by immune cells like macrophages and lymphocytes, cytokines are secreted out exerting their function by engaging respective cell-surface receptors and depending on biological function are labeled as either pro-inflammatory (PIC) or anti-inflammatory (AIC) cytokines (Turner et al. 2014). On the one hand, several cytokines are functionally redundant, and on the other hand, some cells can express receptors for multiple cytokines.

3.2 The Positive-Sense Single-Stranded RNA Viruses

Viruses with positive-sense single-stranded RNA as genome can either be enveloped (Togaviridae, Flaviviridae, and Coronaviridae) or non-enveloped (Astroviridae, Caliciviridae, and Picornaviridae), and several from either group cause severe human pathology (Fields et al. 2013). Entry into human host can be by diverse means including mucosal contact (gut in enteroviruses) or vectorial inoculation (e.g., in dengue and JEV) or parenteral blood transfer (e.g., hepatitis C virus). Immobilization by interaction with extracellular matrix components like glycosaminoglycan is followed by tropism determinant cognate receptor-mediated entry (Chen et al. 1997; Olenina et al. 2005; Tan et al. 2013). In enveloped viruses, the envelope fuses with the endosomal membrane, while non-enveloped viruses breach the membrane of either the cell or the endosome using specific cofactors, ultimately releasing viral genome into the host cytosol (Kumar et al. 2018; Plemper 2011). A culmination of the following steps results in direct translation of the genomic RNA to produce a polyprotein, which is cleaved by virus-derived and host-origin proteases to yield the multiple structural and nonstructural proteins (Fields et al. 2013). The structural features of the genomic RNA facilitating translation can be, e.g., a 5'cap and a poly-A tail (Alphavirus, Togaviridae; Coronavirus, Coronaviridae) or a 5'cap without a poly-A tail (Flavivirus, Flaviviridae) or an internal-ribosome entry site (IRES) serving for ribosome recruitment without a poly-A tail (hepatitis C virus) or an IRES with a poly-A tail (Picornaviridae, Astroviridae, Caliciviridae). After multiple rounds of translation, ribosome loading stops and the genomic RNA is replicated by virus-encoded RNA-dependent RNA polymerase (RdRp), in endoplasmic reticulum (ER) membrane-associated replication complexes (RCs) during which a double-stranded RNA intermediate is produced followed by its asymmetric transcription to produce multiple copies of plus-sense genomic RNA. The new genomic RNAs are packaged into virion particles that exit the cell by either secretory pathway or plasma-membrane budding (for enveloped viruses) or by cell lysis (for non-enveloped viruses) (Bird and Kirkegaard 2015; Pornillos et al. 2002).

3.3 Infection, Intimation, and Initiation of Inflammation

Although viruses can replicate in multiple types of cells, the pathological outcome manifests in only one or a few specific cell/tissue types. Independent of organismal entry site, the likeliest primary encounter of a virus is with mononuclear-phagocytic cells like monocytes, macrophages (Mϕ), and dendritic cells (DCs). Mϕ and DCs are tissue-localized cells constituting the first line of cellular defense, which when infected can undertake antiviral steps in addition to "informing" the other effectors of the innate and adaptive immune system (Pohl et al. 2007; Ginhoux and Jung 2014). Activated DCs shift to lymph nodes, process viral antigen, and "present" or display it for clonal expansion of cognate lymphocytes population. Mϕ, which can be either tissue-resident or differentiated from afferent monocytes postinfection, play a more regulatory role and are important determinants of the outcome of the inflammatory response (Ginhoux and Jung 2014; Mercer and Greber 2013; Hou et al. 2012; Schulz et al. 2012). Tissue-resident Mϕ , which are a distinct population from monocyte-derived ones, include microglial cells in CNS, liver Kupffer cells, skin Langerhans cells, etc. (Davies et al. 2013)

Monocytes/Mϕ (and many other cell types) express molecular detector proteins called pattern recognition receptors (PRRs), specialized for interacting with signature motifs on microbe-derived molecules, termed as pathogen-associated molecular pattern (PAMP). Viral PAMPs include double-stranded (dsRNA) RNA (replication-intermediate formed during replication) and 5′-ppp (uncapped genomic RNA polymerized by de novo replication). Cellular PRRs specific for these include toll-like receptors (TLRs) like TLR3 (dsRNA) and RIG-I-like receptors (RLRs) like RIG-I, MDA5 (dsRNA, 5′-ppp end on RNA) (Jensen and Thomsen 2012). Nod-like receptors or NLRs form another class of cytosolic PRRs that can detect virus infection, albeit in an indirect manner (Takeuchi and Akira 2010; Ichinohe et al. 2013). Physical engagement with PAMPs activates the respective PRRs, stimulating alterations in conformation of these sensors that allow them to interact with adapter molecules mediating the assembly of multi-protein complexes called inflammosome, in parallel to activating the expression of cytokine genes coding for type-1 interferons (IFNs) and NFκB target genes (Kawai et al. 2005; Pichlmair and Reis e Sousa 2007; Chen and Ichinohe 2015; Seth et al. 2005). Secreted type-I IFNs attach specific receptors, in a paracrine or autocrine manner, thereby activating the expression of many interferon-sensitive genes (ISGs) with diverse functions that confer antiviral property to their activity (Schneider et al. 2014; Schoggins and Rice 2011). ISGs include PRR-coding genes producing a feed-forward loop and aggravating inflammation. In parallel, NFκB enhances expression of pro-inflammatory genes like TNF-α, IL-1β, COX2, IL6, IL-12p40, or IL-12 besides components of NLRP3 (Tak and Firestein 2001; Bauernfeind et al. 2009). Upon assembly the NLRP3 inflammosome catalyzes caspase-1 activation, a protease which slices the precursor form of pleiotropic pro-inflammatory cytokines like IL-1β and IL-18 generating their active secreted forms (Garlanda et al. 2013; Biet et al. 2002). IL-1β potentiates the antiviral response by multiple ways in addition to inducing expression of type-I IFNs and ISGs in DCs (Ben-Sasson et al. 2011; Aarreberg et al. 2018). Chemokines (chemotactic cytokines) flag/point to the site of infection by a concentration gradient, attracting leukocytes like neutrophils, monocytes, and lymphocytes, subsequently activating them to release more cytokines thereby amplifying the inflammatory response (Sokol and Luster 2015; Ley 2014). Among these IL-12 and IL-2 (produced predominantly by DCs) have crucial immunomodulatory functions. IL-12 attracts CD4+ T-helper (Th) cells influencing their differentiation into IFN-γ secreting Th_1 cells in addition to augmenting the cytotoxic activity of CD8+ T cells and NK cells (Athie-Morales et al. 2004; Henry et al. 2008). IL-2 on the other hand increases NK-cell sensitivity to IL-12 by receptor upregulation (Wang et al. 2000). IFN-γ which in contrast to type-I IFNs is produced exclusively by immune cells (T and NK cells) has pleiotropic antiviral effect including the capacity to polarize existing or newly recruited Mϕ to M1 phenotype (Hu and Ivashkiv 2009; Verreck et al. 2004). Mϕ either resident or monocyte-derived can acquire either an M1 or an M2 phenotype differing in ontology, phenotype, and secretome, with unidirectional plasticity from M1 to M2 (Halstead et al. 2010; Guiducci et al. 2005; Smith et al. 2016). M1-Mϕ promotes a Th_1 immune response which is necessary for resolution of infection, while the M2-Mϕ endorses tissue repair following inflammation, suggesting that a premature skew in abundance of M2-Mϕ at the expense of M1-Mϕ would limit viral clearance leading to chronic infection and prolonged inflammatory response (Klenerman and Hill 2005). An emerging concept in modulation of inflammation involves the role of bacterial surface components like lipopolysaccharide on concurrent viral infection (Smith et al. 2016; Wilks and Golovkina 2012). Alterations in gut microbiome have been reported and potential influences this might have on disease outcome have been suggested (Preveden et al. 2017; Banks et al. 2015).

Though it is difficult to ascertain the number of asymptomatic infections for any given virus, the percentage of symptomatic infection vis-à-vis asymptomatic ones is often a multivariate variable, being known for only a few. For example, only 1 among 4 individuals infected with DENV shows febrile symptoms. This suggests a success for the antiviral immune mechanisms in the majority of individuals. Animal studies using gene knockout models have given evidence of this efficacy for many viruses (Suthar et al. 2010; Samuel and Diamond 2005; Lazear et al. 2011; Deonarain et al. 2004; Burdeinick-Kerr et al. 2007). In case of humans, these information are complicated by differential efficacy of these pathways, protecting or predisposing individuals under the influence of genotype, environment, etc. (Paalani et al. 2011; Mitchell and Aneshensel 2016; Liu and Taioli 2015) Besides, there are few studies that indicate potential influence of medication or noninfectious ailments or societal stress on the outcome of infection through an influence on the immune system (Mehrbod et al. 2014; Gilbert et al. 2005; Htun et al. 2015; Jean et al. 2007).

3.3.1 Liver Damage Due to Hepatitis C Virus and Dengue Virus Infection

HCV and DENV infection can cause liver damage through a chronic and acute infection regime, respectively (Samanta and Sharma 2015; Axley et al. 2018). Liver as an organ is characterized by a high capacity to regenerate; however, chronic injury/scarring can lead to fibrosis, steatosis, or even hepatocellular carcinoma resulting in liver failure (Forbes and Newsome 2016). Hepatocytes constitute two-thirds of all liver cells and are associated with all major liver functions besides playing a crucial role in innate immune signaling (Kmiec 2001; Zhou et al. 2016). Hepatocytes are permissible to both HCV and DENV, the latter being reported to additionally infect Kupffer cells (Chang et al. 2003; Zehender et al. 1997; Boisvert et al. 2001; Caussin-Schwemling et al. 2001; Goutagny et al. 2003; Marianneau et al. 1999; de Macedo et al. 2006; Huerre et al. 2001). In acute infection, the major damage is through apoptosis following direct infection of these cells, whereas establishment of a chronic infection usually causes a sustained inflammation leading to infiltration of polymorphonuclear cells and lymphocytes (Huerre et al. 2001; Lim et al. 2014; Masalova et al. 2017; Deng et al. 2008; Bala et al. 2012; Sung et al. 2012). Irrespective of the virus, these infections augment PIC levels in the liver with drastic consequences. For example, hepatocyte apoptosis caused by either direct infection or effect of PICs like TNF-α generates apoptotic bodies which when engulfed by Kupffer cells induce the latter to release more PIC providing a positive loop toward inflammation (Canbay et al. 2003a; Burdette et al. 2012; Negash et al. 2013; Shimizu et al. 2005). Cytokines like TGFβ and PDGF thus released can "activate" hepatic stellate cells initiating a metabolic transformation in them to secrete more extracellular matrix that deposits as fibrotic tissue in addition to converting them into smooth muscle fibers (Canbay et al. 2003b; Hernandez-Gea and Friedman 2011). In addition to virus infection-induced changes, bacterial LPS can also potentially "activate" hepatic stellate cells (Brun et al. 2005). HCV infection skews macrophage population to M2 phenotype restraining virus clearance while promoting hepatic stellate cell activation mediated by TGFβ (Saha et al. 2016). Additionally, in case of infection by both of these viruses, immune suppression mediated by AIC like IL10 is implicated for virus persistence and augmented pathology (MacDonald et al. 2002; Sugimoto et al. 2003). In fact higher levels of cytokines like IL10 and IL17 have shown positive correlation with liver damage (Fernando et al. 2016). Liver steatosis, a clinical feature common among HCV patients, is the result of intracellular ROS in hepatocytes (Okuda et al. 2002; Perlemuter et al. 2002). Irrespective of the stimulus, a continuous cycle of injury and repair involving hepatocytes strongly prognoses the growth of hepatocellular carcinoma, DNA damage by augmented levels of ROS and RNS level playing a critical role (Bishayee 2014; Capone et al. 2010).

3.3.2 CNS Damage Due to JEV, WNV, Zika Virus Infection, and Sometimes DENV Too

The central nervous system (CNS) is physiologically isolated from the rest of the body by a specialized selectively permeable barricade called as the blood–brain barrier (BBB), which allows passage to selected metabolites, respiratory gases, and an extremely limited repertoire of circulatory tissue cells. This isolation is necessary for protection of low regeneration capacity neuronal cells from systemic inflammation, which can also upset the structural and functional plasticity of neurons that is dependent on cytokine signaling (Arnett et al. 2001; Gougeon et al. 2013; Mason et al. 2001; Fischer et al. 2011; Brissoni et al. 2006). The CNS can have either neuronal or non-neuronal glial cells; the latter provide vital functional support and include microglia (macrophage-like immune cells), oligodendrocytes (which provide insulation for neurons), and astrocytes (responsible for repair of damaged neuronal tissue). Microglial cells have immunomodulatory function in suppressing a pathogenic inflammation (Seitz et al. 2018). Multiple viruses in the +ve-ssRNA genome group, including Coronavirus, Picornavirus, Flaviviridae, and Togaviridae, cause opportunistic infection of CNS (Bergmann et al. 2006; Koyuncu et al. 2013; Fletcher and McKeating 2012).

In the absence of a direct admission route, these viruses undergo limited replication in peripheral tissue, before entering through either peripheral nerves or BBB microvasculature or CNS infiltrating leukocytes (functioning as the proverbial "Trojan horse") (Koyuncu et al. 2013; Jeha et al. 2003). A feature common here is a breach of the vascular endothelial barrier at varying locations, e.g., BBB for JEV/WNV, blood retinal barrier for ZIKV, and endothelial barriers in lungs/peritoneum for DENV. Breach in BBB is more common for some viruses (e.g., WNV, JEV, ZIKAV, poliovirus) correlating with fatality. Interestingly, WNV and JEV have been suggested to cause BBB disruption from inside the CNS (Li et al. 2015; Verma et al. 2009). Still other reports suggest infected endothelial cells to secrete PICs that disrupt the BBB (Chen et al. 2014; Chang et al. 2017; Roach and Alcendor 2017). The tissue damage is caused from a combination of either direct neuronal infection which activates intrinsic apoptosis or a hyperactive inflammatory response mediated by PICs or CD8+ cytotoxic T cells (CTLs) (Wang et al.

2003; Samuel et al. 2007). Infected neurons secrete chemokines that attract leukocytes like monocytes and lymphocytes (Klein et al. 2005; Shrestha and Diamond 2004; Glass et al. 2005; Kelley et al. 2003; Lim et al. 2011; Bardina et al. 2015; Durrant et al. 2015; Shrestha et al. 2008). The relation between a "good" and "bad" immune response is, however, very tricky when it comes to the CNS. Migration of CTLs expressing receptors for chemokines like CCL2, CCL3, CCL4, CCL5, CXCL9–11, as well as its timing with respect to establishment of infection, seems to play a crucial role in virus eradication and survival (Wang et al. 2003; Shrestha and Diamond 2004; Diamond et al. 2003; Sussman et al. 1989; Getts et al. 2010; Nansen et al. 2000; Chen et al. 2004; Liu et al. 2001; Zink et al. 2001; Winter et al. 2004). The CTLs exert their antiviral role by inducing cell death through either a perforin-dependent or Fas-FasL-mediated mechanism (Rossi et al. 1998; Shrestha and Diamond 2007). In addition to CTLs, other PICs might also induce direct cell death in neurons (Dhanwani et al. 2012; Olmo et al. 2017; Baer et al. 2016; Kumar et al. 2010).

3.3.3 Dengue Infection-Associated Vascular Leakage

Dengue virus causes a febrile illness with can turn fatal after a subsidence of the fever. The severity emanates from leakage of fluid from the blood vessels by a breach of the vascular endothelium. Circulating in four serotypes, severe disease is mostly associated with secondary infection by a serotype different from the one causing primary infection. Neutralizing antibodies generated during primary infection incompletely neutralize the secondary infection virus and instead promote their uptake by monocytes, by a phenomenon called antibody-dependent enhancement or ADE (Katzelnick et al. 2017; Dejnirattisai et al. 2016). Notwithstanding a primary or secondary infection, the pathological symptoms are considered to be the result of an unbridled host immune response (Basu and Chaturvedi 2008; Rothman 2011).

DENV infects a variety of cells including monocytes, dendritic cells (skin Langerhans cells), macrophages (Kupffer cells), and vascular endothelial cells, expectedly leading to PIC secretion (Wu et al. 2000; Jessie et al. 2004; Tolfvenstam et al. 2011). Different studies have reported a positive association of DHF/DSS development with extraordinarily augmented levels of different PICs that include macrophage migration inhibitory factor (MIF), IFN-α, TNF-α (Green et al. 1999; Kurane et al. 1993; Huang et al. 2000; Chen et al. 2006). Although multiple reports have suggested correlation between specific PIC level and plasma leakage, the mechanism is still elusive and limited to association studies (Priyadarshini et al. 2010; Her et al. 2017; Sehrawat et al. 2018; Malavige et al. 2012). Interestingly however, multiple similar association studies have suggested a positive association between levels of IL10 (an AIC) and severe/critical symptoms related to dengue infection (Malavige et al. 2013; Tsai et al. 2013; Flores-Mendoza et al. 2017). IL10, produced by multiple immune cells, suppresses immune response through upregulation of SOCS (suppressor of cytokine signaling) function and downregulation of IFN activity, the result being decreased T-cell cytotoxicity (Halstead et al. 2010;

Katzelnick et al. 2017; Tsai et al. 2013; Azeredo et al. 2001; Brasier et al. 2012). The augmentation of IL10 level has been suggested to emanate from monocytes infected by the ADE route with additional influence from high viremia (Tsai et al. 2014). IL10 is a dominant regulator of the immune system that can prolong pathogen clearance through a subversion of the immune response (Couper et al. 2008).

3.3.4 Lung Infection and Pathology by Coronaviruses

Coronavirus infections are usually benign causing self-limiting mild flu-like symptoms. However, recent outbreaks involving, e.g., severe acute respiratory syndrome coronavirus (SARS-CoV), which jumped species barrier through acquisition of minor genome mutations, have projected them as potentially severe human pathogens (Guan et al. 2004). Spread through aerosols, SARS-CoV primarily infect lung cells triggering an often fatal inflammatory response clinically called acute respiratory distress syndrome (ARDS) that starts with severe hypoxia, pulmonary edema progressing to systemic inflammation, and failure of multiple organs, culminating in high rate of mortality (Peiris et al. 2003; Lew et al. 2003; Tsushima et al. 2009; Farcas et al. 2005). Although evidence suggests that SARS-CoV can infect multiple cell types, lung type-II pneumocytes and ciliated epithelial cells constitute primary sites of virus replication, consequent to which these cells undergo apoptotic and/or necrotic death attracting innate immune cells and activating them to secrete PICs (Sims et al. 2005; Chow et al. 2004; Nicholls et al. 2003). The nature of inflammation following SARS-CoV infection is characterized by a prompt production of PICs through immediate NFκB activation and a delayed expression of type-I IFN genes (Shi et al. 2014; Kong et al. 2009; Wong et al. 2004). Severity of symptoms correlates positively with IL-6 levels while exhibiting negative correlation with that of IL-8 and TGFβ (Zhang et al. 2004). As observed with many other viral pathogenesis models, macrophage polarization culminating in preferential enrichment of M2-macrophages has been suggested to be responsible for SARS-CoV pathogenesis (Page et al. 2012). SARS-CoV infection is also associated with hemophagocytosis or engulfment of different types of blood cells by histiocytes (a class within macrophages), which is a clinical marker of immune system hyper-activation (Usmani et al. 2013).

3.4 Therapeutic Approaches Using Cytokine

Traditionally prophylactic or therapeutic strategies for combating viral pathogenesis are designed using vaccines or directly acting antivirals (DAA), respectively. But for many viruses there is no clinically approved product to serve in either approach. Since the etiology of critical pathogenic symptoms is often associated with an unbridled host inflammatory response, there have been suggestions and attempts to control the harmful effects through modulation of key inflammatory signaling (D'Elia et al. 2013). However, a holistic approach to complete "cure" should probably involve investigations to provide support to both approaches simultaneously. Only ribavirin or the same combined with pegylated IFN-α was the therapeutic strategy for controlling HCV infection, before the advent of high efficacy DAAs. Similarly IFN-λ and glucocorticoids, both of which can consolidate the BBB, have been suggested as therapeutics for combating viral diseases that disrupt this barrier (Rhen and Cidlowski 2005; Daniels et al. 2014; Lazear et al. 2015; Wang et al. 2004; Blecharz et al. 2010; Fabene et al. 2008). Likewise, administration of PICs like CCL7 and IL17A has shown efficacy in increasing survival of mice experimentally infected with WNV (Bardina et al. 2015; Acharya et al. 2016). In dengue patients, however, meddling with either promoter or inhibitor of inflammation has been suggested as possible approaches (Tsai et al. 2013; Goh et al. 2014; Callaway et al. 2015; Ji et al. 2005; Dinarello 2011). Small molecules that can influence the function of the NLRP3 inflammosome have also been projected as potential therapies for CHIKV and can be tested against dengue as well (Chen et al. 2017; Coll et al. 2015; Hottz et al. 2013). Alternative approaches using pharmaceuticals that indirectly mitigate the pathological effect without interfering with inflammation have also been discussed (Olmo et al. 2017; Grip and Janciauskiene 2009; Reynolds and Miller 1989; Thomas and Grossberg 2009; Giguere and Tremblay 2004; Raemer et al. 2009).

3.5 Concluding Remarks

An ability to suppress innate immunity pathways is common among viruses that cause severe human diseases. Nonetheless modulating inflammation needs extreme caution, in order to reduce potential cytotoxicity of the administered therapeutic. Therefore, there is a need to go beyond association studies to generate a clearer picture of the exact role that inflammation plays in viral pathology, which can then assist in developing therapeutic strategies that tinker with inflammation.

Conflict of Interest The author declares that they have no competing interests.

References

Aarreberg LD et al (2018) Interleukin-1beta signaling in dendritic cells induces antiviral interferon responses. MBio 9.). pii: e00342-18.

Acharya D et al (2016) Interleukin-17A promotes CD8+ T cell cytotoxicity to facilitate West Nile virus clearance. J Virol 91:e01529

Arnett HA et al (2001) TNF alpha promotes proliferation of oligodendrocyte progenitors and remyelination. Nat Neurosci 4:1116–1122

Athie-Morales V, Smits HH, Cantrell DA, Hilkens CM (2004) Sustained IL-12 signaling is required for Th1 development. J Immunol 172:61–69

Axley P, Ahmed Z, Ravi S, Singal AK (2018) Hepatitis C virus and hepatocellular carcinoma: a narrative review. J Clin Transl Hepatol 6:79–84

Azeredo EL et al (2001) Characterisation of lymphocyte response and cytokine patterns in patients with dengue fever. Immunobiology 204:494–507

Baer A et al (2016) Venezuelan equine encephalitis virus induces apoptosis through the unfolded protein response activation of EGR1. J Virol 90:3558–3572

Bala S et al (2012) Increased microRNA-155 expression in the serum and peripheral monocytes in chronic HCV infection. J Transl Med 10:151

Banks WA et al (2015) Lipopolysaccharide-induced blood-brain barrier disruption: roles of cyclooxygenase, oxidative stress, neuroinflammation, and elements of the neurovascular unit. J Neuroinflammation 12:223

Bardina SV et al (2015) Differential roles of chemokines CCL2 and CCL7 in monocytosis and leukocyte migration during West Nile virus infection. J Immunol 195:4306–4318

Basu A, Chaturvedi UC (2008) Vascular endothelium: the battlefield of dengue viruses. FEMS Immunol Med Microbiol 53:287–299

Bauernfeind FG et al (2009) Cutting edge: NF-kappaB activating pattern recognition and cytokine receptors license NLRP3 inflammasome activation by regulating NLRP3 expression. J Immunol 183:787–791

Ben-Sasson SZ, Caucheteux S, Crank M, Hu-Li J, Paul WE (2011) IL-1 acts on T cells to enhance the magnitude of in vivo immune responses. Cytokine 56:122–125

Bergmann CC, Lane TE, Stohlman SA (2006) Coronavirus infection of the central nervous system: host-virus stand-off. Nat Rev Microbiol 4:121–132

Biet F, Locht C, Kremer L (2002) Immunoregulatory functions of interleukin 18 and its role in defense against bacterial pathogens. J Mol Med (Berl) 80:147–162

Bird SW, Kirkegaard K (2015) Escape of non-enveloped virus from intact cells. Virology 479–480:444–449

Bishayee A (2014) The role of inflammation and liver cancer. Adv Exp Med Biol 816:401–435

Blecharz KG et al (2010) Glucocorticoid effects on endothelial barrier function in the murine brain endothelial cell line cEND incubated with sera from patients with multiple sclerosis. Mult Scler 16:293–302

Boisvert J et al (2001) Quantitative analysis of hepatitis C virus in peripheral blood and liver: replication detected only in liver. J Infect Dis 184:827–835

Brasier AR et al (2012) A three-component biomarker panel for prediction of dengue hemorrhagic fever. Am J Trop Med Hyg 86:341–348

Brissoni B et al (2006) Intracellular trafficking of interleukin-1 receptor I requires Tollip. Curr Biol 16:2265–2270

Brun P, Castagliuolo I, Pinzani M, Palu G, Martines D (2005) Exposure to bacterial cell wall products triggers an inflammatory phenotype in hepatic stellate cells. Am J Physiol Gastrointest Liver Physiol 289:G571–G578

Burdeinick-Kerr R, Wind J, Griffin DE (2007) Synergistic roles of antibody and interferon in noncytolytic clearance of Sindbis virus from different regions of the central nervous system. J Virol 81:5628–5636

Burdette D et al (2012) Hepatitis C virus activates interleukin-1beta via caspase-1-inflammasome complex. J Gen Virol 93:235–246

Callaway JB et al (2015) Spleen tyrosine kinase (Syk) mediates IL-1beta induction by primary human monocytes during antibody-enhanced dengue virus infection. J Biol Chem 290:17306–17320

Canbay A et al (2003a) Kupffer cell engulfment of apoptotic bodies stimulates death ligand and cytokine expression. Hepatology 38:1188–1198

Canbay A et al (2003b) Apoptotic body engulfment by a human stellate cell line is profibrogenic. Lab Investig 83:655–663

Capone F et al (2010) Serum cytokine levels in patients with hepatocellular carcinoma. Eur Cytokine Netw 21:99–104

Caussin-Schwemling C, Schmitt C, Stoll-Keller F (2001) Study of the infection of human blood derived monocyte/macrophages with hepatitis C virus in vitro. J Med Virol 65:14–22

Chang M et al (2003) Dynamics of hepatitis C virus replication in human liver. Am J Pathol 163:433–444
Chang CY et al (2017) Interplay of inflammatory gene expression in pericytes following Japanese encephalitis virus infection. Brain Behav Immun 66:230–243
Chen IY, Ichinohe T (2015) Response of host inflammasomes to viral infection. Trends Microbiol 23:55–63
Chen Y et al (1997) Dengue virus infectivity depends on envelope protein binding to target cell heparan sulfate. Nat Med 3:866–871
Chen CJ, Chen JH, Chen SY, Liao SL, Raung SL (2004) Upregulation of RANTES gene expression in neuroglia by Japanese encephalitis virus infection. J Virol 78:12107–12119
Chen LC et al (2006) Correlation of serum levels of macrophage migration inhibitory factor with disease severity and clinical outcome in dengue patients. Am J Trop Med Hyg 74:142–147
Chen CJ et al (2014) Infection of pericytes in vitro by Japanese encephalitis virus disrupts the integrity of the endothelial barrier. J Virol 88:1150–1161
Chen W et al (2017) Specific inhibition of NLRP3 in chikungunya disease reveals a role for inflammasomes in alphavirus-induced inflammation. Nat Microbiol 2:1435–1445
Chow KC, Hsiao CH, Lin TY, Chen CL, Chiou SH (2004) Detection of severe acute respiratory syndrome-associated coronavirus in pneumocytes of the lung. Am J Clin Pathol 121:574–580
Coll RC et al (2015) A small-molecule inhibitor of the NLRP3 inflammasome for the treatment of inflammatory diseases. Nat Med 21:248–255
Couper KN, Blount DG, Riley EM (2008) IL-10: the master regulator of immunity to infection. J Immunol 180:5771–5777
Daniels BP et al (2014) Viral pathogen-associated molecular patterns regulate blood-brain barrier integrity via competing innate cytokine signals. MBio 5:e01476–e01414
Davies LC, Jenkins SJ, Allen JE, Taylor PR (2013) Tissue-resident macrophages. Nat Immunol 14:986–995
de Macedo FC et al (2006) Histologic, viral, and molecular correlates of dengue fever infection of the liver using highly sensitive immunohistochemistry. Diagn Mol Pathol 15:223–228
Dejnirattisai W et al (2016) Dengue virus sero-cross-reactivity drives antibody-dependent enhancement of infection with zika virus. Nat Immunol 17:1102–1108
D'Elia RV, Harrison K, Oyston PC, Lukaszewski RA, Clark GC (2013) Targeting the "cytokine storm" for therapeutic benefit. Clin Vaccine Immunol 20:319–327
Deng L et al (2008) Hepatitis C virus infection induces apoptosis through a Bax-triggered, mitochondrion-mediated, caspase 3-dependent pathway. J Virol 82:10375–10385
Deonarain R, Cerullo D, Fuse K, Liu PP, Fish EN (2004) Protective role for interferon-beta in coxsackievirus B3 infection. Circulation 110:3540–3543
Dhanwani R et al (2012) Characterization of Chikungunya virus infection in human neuroblastoma SH-SY5Y cells: role of apoptosis in neuronal cell death. Virus Res 163:563–572
Diamond MS, Shrestha B, Mehlhop E, Sitati E, Engle M (2003) Innate and adaptive immune responses determine protection against disseminated infection by West Nile encephalitis virus. Viral Immunol 16:259–278
Dinarello CA (2011) Blocking interleukin-1beta in acute and chronic autoinflammatory diseases. J Intern Med 269:16–28
Durrant DM, Daniels BP, Pasieka T, Dorsey D, Klein RS (2015) CCR5 limits cortical viral loads during West Nile virus infection of the central nervous system. J Neuroinflammation 12:233
Fabene PF et al (2008) A role for leukocyte-endothelial adhesion mechanisms in epilepsy. Nat Med 14:1377–1383
Farcas GA et al (2005) Fatal severe acute respiratory syndrome is associated with multiorgan involvement by coronavirus. J Infect Dis 191:193–197
Fernando S et al (2016) Patterns and causes of liver involvement in acute dengue infection. BMC Infect Dis 16:319
Fields BN, Knipe DM, Howley PM (2013) In: Knipe DM, Howley PM, Cohen JI et al (eds) Fields virology, 6th edn. Wolters Kluwer Health/Lippincott Williams & Wilkins, Philadelphia

Fischer R et al (2011) Ligand-induced internalization of TNF receptor 2 mediated by a di-leucin motif is dispensable for activation of the NFkappaB pathway. Cell Signal 23:161–170
Fletcher NF, McKeating JA (2012) Hepatitis C virus and the brain. J Viral Hepat 19:301–306
Flores-Mendoza LK et al (2017) IL-10 and socs3 are predictive biomarkers of dengue hemorrhagic fever. Mediat Inflamm 2017(5197592):1
Forbes SJ, Newsome PN (2016) Liver regeneration – mechanisms and models to clinical application. Nat Rev Gastroenterol Hepatol 13:473–485
Garlanda C, Dinarello CA, Mantovani A (2013) The interleukin-1 family: back to the future. Immunity 39:1003–1018
Getts MT, Richards MH, Miller SD (2010) A critical role for virus-specific CD8(+) CTLs in protection from Theiler's virus-induced demyelination in disease-susceptible SJL mice. Virology 402:102–111
Giguere JF, Tremblay MJ (2004) Statin compounds reduce human immunodeficiency virus type 1 replication by preventing the interaction between virion-associated host intercellular adhesion molecule 1 and its natural cell surface ligand LFA-1. J Virol 78:12062–12065
Gilbert C, Bergeron M, Methot S, Giguere JF, Tremblay MJ (2005) Statins could be used to control replication of some viruses, including HIV-1. Viral Immunol 18:474–489
Ginhoux F, Jung S (2014) Monocytes and macrophages: developmental pathways and tissue homeostasis. Nat Rev Immunol 14:392–404
Glass WG et al (2005) Chemokine receptor CCR5 promotes leukocyte trafficking to the brain and survival in West Nile virus infection. J Exp Med 202:1087–1098
Goh AX et al (2014) A novel human anti-interleukin-1beta neutralizing monoclonal antibody showing in vivo efficacy. MAbs 6:765–773
Gougeon PY et al (2013) The pro-inflammatory cytokines IL-1beta and TNFalpha are neurotrophic for enteric neurons. J Neurosci 33:3339–3351
Goutagny N et al (2003) Evidence of viral replication in circulating dendritic cells during hepatitis C virus infection. J Infect Dis 187:1951–1958
Green S et al (1999) Early immune activation in acute dengue illness is related to development of plasma leakage and disease severity. J Infect Dis 179:755–762
Grip O, Janciauskiene S (2009) Atorvastatin reduces plasma levels of chemokine (CXCL10) in patients with Crohn's disease. PLoS One 4:e5263
Guan Y et al (2004) Molecular epidemiology of the novel coronavirus that causes severe acute respiratory syndrome. Lancet 363:99–104
Guiducci C, Vicari AP, Sangaletti S, Trinchieri G, Colombo MP (2005) Redirecting in vivo elicited tumor infiltrating macrophages and dendritic cells towards tumor rejection. Cancer Res 65:3437–3446
Halstead SB, Mahalingam S, Marovich MA, Ubol S, Mosser DM (2010) Intrinsic antibody-dependent enhancement of microbial infection in macrophages: disease regulation by immune complexes. Lancet Infect Dis 10:712–722
Henry CJ, Ornelles DA, Mitchell LM, Brzoza-Lewis KL, Hiltbold EM (2008) IL-12 produced by dendritic cells augments CD8+ T cell activation through the production of the chemokines CCL1 and CCL17. J Immunol 181:8576–8584
Her Z et al (2017) Severity of plasma leakage is associated with high levels of interferon gamma-inducible protein 10, hepatocyte growth factor, matrix metalloproteinase 2 (MMP-2), and MMP-9 during dengue virus infection. J Infect Dis 215:42–51
Hernandez-Gea V, Friedman SL (2011) Pathogenesis of liver fibrosis. Annu Rev Pathol 6:425–456
Hottz ED et al (2013) Platelets mediate increased endothelium permeability in dengue through NLRP3-inflammasome activation. Blood 122:3405–3414
Hou W et al (2012) Viral infection triggers rapid differentiation of human blood monocytes into dendritic cells. Blood 119:3128–3131
Htun NS et al (2015) Is diabetes a risk factor for a severe clinical presentation of dengue?--review and meta-analysis. PLoS Negl Trop Dis 9:e0003741
Hu X, Ivashkiv LB (2009) Cross-regulation of signaling pathways by interferon-gamma: implications for immune responses and autoimmune diseases. Immunity 31:539–550

Huang YH et al (2000) Dengue virus infects human endothelial cells and induces IL-6 and IL-8 production. Am J Trop Med Hyg 63:71–75

Huerre MR et al (2001) Liver histopathology and biological correlates in five cases of fatal dengue fever in Vietnamese children. Virchows Arch 438:107–115

Ichinohe T, Yamazaki T, Koshiba T, Yanagi Y (2013) Mitochondrial protein mitofusin 2 is required for NLRP3 inflammasome activation after RNA virus infection. Proc Natl Acad Sci U S A 110:17963–17968

Jean CM, Honarmand S, Louie JK, Glaser CA (2007) Risk factors for West Nile virus neuroinvasive disease, California, 2005. Emerg Infect Dis 13:1918–1920

Jeha LE et al (2003) West Nile virus infection: a new acute paralytic illness. Neurology 61:55–59

Jensen S, Thomsen AR (2012) Sensing of RNA viruses: a review of innate immune receptors involved in recognizing RNA virus invasion. J Virol 86:2900–2910

Jessie K, Fong MY, Devi S, Lam SK, Wong KT (2004) Localization of dengue virus in naturally infected human tissues, by immunohistochemistry and in situ hybridization. J Infect Dis 189:1411–1418

Ji JD et al (2005) Inhibition of IL-10-induced STAT3 activation by 15-deoxy-Delta12,14-prostaglandin J2. Rheumatology (Oxford) 44:983–988

Katzelnick LC et al (2017) Antibody-dependent enhancement of severe dengue disease in humans. Science 358:929–932

Kawai T et al (2005) IPS-1, an adaptor triggering RIG-I- and Mda5-mediated type I interferon induction. Nat Immunol 6:981–988

Kelley TW, Prayson RA, Ruiz AI, Isada CM, Gordon SM (2003) The neuropathology of West Nile virus meningoencephalitis. A report of two cases and review of the literature. Am J Clin Pathol 119:749–753

Klein RS et al (2005) Neuronal CXCL10 directs CD8+ T-cell recruitment and control of West Nile virus encephalitis. J Virol 79:11457–11466

Klenerman P, Hill A (2005) T cells and viral persistence: lessons from diverse infections. Nat Immunol 6:873–879

Kmiec Z (2001) Cooperation of liver cells in health and disease. In: Advances in anatomy, embryology and cell biology, vol 161. Springer, Berlin, pp 1–151

Kong SL, Chui P, Lim B, Salto-Tellez M (2009) Elucidating the molecular physiopathology of acute respiratory distress syndrome in severe acute respiratory syndrome patients. Virus Res 145:260–269

Koyuncu OO, Hogue IB, Enquist LW (2013) Virus infections in the nervous system. Cell Host Microbe 13:379–393

Kumar M, Verma S, Nerurkar VR (2010) Pro-inflammatory cytokines derived from West Nile virus (WNV)-infected SK-N-SH cells mediate neuroinflammatory markers and neuronal death. J Neuroinflammation 7:73

Kumar CS, Dey D, Ghosh S, Banerjee M (2018) Breach: host membrane penetration and entry by nonenveloped viruses. Trends Microbiol 26:525–537

Kurane I et al (1993) High levels of interferon alpha in the sera of children with dengue virus infection. Am J Trop Med Hyg 48:222–229

Lazear HM, Pinto AK, Vogt MR, Gale M Jr, Diamond MS (2011) Beta interferon controls West Nile virus infection and pathogenesis in mice. J Virol 85:7186–7194

Lazear HM et al (2015) Interferon-lambda restricts West Nile virus neuroinvasion by tightening the blood-brain barrier. Sci Transl Med 7:284ra59

Lew TW et al (2003) Acute respiratory distress syndrome in critically ill patients with severe acute respiratory syndrome. JAMA 290:374–380

Ley K (2014) The second touch hypothesis: T cell activation, homing and polarization. F1000Res 3:37

Li F et al (2015) Viral infection of the central nervous system and neuroinflammation precede blood-brain barrier disruption during Japanese encephalitis virus infection. J Virol 89:5602–5614

Lim JK et al (2011) Chemokine receptor Ccr2 is critical for monocyte accumulation and survival in West Nile virus encephalitis. J Immunol 186:471–478

Lim EJ et al (2014) Hepatitis C virus-induced hepatocyte cell death and protection by inhibition of apoptosis. J Gen Virol 95:2204–2215
Liu B, Taioli E (2015) Seasonal variations of complete blood count and inflammatory biomarkers in the US population – analysis of NHANES data. PLoS One 10:e0142382
Liu MT, Armstrong D, Hamilton TA, Lane TE (2001) Expression of Mig (monokine induced by interferon-gamma) is important in T lymphocyte recruitment and host defense following viral infection of the central nervous system. J Immunol 166:1790–1795
MacDonald AJ et al (2002) CD4 T helper type 1 and regulatory T cells induced against the same epitopes on the core protein in hepatitis C virus-infected persons. J Infect Dis 185:720–727
Malavige GN et al (2012) Cellular and cytokine correlates of severe dengue infection. PLoS One 7:e50387
Malavige GN et al (2013) Serum IL-10 as a marker of severe dengue infection. BMC Infect Dis 13:341
Marianneau P et al (1999) Infection of primary cultures of human Kupffer cells by dengue virus: no viral progeny synthesis, but cytokine production is evident. J Virol 73:5201–5206
Masalova OV et al (2017) Modulation of Cell Death Pathways by Hepatitis C Virus Proteins in Huh7.5 Hepatoma Cells. Int J Mol Sci 18:E2346
Mason JL, Suzuki K, Chaplin DD, Matsushima GK (2001) Interleukin-1beta promotes repair of the CNS. J Neurosci 21:7046–7052
Mehrbod P, Omar AR, Hair-Bejo M, Haghani A, Ideris A (2014) Mechanisms of action and efficacy of statins against influenza. Biomed Res Int 2014(872370):1
Mercer J, Greber UF (2013) Virus interactions with endocytic pathways in macrophages and dendritic cells. Trends Microbiol 21:380–388
Mitchell UA, Aneshensel CS (2016) Social inequalities in inflammation. J Aging Health 29:769–787
Nansen A, Marker O, Bartholdy C, Thomsen AR (2000) CCR2+ and CCR5+ CD8+ T cells increase during viral infection and migrate to sites of infection. Eur J Immunol 30:1797–1806
Negash AA et al (2013) IL-1beta production through the NLRP3 inflammasome by hepatic macrophages links hepatitis C virus infection with liver inflammation and disease. PLoS Pathog 9:e1003330
Nicholls JM et al (2003) Lung pathology of fatal severe acute respiratory syndrome. Lancet 361:1773–1778
Okuda M et al (2002) Mitochondrial injury, oxidative stress, and antioxidant gene expression are induced by hepatitis C virus core protein. Gastroenterology 122:366–375
Olenina LV et al (2005) Identification of glycosaminoglycan-binding sites within hepatitis C virus envelope glycoprotein E2*. J Viral Hepat 12:584–593
Olmo IG et al (2017) Zika virus promotes neuronal cell death in a non-cell autonomous manner by triggering the release of neurotoxic factors. Front Immunol 8:1016
Paalani M, Lee JW, Haddad E, Tonstad S (2011) Determinants of inflammatory markers in a bi-ethnic population. Ethn Dis 21:142–149
Page C et al (2012) Induction of alternatively activated macrophages enhances pathogenesis during severe acute respiratory syndrome coronavirus infection. J Virol 86:13334–13349
Peiris JS, Yuen KY, Osterhaus AD, Stohr K (2003) The severe acute respiratory syndrome. N Engl J Med 349:2431–2441
Perlemuter G et al (2002) Hepatitis C virus core protein inhibits microsomal triglyceride transfer protein activity and very low density lipoprotein secretion: a model of viral-related steatosis. FASEB J 16:185–194
Pichlmair A, Reis e Sousa C (2007) Innate recognition of viruses. Immunity 27:370–383
Plemper RK (2011) Cell entry of enveloped viruses. Curr Opin Virol 1:92–100
Pohl C, Shishkova J, Schneider-Schaulies S (2007) Viruses and dendritic cells: enemy mine. Cell Microbiol 9:279–289
Pornillos O, Garrus JE, Sundquist WI (2002) Mechanisms of enveloped RNA virus budding. Trends Cell Biol 12:569–579

Preveden T, Scarpellini E, Milic N, Luzza F, Abenavoli L (2017) Gut microbiota changes and chronic hepatitis C virus infection. Expert Rev Gastroenterol Hepatol 11:813–819
Priyadarshini D et al (2010) Clinical findings and pro-inflammatory cytokines in dengue patients in Western India: a facility-based study. PLoS One 5:e8709
Raemer PC, Kohl K, Watzl C (2009) Statins inhibit NK-cell cytotoxicity by interfering with LFA-1-mediated conjugate formation. Eur J Immunol 39:1456–1465
Reynolds IJ, Miller RJ (1989) Ifenprodil is a novel type of N-methyl-D-aspartate receptor antagonist: interaction with polyamines. Mol Pharmacol 36:758–765
Rhen T, Cidlowski JA (2005) Antiinflammatory action of glucocorticoids--new mechanisms for old drugs. N Engl J Med 353:1711–1723
Roach T, Alcendor DJ (2017) Zika virus infection of cellular components of the blood-retinal barriers: implications for viral associated congenital ocular disease. J Neuroinflammation 14:43
Rossi CP, McAllister A, Tanguy M, Kagi D, Brahic M (1998) Theiler's virus infection of perforin-deficient mice. J Virol 72:4515–4519
Rothman AL (2011) Immunity to dengue virus: a tale of original antigenic sin and tropical cytokine storms. Nat Rev Immunol 11:532–543
Saha B, Kodys K, Szabo G (2016) Hepatitis C virus-induced monocyte differentiation into polarized M2 macrophages promotes stellate cell activation via TGF-beta. Cell Mol Gastroenterol Hepatol 2:302–316.e8
Samanta J, Sharma V (2015) Dengue and its effects on liver. World J Clin Cases 3:125–131
Samuel MA, Diamond MS (2005) Alpha/beta interferon protects against lethal West Nile virus infection by restricting cellular tropism and enhancing neuronal survival. J Virol 79:13350–13361
Samuel MA, Morrey JD, Diamond MS (2007) Caspase 3-dependent cell death of neurons contributes to the pathogenesis of West Nile virus encephalitis. J Virol 81:2614–2623
Scarpellini E et al (2015) The human gut microbiota and virome: potential therapeutic implications. Dig Liver Dis 47:1007–1012
Schneider WM, Chevillotte MD, Rice CM (2014) Interferon-stimulated genes: a complex web of host defenses. Annu Rev Immunol 32:513–545
Schoggins JW, Rice CM (2011) Interferon-stimulated genes and their antiviral effector functions. Curr Opin Virol 1:519–525
Schulz C et al (2012) A lineage of myeloid cells independent of Myb and hematopoietic stem cells. Science 336:86–90
Sehrawat P et al (2018) Role of cytokines as molecular marker of dengue severity. Mediterr J Hematol Infect Dis 10:e2018023
Seitz S, Clarke P, Tyler KL (2018) Pharmacologic depletion of microglia increases viral load in the brain and enhances mortality in murine models of flavivirus-induced encephalitis. J Virol 92:e00525
Seth RB, Sun L, Ea CK, Chen ZJ (2005) Identification and characterization of MAVS, a mitochondrial antiviral signaling protein that activates NF-kappaB and IRF 3. Cell 122:669–682
Shi CS et al (2014) SARS-coronavirus open reading frame-9b suppresses innate immunity by targeting mitochondria and the MAVS/TRAF3/TRAF6 signalosome. J Immunol 193:3080–3089
Shimizu S et al (2005) Liver injury induced by lipopolysaccharide is mediated by TNFR-1 but not by TNFR-2 or Fas in mice. Hepatol Res 31:136–142
Shrestha B, Diamond MS (2004) Role of CD8+ T cells in control of West Nile virus infection. J Virol 78:8312–8321
Shrestha B, Diamond MS (2007) Fas ligand interactions contribute to CD8+ T-cell-mediated control of West Nile virus infection in the central nervous system. J Virol 81:11749–11757
Shrestha B, Zhang B, Purtha WE, Klein RS, Diamond MS (2008) Tumor necrosis factor alpha protects against lethal West Nile virus infection by promoting trafficking of mononuclear leukocytes into the central nervous system. J Virol 82:8956–8964
Sims AC et al (2005) Severe acute respiratory syndrome coronavirus infection of human ciliated airway epithelia: role of ciliated cells in viral spread in the conducting airways of the lungs. J Virol 79:15511–15524

Smith TD, Tse MJ, Read EL, Liu WF (2016) Regulation of macrophage polarization and plasticity by complex activation signals. Integr Biol (Camb) 8:946–955

Sokol CL, Luster AD (2015) The chemokine system in innate immunity. Cold Spring Harb Perspect Biol 7. pii: a016303.

Sugimoto K et al (2003) Suppression of HCV-specific T cells without differential hierarchy demonstrated ex vivo in persistent HCV infection. Hepatology 38:1437–1448

Sung JM, Lee CK, Wu-Hsieh BA (2012) Intrahepatic infiltrating NK and CD8 T cells cause liver cell death in different phases of dengue virus infection. PLoS One 7:e46292

Sussman MA, Shubin RA, Kyuwa S, Stohlman SA (1989) T-cell-mediated clearance of mouse hepatitis virus strain JHM from the central nervous system. J Virol 63:3051–3056

Suthar MS et al (2010) IPS-1 is essential for the control of West Nile virus infection and immunity. PLoS Pathog 6:e1000757

Tak PP, Firestein GS (2001) NF-kappaB: a key role in inflammatory diseases. J Clin Invest 107:7–11

Takeuchi O, Akira S (2010) Pattern recognition receptors and inflammation. Cell 140:805–820

Tan CW, Poh CL, Sam IC, Chan YF (2013) Enterovirus 71 uses cell surface heparan sulfate glycosaminoglycan as an attachment receptor. J Virol 87:611–620

Thomas SJ, Grossberg GT (2009) Memantine: a review of studies into its safety and efficacy in treating Alzheimer's disease and other dementias. Clin Interv Aging 4:367–377

Tolfvenstam T et al (2011) Characterization of early host responses in adults with dengue disease. BMC Infect Dis 11:209

Tsai TT et al (2013) An emerging role for the anti-inflammatory cytokine interleukin-10 in dengue virus infection. J Biomed Sci 20:40

Tsai TT et al (2014) Antibody-dependent enhancement infection facilitates dengue virus-regulated signaling of IL-10 production in monocytes. PLoS Negl Trop Dis 8:e3320

Tsushima K et al (2009) Acute lung injury review. Intern Med 48:621–630

Turner MD, Nedjai B, Hurst T, Pennington DJ (2014) Cytokines and chemokines: at the crossroads of cell signalling and inflammatory disease. Biochim Biophys Acta 1843:2563–2582

Usmani GN, Woda BA, Newburger PE (2013) Advances in understanding the pathogenesis of HLH. Br J Haematol 161:609–622

Verma S et al (2009) West Nile virus infection modulates human brain microvascular endothelial cells tight junction proteins and cell adhesion molecules: transmigration across the in vitro blood-brain barrier. Virology 385:425–433

Verreck FA et al (2004) Human IL-23-producing type 1 macrophages promote but IL-10-producing type 2 macrophages subvert immunity to (myco)bacteria. Proc Natl Acad Sci U S A 101:4560–4565

Wang KS, Frank DA, Ritz J (2000) Interleukin-2 enhances the response of natural killer cells to interleukin-12 through up-regulation of the interleukin-12 receptor and STAT4. Blood 95:3183–3190

Wang Y, Lobigs M, Lee E, Mullbacher A (2003) CD8+ T cells mediate recovery and immunopathology in West Nile virus encephalitis. J Virol 77:13323–13334

Wang T et al (2004) Toll-like receptor 3 mediates West Nile virus entry into the brain causing lethal encephalitis. Nat Med 10:1366–1373

Wilks J, Golovkina T (2012) Influence of microbiota on viral infections. PLoS Pathog 8:e1002681

Winter PM et al (2004) Proinflammatory cytokines and chemokines in humans with Japanese encephalitis. J Infect Dis 190:1618–1626

Wong CK et al (2004) Plasma inflammatory cytokines and chemokines in severe acute respiratory syndrome. Clin Exp Immunol 136:95–103

Wu SJ et al (2000) Human skin Langerhans cells are targets of dengue virus infection. Nat Med 6:816–820

Zehender G et al (1997) Detection of hepatitis C virus RNA in CD19 peripheral blood mononuclear cells of chronically infected patients. J Infect Dis 176:1209–1214

Zhang Y et al (2004) Analysis of serum cytokines in patients with severe acute respiratory syndrome. Infect Immun 72:4410–4415

Zhou Z, Xu MJ, Gao B (2016) Hepatocytes: a key cell type for innate immunity. Cell Mol Immunol 13:301–315

Zink MC et al (2001) Increased macrophage chemoattractant protein-1 in cerebrospinal fluid precedes and predicts simian immunodeficiency virus encephalitis. J Infect Dis 184:1015–1021

Interferons and Their Role in Viral Infection

4

Suji George, Gururaj Rao Deshpande,
and Gajanan N. Sapkal

Abstract

Interferons (IFNs) are a large family of naturally occurring cytokines, discovered in 1957. They are secreted by host immune cells in response to virus infections, tumors, and other biological inducers as diverse countermeasures against stimulation. They elicit and regulate inter- and intracellular networks of immune system by producing interferon-induced proteins. These proteins play a vital role in inhibiting any vulnerable step in viral replication cycle starting from viral entry/uncoating to maturation and release of the virus. Presently many interferons are approved as therapeutic agents, and tremendous progress has been made in understanding their role and usage as antiviral agents. This chapter explores the essential link of IFNs with the innate and adaptive immune system during viral infections. Further it discusses in detail the types of IFNs, genes induced by IFNs in the host, IFN-induced proteins and their antiviral roles, and the strategies used by viruses to counter the effects of antiviral action by IFNs. Finally, the chapter concludes by summarizing the therapeutic application of IFNs to fight viral diseases.

Keywords

Interferons · Antiviral activity · Immune system

4.1 Introduction

Viruses cause infectious diseases by invading living cells to multiply and produce progeny viruses. The host immune system plays a key role in defending against the infection and this defense barrier consists of innate immune cells such as dendritic

S. George · G. R. Deshpande · G. N. Sapkal (✉)
National Institute of Virology, Pune, Maharashtra, India

P. V. Bramhachari (ed.), *Dynamics of immune activation in viral diseases*,
https://doi.org/10.1007/978-981-15-1045-8_4

cells, macrophages, and natural killer cells. These cells are capable of mounting an immune response by recognizing a wide range of viral products. Activation of these cells can result in rapid viral sensing and release of key antiviral cytokines, which can inhibit viral replication and can further activate other components of innate and adaptive immunity. Interferons (IFNs) are a group of highly specialized cytokines capable of mediating antiviral, antigrowth, and immune modulation responses. It was discovered by Isaacs and Lindenmann in 1957 as a molecule secreted by virus-infected cells and was able to prevent further infection of cells that were exposed to it (Isaacs and Lindenmann 1957). IFN activity includes the induction of the transcription of IFN-stimulated gene (ISG) that actively participates in the inhibition of crucial steps in viral replication.

4.2 Classification of Interferons

IFNs are classified into three types (I, II, and III) based on the structural homology, chromosomal location, and cell surface receptors to which they bind to induce intracellular signaling pathways (de Weerd and Nguyen 2012). There are more than 10 IFN I subtypes of IFN-α, named IFN-α_1, IFN-α_2, etc., IFN-β, IFN-κ, IFN-ω, IFN-ε, and limitin. Type I IFNs signal by binding to heterodimeric receptor composed of low- (IFNAR1) and high-affinity (IFNAR2). While IFN II signals through transmembrane tetrameric IFN-γ receptor (IFNGR), comprising two units of IFNGR1 and IFNGR2. Type III IFNs (IFN-λ1, -λ1, -λ2, -λ3, and -λ4) signal through IFN-λ receptor (IFNLR), a heterodimer of IFNLR1 and IL10RB (Fensterl et al. 2015).

The functions of three IFNs in some instances overlap, for example, their ability to inhibit virus replication; at the same time many effects of signaling by type I, II, III IFNs are distinct. These differences may be attributed to two properties. First, IFN and its cognate receptors show differential expression as there are cell- and tissue-specific differences. Although most of the virally infected cells can produce IFN-α/β, IFN-γ is synthesized by certain cells of the immune system such as natural killer T cells, B cells, and antigen-presenting cells (Samuel 2001). Type I, type II, and IL10RB receptors are expressed in almost all cell types, except that IFNGR2 is not found on the surface of Th1 cells and low expression of IFNAR2 is seen in the human brain (de Weerd and Nguyen 2012). Unlike type I and II IFN receptors, the distribution of type III IFN receptor (IFNLR1) is more restricted. IFNAR is abundantly expressed only in cells of epithelial origin and hepatocytes and responds to type III IFNs (IFN-λ) (Mordstein et al. 2010; Sheahan et al. 2014; Sommereyns et al. 2008) Second, downstream targets of IFNs are different and this leads to induction of different IFN-stimulated genes (Der et al. 1998), and there is only partial overlap between IFNAR, IFNLR, and IFNGR. The set of genes induced by IFNAR and IFNLR is similar and they do so by activating intracellular signaling pathway through transcription factor complex ISG factor 3 (ISGF3) (Doyle et al. 2006). In contrast, activation of the transcription factor gamma-activated factor (GAF) by IFNGR induces different set of genes (Decker et al. 1997; Fensterl et al. 2015) (Fig. 4.1).

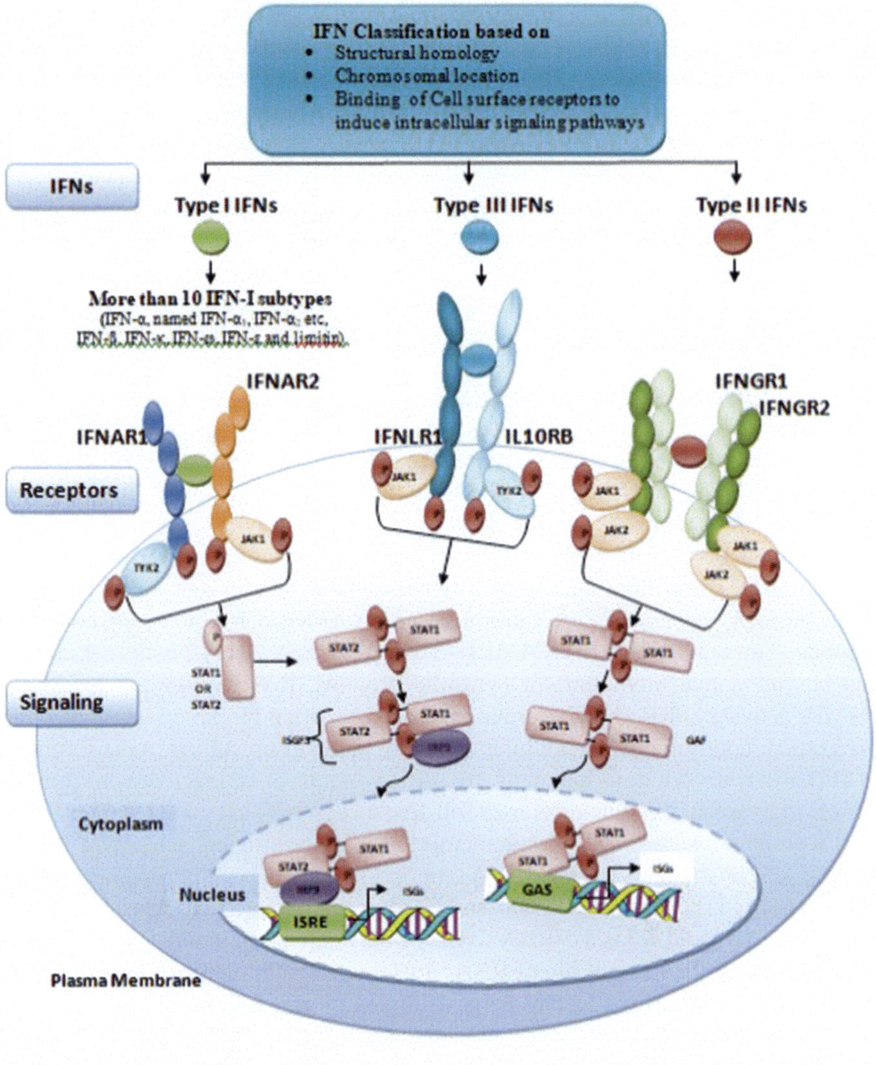

Fig. 4.1 Summary of Type I, Type II, and Type III interferon signaling: Different cell surface receptors are involved in signaling of Type I, Type II, and Type III interferons. Binding of IFNs to cognate receptor triggers phosphorylation of STAT through kinases (JAK-1, JAK-2, and TYK-2). Further, phosphorylated STAT undergoes oligomerization and interacts with IRF-9 in case of Type I and III IFNs to form a complex of ISGF3. Similarly, in the case of Type II IFNs STAT undergoes dimerization and associates with GAF to form a complex which then triggers the transcriptional activity

Type I and type III IFNs through IFNAR1/2 or IL-10RB/IFNLR1complexes mediate phosphorylation of pre-associated Janus kinase 1 (JAK1) and tyrosine kinase2 (TYK2), which then phosphorylate the receptors at specific intracellular tyrosine residues. The phosphorylated JAK1 and TYK2 then recruit and phosphorylate signal transducers and activators of transcription 1 and 2 (STAT1and 2) which further associate to form a heterodimer. The STAT heterodimer then interacts with IFN-regulatory factor 9 (IRF9) to form the IFN-stimulated gene factor 3 (ISGF3). Type II IFN dimers binding to the IFNGR1/2complex activate the cascade by phosphorylation of pre-associated JAK1 and JAK2 tyrosine kinases and transphosphorylation of the receptor chains leading to the recruitment and phosphorylation of STAT1. Phosphorylated STAT1 homodimers form IFN-γ activation factor (GAF) and translocate to the nucleus and bind to the gamma-activated sequence (GAS) promoter elements, respectively, resulting in expression of antiviral genes. Similarly the ISGF3 translocates to the nucleus to induce genes regulated by IFN-stimulated response elements (ISRE), resulting in expression of antiviral genes.

4.3 Synthesis of Interferons

Although IFNs are not expressed in resting cells, in virus-infected cells there is a need to accelerate the synthesis of IFNs as they are effective in limiting the virus replication and spread. As all the viruses replicate inside the cells, detecting viral products can be an effective strategy for inducing an immune response through production of IFNs. Viral nucleic acid (RNA or DNA) can trigger the synthesis of IFNs through the pattern recognition receptors (PRRs) by engaging pathogen-associated molecular patterns (PAMPs) (Thompson et al. 2011). Nucleic acid PRRs can be endosomal membrane or cytosolic proteins. Recognition and binding of PAMPs activates PRRs by conformational change, often by dimerization or oligomerization, and initiates intracellular signaling pathways resulting in the formation of IFNs. In uninfected normal cells, the PRRs remain in an inactive conformation unable to evoke an immune response. Toll-like receptors (TLRs) are receptors localized within endosomes and plasma membrane and are efficient sensors of PAMPs such as single-stranded RNA (ssRNA), lipopolysaccharides, lipoproteins, and flagella. TLRs (TLR3, TLR7, TLR8, and TLR9) are nucleic acid sensing receptor (Gay et al. 2014). TLR3 is a dsRNA sensor widely expressed in innate immune cells and involved in the recognition of RNA viruses such as influenza A virus, picornaviruses, and DNA viruses such as herpes simplex virus-1 (Alexopoulou et al. 2001; Le Goffic et al. 2006; Hardarson et al. 2007; Zhang et al. 2007). TLR-7 and TLR-8 are activated by ssRNA from many viruses such as vesicular stomatitis virus and influenza A virus (Lee et al. 2007). DNA viruses are recognized by endosomal TLR9 and several other cytoplasmic DNA sensors such as IFN-induced 16-kDA protein (IFI16), cyclic GMP-AMP synthase (cGAS), DEAD-Box RNA Helicase (DDX41), and DNA-dependent activator of IRFs (DAI) (Pandey et al. 2015; Wu and Chen 2014). TLR9 can recognize unmethylated deoxycytidylate-phosphate-deoxyguanylate (CpG) motif in viral DNA in plasmacytoid dendritic cells (Krug et al. 2004). RIG-I-like receptor family (RLRs) is another closely related family consisting of retinoic acid-inducible gene (RIG-I), melanoma differentiation-associated gene 5(MDA-5), and laboratory of genetics and physiology-2 (LGP2). RLRs are cytosolic RNA helicases involved in recognition of RNA viruses (Yoneyama et al. 2015). RLRs play a crucial role in antiviral defense pathway in cell types such as fibroblasts, epithelial cells, and dendrite cells (Thompson et al. 2011).

RIG-I uses 5′ end of the ssRNA as a strategy to distinguish between viral and host RNA. In addition, RIG-I can also recognize short dsRNA product produced during viral replication (Goubau et al. 2014; Weber et al. 2013; Yoneyama et al. 2015). MDA-5 differentiates viral and host RNA based on the length of the sequences as long as RNA is not common in host cells (Kato et al. 2008; Züst et al. 2011).

RIG-I recognizes Vesicular Stomatitis virus (VSV), rabies virus, Newcastle disease virus, measles virus, influenza A and B, hepatitis C virus, Japanese encephalitis virus, and Ebola virus (Fensterl et al. 2015; Thompson et al. 2011). LGP-2 can act via two different pathways during cytoplasmic recognition. LGP2 binds to dsRNA in tandem with MDA5 and thereby activates MDA5 (Bruns et al. 2014; Childs et al. 2013). Second, it can inhibit RIG-I activation by competing for dsRNA and by direct interaction with RIG I (Saito et al. 2007).

Detection of PAMPs results in the activation of downstream signaling pathways leading to the induction of IFNs. IFN-α subtypes use transcription factor IRF-7 and IFN-λs utilize IRF-3, IRF-7, and NF-κB (Génin et al. 2009; Iversen and Paludan 2010; Onoguchi et al. 2007; Osterlund et al. 2007). IFN-β is induced by the synergistic promoter binding of transcription factors IRF-3, NF-κB, and API (Ford and Thanos 2010; Panne et al. 2007). Once activated by the PAMP, PRRs interact with respective adapter proteins; TLR3 uses the TRIF and TLR7, TLR8, and TLR9 use My88, while RLRs use mitochondrial and peroxisome-bound MAVS (Fensterl et al. 2015). Cytoplasmic DNA sensors use STING (stimulator of interferon genes) for their signaling. Further the TRIF and MAVS interact with signaling proteins of the TRAF family of ubiquitin ligases, as well as IKKα/β and TBK1 kinases which phosphorylate and activate downstream transcription factors (Fensterl et al. 2015). Phosphorylation of IRF3 at specific serine and threonine residues results in opening of its conformation and subsequent dimerization and translocation to nucleus where it binds to IFN-stimulated response elements (ISREs) in the promoter region of IFN-β and IFN-γ (Pandey et al. 2015). NF-κB another transcription factor upon phosphorylation is released from its inhibitor IκB to induce IFN genes (Fig 4.2).

4.4 Interferon-Mediated Induction of Gene Expression

Evidence suggests that multiple signaling pathways are activated and cooperation as well as coordinated response by these pathways is required for the generation of response to IFNs. Some of the major pathways involved in IFN signaling are dealt in sufficient detail below. In addition to the signaling pathway described here, there are many other pathways whose involvement in IFN signaling is documented; in fact the list of newly identified IFN signaling pathway is growing and it seems probable that more than one signaling pathway is involved in IFN-mediated response.

4.4.1 Jak-Stat Pathway in Interferon Signaling

IFNs are synthesized and secreted by virus-infected cells and it can elicit biological responses by binding to specific receptors on the surface of the target cells. Binding

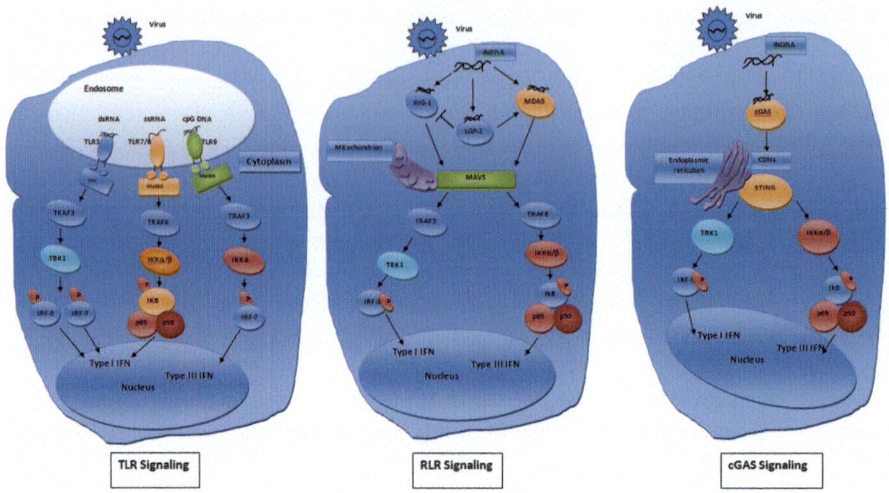

Fig. 4.2 Signaling pathways for Type I and III interferons. Receptor proteins (toll-like receptors (TLRs), RIG-I-like receptors (RLRs), and cyclic GMP-AMP synthase (cGAS) in endosomes or cytoplasm recognize viral RNA or DNA and trigger signaling via adapter proteins such as TRIF, My88, MAVS, and STING, which further activate the transcription factors of the IRF family and NF-κB resulting in the transcription of Type I and Type III interferons

of IFNs to the respective receptors results in change in conformation of the receptor intracellular domain and activates a cascade of intracellular signaling events that culminates in the induction of target genes. The effect of IFNs is mediated by the Jak/STAT signaling cascade. The Jak/STAT pathway has three components: (1) cell receptor, (2) JAK proteins, and (3) STAT proteins. There are four members in the JAK family [JAK 1, 2, 3 and tyrosine kinase 2 (TYK2)] and seven STATs (1–4, 5a, 5b, and 6) (Aaronson 2002). The subunit for receptor for type I IFNs and type II IFNs interacts with member of the Janus-activated kinase (JAK) family. The IFNAR1 subunit is constitutively associated with tyrosine kinase 2 (TYK2), whereas IFNAR2 is associated with JAK1 unit of type I IFNs. While in type II IFN receptor the IFNGR1 subunit associates with JAK1, whereas IFNGR2 is constitutively associated with JAK2 (Raftery and Stevenson 2017). The first step in type I and type II associated signaling is the activation of these receptor-associated JAKs. This happens in response to ligand binding and dimerization of the receptor subunits, subsequent autophosphorylation, and activation of the associated JAKs, TYK2, and JAK1 which in turn regulate the phosphorylation and activation of STATs (Raftery and Stevenson 2017). The type of STATs activated by IFNs depends on the cell type, while Type I activate STAT1, STAT2 and STAT3 in most cell types, in certain other cell types acts via activation of STAT4, STAT5, STAT6 (Wen and Darnell 1997; Wen et al. 1995; Platanias 2005). IFN-α activates STAT4 and STAT6; however, such activation is seen only in certain cell types such as endothelial cells or cells of lymphoid origin (Platanias 2005). After activation by phosphorylation STATs can associate to form homodimers or heterodimers. Then cytoplasmic IRF9 associates with the STAT dimer to form ISG factor 3 complex, which subsequently translocates to the nucleus and initiates transcription by binding specific sites in the promoters (IFN-stimulated response elements) of IFN-stimulated genes (ISGs) (Ivashkiv and Donlin 2014; Platanias 2005). In addition, IFN-induced complex can also bind to another site known as IFN-γ-activated site (GAS) element (Platanias 2005). IFNs induce the expression of hundreds of genes, of these some have only ISREs or only GAS elements in their promoter, whereas others have both elements. Thus, for optimal transcriptional activation a combination of different STATS complex is required to produce functionally distinct genes (Platanias 2005). STAT1 is activated by JAK1 and JAK2 in response to IFN-γ engagement of type II IFN receptors. Phosphorylation of STAT1 on the tyrosine residue at position 701 (Tyr701) results in the formation of STAT1-STAT1 homodimers which translocate to the nucleus to bring about transcription by binding to GAS elements (Aaronson 2002; Boehm et al. 1997; Stark et al. 1998). Type II IFNs are different from type I as they do not form ISGF3 complexes and thereby cannot induce the formation of genes that have only ISREs in their promoter.

4.4.2 Crk-Signaling Pathway in Interferon Signaling

In addition to the Jak STAT pathway, IFNs activates many other signaling pathways. Such a pathway includes the Crk family of adaptor proteins which includes three members—CrkL, CrkI, and CrkII, all of which are cellular homologues of the v-Crk gene (Platanias and Fish 1999; Sattler and Salgia 1998). CrkL is present in the latent form in cytoplasm and is known to constitutively associate with the guanine nucleotide exchange factor (GEF) C3G. Binding of the IFN to the respective receptor results in the association of CrkL-C3G with TYK2 and is activated by phosphorylation of the tyrosine. The active Crkl now forms a complex with STAT-5 to form Crkl-STAT5 complex, which subsequently translocates to the nucleus and binds to GAS elements (TTCTAGGAA) located at the promoters of ISGs. Activation of Crkl can also result in the GEF activity of the C3G, thereby resulting in the activation of GTPase activity of RAPI, a member of the Ras family of small GTPases which mediate induction of the growth suppressive effects by IFNs (Platanias 2005; Schmitt and Stork 2001). Both type I and type II IFNs activate RAPI in a Crkl-dependent manner (Lekmine et al. 2002). RAPI has also been shown to promote cellular proliferation; thus, the effects on cell growth depend on the stimuli and cell type that are involved (Hattori 2003).

4.4.3 PI-3-Kinase Pathway in Interferon Signaling

Phosphatidyl-inositol 3 kinases (PI3K) are enzymes known to regulate an array of protein kinase signaling cascades involved in cellular processes like cell survival, metabolism, proliferation, and inflammation/immunity. The class Ia PI3Ks are dimeric enzymes composed of p110 catalytic subunit attached to a smaller, noncatalytic, regulatory subunit (p85). The p85 subunit contains the SRC homology 2

(SH2) and SH3 domains which interact with other signaling proteins, while the p110 regulates the phosphorylation of the D3 position of the phosphatidylinositol lipids of the inositol ring. Several type I IFNs (IFNs-α, IFNs-β, IFNs-ω) have been shown to induce tyrosine phosphorylation of IRS1 and the p85 subunit of PI3K associates with IRS1 through its SH2 domains in an IFN-dependent manner resulting in activation of catalytic subunit of PI3K (p110) (Uddin et al. 1995). Similar to IRS1, the IRS2 belonging to the same family was subsequently shown to undergo phosphorylation in type I IFN-dependent manner (Burfoot et al. 1997; Platanias et al. 1996). Thus, type I IFN can activate the PI3K signaling pathway in an IRS-dependent and STAT independent manner (Ivashkiv and Donlin 2014).

Similar to type I IFNs, type II IFNs activate PI3K pathway but in an IRS independent manner. The protein that seems to play a role in type II IFN signaling is CBL which has binding sites for the SH3 domain of p85 and is shown to be phosphorylated in an IFN-γ-dependent manner (Alsayed et al. 2000). As mentioned above JAK1 and JAK2 are activated in response to IFN-γ engagement of type II IFN receptors. The activated JAK through an intermediate can also activate the p110 of PI3K, which in turn activates protein kinase C-δ (PKC-δ). Finally, the PKC-δ regulates the phosphorylation of STAT1 on the Ser727 (Nguyen et al. 2001). Although the modification of STAT1 on the Ser727 is not required for its translocation to nucleus, it is essential for transcriptional activation. Many downstream effectors of PI3K signaling pathway have been identified such as 3-phosphatidylinositol-dependent protein kinase 1 (PDK1), serine/threonine kinases (AKT), various isoforms of protein kinase C (PKC), and members of the TEC family of tyrosine kinases (Platanias 2005).

4.4.4 MAPKs in Interferon Signaling

Mitogen-activated protein (MAP) kinases are serine-threonine kinases that play crucial roles in many signal transduction pathways including IFN signaling in mammalian cells (Uddin et al. 1995). There are three types of MAPK in cells: (1) p38 Map kinases, (2) the extracellular signal-regulated kinase family, and (3) the c-Jun NH2-terminal kinase phosphatidylinositol 3′ kinase family. The p38-kinase pathway has been shown to be involved in type I, type II, and type III IFN receptors and acts as an auxiliary cascade to the Jak-Stat pathways and is required for IFN-inducible expression. There are several isoforms of p38 (p38α, p38β, p38γ, and p38δ) encoded by distinct genes, but show structural homology (Fensterl et al. 2015).

Activation of Janus-activated kinase through type I IFN receptor regulates the phosphorylation of VAV proto-oncogene and/or other guanine nucleotide exchange factors (GEFs). This subsequently activates the RAC1 and the small G-protein (SGPs). MAPKKK a MAPK kinase downstream to RAC1 and SGPs is activated, which then regulates MAPK kinases MAPKK3 and MAPKK6, which phosphorylate p38 and activate it. Several downstream effector kinases of p38 including MAPK-activated protein kinase 2 (MAPKAPK2) (MAPKAPK3), mitogen and stress activated kinase 1 (MSK1), and MAPK interacting protein kinase 1 (MNK1) then mediate gene transcription of ISG protein products. p38 MAPK pathway is also activated by the type II IFN and type III receptor and plays important roles in the generation of type II IFN-biological responses in several different cell types (Fensterl et al. 2015).

4.5 Antiviral Action of Interferons

Binding of IFNs to its cognate receptors activates signaling cascades which culminates into the transcriptional induction of a number of antiviral genes called interferon-stimulated genes (ISGs). These genes encode direct antiviral molecules with potential to regulate IFNs signaling positively or negatively. Hundreds of ISGs have been identified by genome-wide transcriptional profiling and most of the ISGs mainly target conserved aspects of viral infection. Specific mechanisms of the ISGs are summarized below (Schneider et al. 2014; Schoggins and Rice 2011).

4.5.1 Classical ISGs

The best characterized IFN-induced ISGs are the "classical ISGs," interferon-induced, double-stranded RNA-activated protein kinase encoded by Eif2ak2 (PKR), MX1 (the myxovirus (influenza virus) resistance 1), and 2′-5′ oligoadenylate synthetase/RNase L system (OAS1, OAS2, OAS3, encoded by Oas1, Oas2, Oas3). PKR is a serine/threonine protein kinase and has been shown to be an inhibitor of cellular and viral mRNA translation by phosphorylation of the protein synthesis initiation factor eIF2. OAS enzyme system catalyzes the formation of 2′-5′ oligoadenylates, which then activate RNase L to degrade viral genomes. The MX1 protein is a dynamin-like large guanosine triphosphatase (GTPase) and it acts by trapping the viral components such as the nucleocapsids and prevents them from reaching their destination by limiting the viral growth.

4.5.2 Cholesterol-25-Hydroxylase (CH25H)

CH25H is an IFN-induced enzyme that converts cholesterol into 25-hydroxycholesterol (25HC), an oxysterol. Recent reports suggest that a change in the physical properties of cell membrane as a result of the formation of high levels of 25HC prevents virus–host membrane fusion. As an enzyme, CH25H modulates the cellular lipid composition and interferes in virus–host membrane fusion, especially among enveloped viruses (e.g., hepatitis C virus, herpes simplex virus).

Besides this direct action, the antiviral activity of 25 HC partly may be due to its ability to regulate sterol biosynthesis pathway. 25 HC can freely permeate membranes and can inhibit sterol biosynthesis in both an autocrine and paracrine manner (Liu et al. 2013).

4.5.3 The Promyelocytic Leukemia (PML) Protein

Around 20 years ago, the expression of PML was detected in 98% of acute promyelocytic leukemia (APL) cases due to chromosomal translocation; later it was identified as INF-induced protein. The PML is known for inhibition of replication of RNA and DNA viruses. The induction of PML enhances the expression of distinct PML isoforms and nuclear bodies (NBs). Several PML isoforms have a key role in transcription regulation and posttranslational modifications such as ubiquitinylation, SUMOylation, acetylation, and phosphorylation, which leads to consolidated antiviral activity against specific viruses (Geoffroy and Chelbi-Alix 2011).

4.5.4 Interferon-Stimulated Gene 15 (ISG15)

Interferon-stimulated gene 15 is a 15 kDa ubiquitin-like protein induced by type 1 interferon. These proteins are responsible for ISGylation, a type of covalent addition to cytoplasmic and nuclear proteins, similar to ubiquitination. ISGylation involves three main enzymes: UBE1L; E1 (ubiquitin-activating enzyme-E1-like protein), E2 (ubiquitin-conjugating enzyme 8), and E3 (estrogen-responsive finger protein and UBP43 or USP18 (ubiquitin-specific protease 18). ISG15 has numerous antiviral functions including inhibition of virus release (e.g., retroviruses HIV-1, ASLV, and Ebola virus) (Morales and Lenschow 2013).

4.5.5 The Interferon-Induced Protein with Tetratricopeptide Repeats (IFIT) Family and IFN-Induced Transmembrane Protein Family (IFITM Families)

Type-1 and type III interferons chiefly induce IFIT family proteins. The genes for these proteins are mainly conserved in mammals, amphibians, and fish but not in lower animals like yeasts, fruit flies, etc. Presence of tetratricopeptide (TPR) multiple repeats is a unique property of these proteins. Some members of the IFIT family act by binding to the subunits of the eIF3 translation initiation complex and thereby reduce the efficiency of cellular cap-dependent protein translation. Besides, IFIT can also function as viral RNA sensor by binding to uncapped 5′-ppp RNA and sequestering it from the actively replicating pool. Studies have also shown that IFIT can bind other proteins such as viral E1 helicase, a protein required for replication, and prevent virus infection.

Humans consist of four IFITM proteins (IFITM1, IFITM2, IFITM3, and IFITM5). Besides antiviral function, IFTM has been shown to play a role in many biological processes such as cell signaling, germ cell homing and maturation, and bone mineralization. IFN-induced transmembrane (IFITM) protein members inhibit endocytic fusion events of a broad spectrum of viruses. However, further work has to be done to establish how IFITM differentially restricts different viruses (Diamond and Farzan 2013; Zhou et al. 2013).

4.5.6 Radical SAM Domain-Containing Molecule/Viperin

Radical SAM domain-containing molecule (RSAD2) known as viperin is induced by JAK-STAT signaling or by direct activation of IRF1/3. Viperin interacts with mitochondrial trifunctional protein after relocalization from endoplasmic reticulum or in the form of ER-derived lipid droplets. Viperin inhibits many viruses which are enveloped, with diverse mechanisms. One of the mechanism identified for influenza A virus is by decreasing the farnesyl diphosphate synthase activity (an enzyme involved in isoprenoid biosynthesis) which alters membrane fluidity and hence interferes in viral budding (e.g., HIV-1). In HCV, viperin has been shown to inhibit viral RNA replication (Duschene and Broderick 2012).

4.5.7 Retroviral Restriction Factors: APOBEC3

Hosts of higher animals including mammals have evolved a mechanism to synthesize virus restriction factors to restrict the infection. Among them apolipoprotein B editing complex (APOBEC3) family members are well characterized for inducing hypermutations in cDNA by interfering in the process of reverse transcription among retroviruses (HIV-1). Overexpression of APOBEC3 leads to cytidine deamination during reverse transcription and that results in lethal mutations. APOBEC3 gene family among humans has A, B, C, D, E, F, G, and H polymorphic forms. The polymorphism in coding and regulatory regions of these genes has deleterious effect on survival of Parvo, Herpes, Papilloma, and hepatitis B viruses in infected host cells. Also APOBEC3 has a key role in boosting innate immunity against many viruses (Harris and Dudley 2015).

4.5.8 SAMHD1

Sterile α motif and histidine-aspartic acid (HD) domain-containing protein 1 (SAMHD1) is another well-characterized retroviral restriction factor. SAMHD1 is a deoxynucleoside triphosphate (dTNP) hydrolase which restricts viral replication (HIV-1) by hydrolyzing deoxynucleotide triphosphates (dNTPs) present in virus-infected cells (CD+T and myeloid cells) (Li et al. 2017).

4.5.9 Bone Marrow Stromal Cell Antigen-2 (BST2)/Tetherin

Tetherin (bone marrow stromal cell antigen 2—BST-2, CD317) is a membrane-bound protein, which inhibits viral budding, release of viral particles from cell surface in retrovirus (all classes), and many other viruses such as filoviruses (Ebola virus and Marburg virus) and paramyxovirus (Nipah virus). Evidence suggests a direct tethering mechanism of virus restriction where the tetherin dimers physically crosslink virion and cellular/other virion membranes, thereby leading to the retention of mature virions in protease-sensitive layers on the plasma membrane. Many viral proteins (glycoproteins) subvert tetherin function (e.g., HIV-1 VPU, HIV-2 ENV, Ebola virus surface glycoprotein-VP-40) by ubiquitination of tetherin resulting in tetherin degradation (Le Tortorec et al. 2011).

4.6 miRNA and IFNs

IFN is known to regulate the expression of a number of noncoding RNA genes, especially micro RNA (miRNA). These miRNAs along with IFN-induced proteins form the innate response system against the invading virus. Each type of IFN (I, II, III) is capable of inducing the expression of miRNA capable of restricting the replication of virus by different mechanisms. Of the several miRNA induced by IFNβ, eight miRNAs show sequence-predicted targets within the HCV genomic RNA and three among these miR-351, miR-431, and miR-488 can inhibit HCV replication. For a more comprehensive review, refer to Pedersen et al. (2007).

4.7 Evasion of the IFN System by Viruses

The IFN system is one of the first lines of defense mechanism mounted by the host against the virus. Virus in due course of evolution has acquired mechanisms to antagonize the IFN-mediated responses by inhibiting the IFN-activated signaling cascade or else by suppressing the antiviral pathways. In the following section of this article, a brief account of how viruses evade the IFN system is described. For a more detailed review, refer to García-Sastre (2017).

Viruses such as flaviviruses form replication complex inside membrane vesicles also termed as viroplasm-like structures (VLS) that protect the viral RNA from interacting with RLRs during RNA synthesis (den Boon and Ahlquist 2010). While in case of influenza virus, the replication occurs in nucleus far away from cytosolic RLRs and TLRs, it has been shown that RIGI can recognize the influenza virus RNA genome during the transit from cytoplasm to the nucleus (Jackson et al. 1982; Weber et al. 2015). The viral RNA with unmethylated 5′ tri- or diphosphate ends is recognized by the RLRs RIG-I and MDA 5, and this in turn activates and induces IFNs. In order to mimic the host, some RNA viruses post-translationally modify RNA by adding 5′ cap structures similar to eukaryotic using viral coded phosphatases and methyltransferases. For some viruses that lack enzymes for posttranslational addition of their own RNA cap, they utilize the host cellular mRNA expressing viral endonucleases and utilize these caps to mask viral RNA (Reguera et al. 2010).

Many DNA viruses such as herpes virus due to their large coding sequences encode viral protein capable of interacting with key mediators of IFN signaling and thereby suppress IFN response. For example, inhibition by herpesviruses of the STING, a signaling molecule which is downstream of cGAS, has been reported (Chen et al. 2016). Similar to DNA virus, some RNA viruses such as arenavirus Z protein (Fan et al. 2010) and paramyxovirus V proteins (Andrejeva et al. 2004) bind to and inhibit members of RIG-I like receptors. The influenza virus PB1-F2 protein is shown to interact with the downstream adaptor mitochondrial protein MAVS (Varga et al. 2012). Viral proteins are also shown to interact with a number of transcription factors involved in IFN induction. The binding of ORF36 protein of murine gamma-herpesvirus to activated IRF3 in the nucleus prevents induction of IFN mRNA synthesis (Hwang et al. 2009) The NS5 protein of yellow fever virus binds to STAT2 and this affects its antiviral action by inhibiting binding of this transcription factor to the IFN responsive promoter elements of the ISGs (Laurent-Rolle et al. 2014). Similarly a number of viral encoded proteins have been shown to bind to STAT1 and STAT2 such as the P proteins of rabies (Vidy et al. 2005), the C6 protein of vaccinia virus (Stuart et al. 2016), and the V and W proteins of paramyxoviruses (Rodriguez et al. 2002; Shaw et al. 2004).

Direct regulation of phosphorylation events involved in IFN induction is another attractive strategy used by virus. These kinases are target of many viral encoded proteins, for example, the IKKε kinase inhibition by the N protein of arenaviruses (Pythoud et al. 2012). The NS4B protein of flaviviruses binds to and inhibits tank binding kinase 1(TBK1) (Dalrymple et al. 2015), VP40 of Ebola virus inhibits the JAK1 (Valmas et al. 2010), and TYK2 inhibition by LMP1 of Epstein-Barr virus (Geiger and Martin 2006). IFN signaling is yet another pathway targeted by the viruses to evade the host. The IFNAR1 chain of the IFN receptor is the preferred target for degradation by phosphorylation-dependent ubiquitination, and this affects type I signaling and antiviral defense (Liu et al. 2009).

Many viral proteases that are involved in the processing of the viral polyprotein are also shown to cleave and degrade proteins necessary for IFN response. This is most commonly seen in positive-strand RNA viruses. For example, the hepatitis C virus protease cleaves MAVS (Li et al. 2005; Meylan et al. 2005), while the dengue virus protease cleaves STING (Aguirre et al. 2012; Yu et al. 2012). Another mechanism by which virus degrades host proteins is by marking the protein by addition of K48 polyubiquitin chains makes it susceptible to proteosome degradation. This is exemplified by NS5 of Zika virus that targets human STAT2 for proteosomal degradation in infected cells (Grant et al. 2016).

Viruses have been shown to prevent gene expression by epigenetic regulation, thereby interfering with transcription mechanism of the host. Adenovirus E1A protein has been shown to bind to and dissociate the host nuclear complex that is required for histone monoubiquitination at H2B lysine 120 (Fonseca et al. 2012). This modification prevents the opening of the chromatin to allow transcription, and therefore IFN is unable to activate transcription of the ISGs. In addition to transcriptional shutoff, viruses can suppress host gene expression by other mechanisms such as the posttranscriptional inhibition of cellular RNA processing, trafficking, and translational shutoff. The NS1 protein of influenza virus has been shown to prevent the termination as well as polyadenylation of cellular mRNA (Nemeroff et al. 1998). The M protein of VSV targets the nuclear pore components and thereby inhibits the RNA export from the nucleus (von Kobbe et al. 2000). Similarly the ORF10 of the Kaposi's sarcoma-associated herpesvirus inhibits the translocation of mRNA of genes coding for mitosis, gene silencing, DNA metabolic process, chromosome organization, cell cycle, and transcription regulation (Gong et al. 2016). These genes although not involved in direct IFN response, may be detrimental for survival of virus. Translational shutdown of host gene expression is another strategy used by many viruses. This is best exemplified by the hepatitis C virus infection during which the PKR mediated phosphorylation of the translation factor eiF2α results in the inhibition of cellular translation, while the hepatitis mRNA translation continues as it depends on a different mechanism of 5′ ribosomal entry site that is insensitive to PKR mediated translation inhibition (Garaigorta and Chisari 2009). Many viruses recruit decoy proteins to sequester the host proteins and prevent its binding to specific targets. The B18R protein of vaccinia virus, which is a poxvirus-encoded soluble IFN receptor, can sequester IFN before it can bind to the IFN receptor (Symons et al. 1995). Similarly the K3L protein of the poxvirus acts as a decoy for PKR substrates (Beattie et al. 1991). Thus, a virus may use any of the above mechanism or multiple mechanisms as a strategy to suppress the effect of IFN-mediated host response.

4.8 Therapeutic Intervention of Viral Infection Using Interferons

As the virus hijacks the host immune response and causes chronic infection, for elimination and controlling the virus replication a therapeutic intervention is needed. The strong antiviral state conferred by the IFNs to the host cells combined with its ability to modulate the immune system has made it an ideal candidate to be used in antiviral therapy. IFN-α was initially used for the treatment of hepatitis C virus (HCV) infection, but due to its short half-life and rapid elimination from the body was unable to achieve the desired effect (Lin and Young 2014). In order to maintain stable level in the blood, pegylated IFN was used for the treatment of chronic HCV. IFN-α is also found to be effective against hepatitis B virus infection. Besides the antiviral activity IFN-α is a powerful immune modulator and the studies related to its efficiency as an adjuvant have met with limited success (Rizza et al. 2008). IFN-α treatment is accompanied by systemic side effects such as neutropenia and thrombocytopenia and as a remedy other IFN molecules such as albumin-IFN-αand consensus IFN (CIFN) have been developed to replace PEG-IFN-α. CIFN with 10–100 times higher antiviral activity compared to standard IFN-α is a recombinant protein and is found to be effective at a lower dose with limited side effects (Sjogren et al. 2007). CIFN is a recombinant IFN with most commonly observed amino acids from several type I IFN subtypes. Similar to type I IFNs, other classes of IFNs have been shown to be useful in therapeutic intervention during viral infections. IFN-γ has been shown to be protective against HIV-associated opportunistic infection with or without highly active antiretroviral therapy (Jarvis et al. 2012). A comparison of treatment with PEG-IFN-α and PEG-IFN-λ along with ribavirin showed higher rates of rapid virologic response regardless of the HCV genotypes. Unlike IFN-α, treatment of HCV with IFN-λ is much safer and less toxic as the restricted tissue expression of the IFN-λR1 chain ensures a localized reaction (Lin and Young 2014).

References

Aaronson DS (2002) A road map for those who don't know JAK-STAT. Science 296(5573):1653–1655. https://doi.org/10.1126/science.1071545

Aguirre S, Maestre AM, Pagni S, Patel JR, Savage T, Gutman D et al (2012) DENV inhibits type I IFN production in infected cells by cleaving human STING. PLoS Pathog 8(10):e1002934. https://doi.org/10.1371/journal.ppat.1002934

Alexopoulou L, Holt AC, Medzhitov R, Flavell RA (2001) Recognition of double-stranded RNA and activation of NF-kappaB by toll-like receptor 3. Nature 413(6857):732–738. https://doi.org/10.1038/35099560

Alsayed Y, Uddin S, Ahmad S, Majchrzak B, Druker BJ, Fish EN, Platanias LC (2000) IFN-gamma activates the C3G/Rap1 signaling pathway. Journal of Immunology (Baltimore, Md. 1950) 164(4):1800–1806. http://www.ncbi.nlm.nih.gov/pubmed/10657627

Andrejeva J, Childs KS, Young DF, Carlos TS, Stock N, Goodbourn S, Randall RE (2004) The V proteins of paramyxoviruses bind the IFN-inducible RNA helicase, mda-5, and inhibit its activation of the IFN-beta promoter. Proc Natl Acad Sci U S A 101(49):17264–17269. https://doi.org/10.1073/pnas.0407639101

Beattie E, Tartaglia J, Paoletti E (1991) Vaccinia virus-encoded eIF-2 alpha homolog abrogates the antiviral effect of interferon. Virology 183(1):419–422. http://www.ncbi.nlm.nih.gov/pubmed/1711259

Boehm U, Klamp T, Groot M, Howard JC (1997) Cellular responses to interferon-γ. Annu Rev Immunol 15(1):749–795. https://doi.org/10.1146/annurev.immunol.15.1.749

Bruns AM, Leser GP, Lamb RA, Horvath CM (2014) The innate immune sensor LGP2 activates antiviral signaling by regulating MDA5-RNA interaction and filament assembly. Mol Cell 55(5):771–781. https://doi.org/10.1016/j.molcel.2014.07.003

Burfoot MS, Rogers NC, Watling D, Smith JM, Pons S, Paonessaw G et al (1997) Janus kinase-dependent activation of insulin receptor substrate 1 in response to interleukin-4, oncostatin M, and the interferons. The Journal of Biological Chemistry 272(39):24183–24190. http://www.ncbi.nlm.nih.gov/pubmed/9305869

Chen Q, Sun L, Chen ZJ (2016) Regulation and function of the cGAS-STING pathway of cytosolic DNA sensing. Nat Immunol 17(10):1142–1149. https://doi.org/10.1038/ni.3558

Childs KS, Randall RE, Goodbourn S (2013) LGP2 plays a critical role in sensitizing mda-5 to activation by double-stranded RNA. PLoS One 8(5):e64202. https://doi.org/10.1371/journal.pone.0064202

Dalrymple NA, Cimica V, Mackow ER (2015) Dengue virus NS proteins inhibit RIG-I/MAVS signaling by blocking TBK1/IRF3 phosphorylation: dengue virus serotype 1 NS4A is a unique interferon-regulating virulence determinant. MBio 6(3):e00553–e00515. https://doi.org/10.1128/mBio.00553-15

de Weerd NA, Nguyen T (2012) The interferons and their receptors – distribution and regulation. Immunol Cell Biol 90(5):483–491. https://doi.org/10.1038/icb.2012.9

Decker T, Kovarik P, Meinke A (1997) GAS elements: a few nucleotides with a major impact on cytokine-induced gene expression. J Interf Cytokine Res 17(3):121–134. https://doi.org/10.1089/jir.1997.17.121

den Boon JA, Ahlquist P (2010) Organelle-like membrane compartmentalization of positive-strand RNA virus replication factories. Annu Rev Microbiol 64:241–256. https://doi.org/10.1146/annurev.micro.112408.134012

Der SD, Zhou A, Williams BR, Silverman RH (1998) Identification of genes differentially regulated by interferon alpha, beta, or gamma using oligonucleotide arrays. Proceedings of the National Academy of Sciences of the United States of America 95(26):15623–15628. http://www.ncbi.nlm.nih.gov/pubmed/9861020

Diamond MS, Farzan M (2013) The broad-spectrum antiviral functions of IFIT and IFITM proteins. Nat Rev Immunol 13(1):46–57. https://doi.org/10.1038/nri3344

Doyle SE, Schreckhise H, Khuu-Duong K, Henderson K, Rosler R, Storey H et al (2006) Interleukin-29 uses a type 1 interferon-like program to promote antiviral responses in human hepatocytes. Hepatology 44(4):896–906. https://doi.org/10.1002/hep.21312

Duschene KS, Broderick JB (2012) Viperin: a radical response to viral infection. Biomol Concepts 3(3):255. https://doi.org/10.1515/bmc-2011-0057

Fan L, Briese T, Lipkin WI (2010) Z proteins of New World arenaviruses bind RIG-I and interfere with type I interferon induction. J Virol 84(4):1785–1791. https://doi.org/10.1128/JVI.01362-09

Fensterl V, Chattopadhyay S, Sen GC (2015) No love lost between viruses and interferons. Ann Rev Virol 2(1):549–572. https://doi.org/10.1146/annurev-virology-100114-055249

Fonseca GJ, Thillainadesan G, Yousef AF, Ablack JN, Mossman KL, Torchia J, Mymryk JS (2012) Adenovirus evasion of interferon-mediated innate immunity by direct antagonism of a cellular histone posttranslational modification. Cell Host Microbe 11(6):597–606. https://doi.org/10.1016/j.chom.2012.05.005

Ford E, Thanos D (2010) The transcriptional code of human IFN-β gene expression. Biochim Biophys Acta (BBA) - Gene Regul Mech 1799(3–4):328–336. https://doi.org/10.1016/j.bbagrm.2010.01.010

Garaigorta U, Chisari FV (2009) Hepatitis C virus blocks interferon effector function by inducing protein kinase R phosphorylation. Cell Host Microbe 6(6):513–522. https://doi.org/10.1016/j.chom.2009.11.004

García-Sastre A (2017) Ten strategies of interferon evasion by viruses. Cell Host Microbe 22(2):176–184

Gay NJ, Symmons MF, Gangloff M, Bryant CE (2014) Assembly and localization of toll-like receptor signalling complexes. Nat Rev Immunol 14(8):546–558. https://doi.org/10.1038/nri3713

Geiger TR, Martin JM (2006) The Epstein-Barr virus-encoded LMP-1 oncoprotein negatively affects Tyk2 phosphorylation and interferon signaling in human B cells. J Virol 80(23):11638–11650. https://doi.org/10.1128/JVI.01570-06

Génin P, Vaccaro A, Civas A (2009) The role of differential expression of human interferon-A genes in antiviral immunity. Cytokine Growth Factor Rev 20(4):283–295. https://doi.org/10.1016/j.cytogfr.2009.07.005

Geoffroy M-C, Chelbi-Alix MK (2011) Role of promyelocytic leukemia protein in host antiviral defense. J Interfer Cytokine Res 31(1):145–158. https://doi.org/10.1089/jir.2010.0111

Gong D, Kim YH, Xiao Y, Du Y, Xie Y, Lee KK et al (2016) A herpesvirus protein selectively inhibits cellular mRNA nuclear export. Cell Host Microbe 20(5):642–653. https://doi.org/10.1016/j.chom.2016.10.004

Goubau D, Schlee M, Deddouche S, Pruijssers AJ, Zillinger T, Goldeck M et al (2014) Antiviral immunity via RIG-I-mediated recognition of RNA bearing 5′-diphosphates. Nature 514(7522):372–375. https://doi.org/10.1038/nature13590

Grant A, Ponia SS, Tripathi S, Balasubramaniam V, Miorin L, Sourisseau M et al (2016) Zika virus targets human STAT2 to inhibit type I interferon signaling. Cell Host Microbe 19(6):882–890. https://doi.org/10.1016/j.chom.2016.05.009

Hardarson HS, Baker JS, Yang Z, Purevjav E, Huang C-H, Alexopoulou L et al (2007) Toll-like receptor 3 is an essential component of the innate stress response in virus-induced cardiac injury. Am J Phys Heart Circ Phys 292(1):H251–H258. https://doi.org/10.1152/ajpheart.00398.2006

Harris RS, Dudley JP (2015) APOBECs and virus restriction. Virology 479–480:131–145. https://doi.org/10.1016/j.virol.2015.03.012

Hattori M (2003) Rap1 GTPase: functions, regulation, and malignancy. J Biochem 134(4):479–484. https://doi.org/10.1093/jb/mvg180

Hwang S, Kim KS, Flano E, Wu T-T, Tong LM, Park AN et al (2009) Conserved herpesviral kinase promotes viral persistence by inhibiting the IRF-3-mediated type I interferon response. Cell Host Microbe 5(2):166–178. https://doi.org/10.1016/j.chom.2008.12.013

Isaacs A, Lindenmann J (1957) Virus interference. I. The interferon. Proc R Soc Lond Ser B Biol Sci 147(927):258–267. http://www.ncbi.nlm.nih.gov/pubmed/13465720

Ivashkiv LB, Donlin LT (2014) Regulation of type I interferon responses. Nat Rev Immunol 14(1):36–49. https://doi.org/10.1038/nri3581
Iversen MB, Paludan SR (2010) Mechanisms of type III interferon expression. J Interf Cytokine Res 30(8):573–578. https://doi.org/10.1089/jir.2010.0063
Jackson DA, Caton AJ, McCready SJ, Cook PR (1982) Influenza virus RNA is synthesized at fixed sites in the nucleus. Nature 296(5855):366–368. http://www.ncbi.nlm.nih.gov/pubmed/7063035
Jarvis JN, Meintjes G, Rebe K, Williams GN, Bicanic T, Williams A et al (2012) Adjunctive interferon-γ immunotherapy for the treatment of HIV-associated cryptococcal meningitis: a randomized controlled trial. AIDS (London, England) 26(9):1105–1113. https://doi.org/10.1097/QAD.0b013e3283536a93
Kato H, Takeuchi O, Mikamo-Satoh E, Hirai R, Kawai T, Matsushita K et al (2008) Length-dependent recognition of double-stranded ribonucleic acids by retinoic acid–inducible gene-I and melanoma differentiation–associated gene 5. J Exp Med 205(7):1601–1610. https://doi.org/10.1084/jem.20080091
Krug A, French AR, Barchet W, Fischer JAA, Dzionek A, Pingel JT et al (2004) TLR9-dependent recognition of MCMV by IPC and DC generates coordinated cytokine responses that activate antiviral NK cell function. Immunity 21(1):107–119. https://doi.org/10.1016/j.immuni.2004.06.007
Laurent-Rolle M, Morrison J, Rajsbaum R, Macleod JML, Pisanelli G, Pham A et al (2014) The interferon signaling antagonist function of yellow fever virus NS5 protein is activated by type I interferon. Cell Host Microbe 16(3):314–327. https://doi.org/10.1016/j.chom.2014.07.015
Le Goffic R, Balloy V, Lagranderie M, Alexopoulou L, Escriou N, Flavell R et al (2006) Detrimental contribution of the toll-like receptor (TLR)3 to influenza a virus–induced acute pneumonia. PLoS Pathog 2(6):e53. https://doi.org/10.1371/journal.ppat.0020053
Le Tortorec A, Willey S, Neil SJD (2011) Antiviral inhibition of enveloped virus release by tetherin/BST-2: action and counteraction. Viruses 3(5):520–540. https://doi.org/10.3390/v3050520
Lee HK, Lund JM, Ramanathan B, Mizushima N, Iwasaki A (2007) Autophagy-dependent viral recognition by plasmacytoid dendritic cells. Science 315(5817):1398–1401. https://doi.org/10.1126/science.1136880
Lekmine F, Sassano A, Uddin S, Majchrzak B, Miura O, Druker BJ et al (2002) The CrkL adapter protein is required for type I interferon-dependent gene transcription and activation of the small G-protein Rap1. Biochem Biophys Res Commun 291(4):744–750. https://doi.org/10.1006/bbrc.2002.6516
Li X-D, Sun L, Seth RB, Pineda G, Chen ZJ (2005) Hepatitis C virus protease NS3/4A cleaves mitochondrial antiviral signaling protein off the mitochondria to evade innate immunity. Proc Natl Acad Sci U S A 102(49):17717–17722. https://doi.org/10.1073/pnas.0508531102
Li M, Zhang D, Zhu M, Shen Y, Wei W, Ying S et al (2017) Roles of SAMHD1 in antiviral defense, autoimmunity and cancer. Rev Med Virol 27(4). https://doi.org/10.1002/rmv.1931
Lin F, Young HA (2014) Interferons: success in anti-viral immunotherapy. Cytokine Growth Factor Rev 25(4):369–376. https://doi.org/10.1016/j.cytogfr.2014.07.015
Liu J, HuangFu W-C, Kumar KGS, Qian J, Casey JP, Hamanaka RB et al (2009) Virus-induced unfolded protein response attenuates antiviral defenses via phosphorylation-dependent degradation of the type I interferon receptor. Cell Host Microbe 5(1):72–83. https://doi.org/10.1016/j.chom.2008.11.008
Liu S-Y, Aliyari R, Chikere K, Li G, Marsden MD, Smith JK et al (2013) Interferon-inducible cholesterol-25-hydroxylase broadly inhibits viral entry by production of 25-hydroxycholesterol. Immunity 38(1):92–105. https://doi.org/10.1016/j.immuni.2012.11.005
Meylan E, Curran J, Hofmann K, Moradpour D, Binder M, Bartenschlager R, Tschopp J (2005) Cardif is an adaptor protein in the RIG-I antiviral pathway and is targeted by hepatitis C virus. Nature 437(7062):1167–1172. https://doi.org/10.1038/nature04193
Morales DJ, Lenschow DJ (2013) The antiviral activities of ISG15. J Mol Biol 425(24):4995–5008. https://doi.org/10.1016/j.jmb.2013.09.041

Mordstein M, Neugebauer E, Ditt V, Jessen B, Rieger T, Falcone V et al (2010) Lambda interferon renders epithelial cells of the respiratory and gastrointestinal tracts resistant to viral infections. J Virol 84(11):5670–5677. https://doi.org/10.1128/JVI.00272-10

Nemeroff ME, Barabino SM, Li Y, Keller W, Krug RM (1998) Influenza virus NS1 protein interacts with the cellular 30 kDa subunit of CPSF and inhibits 3'end formation of cellular pre-mRNAs. Mol Cell 1(7):991–1000. http://www.ncbi.nlm.nih.gov/pubmed/9651582

Nguyen VQ, Co C, Li JJ (2001) Cyclin-dependent kinases prevent DNA re-replication through multiple mechanisms. Nature 411(6841):1068–1073. https://doi.org/10.1038/35082600

Onoguchi K, Yoneyama M, Takemura A, Akira S, Taniguchi T, Namiki H, Fujita T (2007) Viral infections activate types I and III interferon genes through a common mechanism. J Biol Chem 282(10):7576–7581. https://doi.org/10.1074/jbc.M608618200

Osterlund PI, Pietilä TE, Veckman V, Kotenko SV, Julkunen I (2007) IFN regulatory factor family members differentially regulate the expression of type III IFN (IFN-lambda) genes. Journal of Immunology (Baltimore, Md. : 1950) 179(6):3434–3442. http://www.ncbi.nlm.nih.gov/pubmed/17785777

Pandey S, Kawai T, Akira S (2015) Microbial sensing by toll-like receptors and intracellular nucleic acid sensors. Cold Spring Harb Perspect Biol 7(1):a016246. https://doi.org/10.1101/cshperspect.a016246

Panne D, Maniatis T, Harrison SC (2007) An atomic model of the interferon-β enhanceosome. Cell 129(6):1111–1123. https://doi.org/10.1016/j.cell.2007.05.019

Pedersen IM, Cheng G, Wieland S, Volinia S, Croce CM, Chisari FV, David M (2007) Interferon modulation of cellular microRNAs as an antiviral mechanism. Nature 449(7164):919–922. https://doi.org/10.1038/nature06205

Platanias LC (2005) Mechanisms of type-I- and type-II-interferon-mediated signalling. Nat Rev Immunol 5(5):375–386. https://doi.org/10.1038/nri1604

Platanias LC, Fish EN (1999) Signaling pathways activated by interferons. Experimental Hematol 27(11):1583–1592. http://www.ncbi.nlm.nih.gov/pubmed/10560905

Platanias LC, Uddin S, Yetter A, Sun XJ, White MF (1996) The type I interferon receptor mediates tyrosine phosphorylation of insulin receptor substrate 2. J Biol Chem 271(1):278–282. http://www.ncbi.nlm.nih.gov/pubmed/8550573

Pythoud C, Rodrigo WWSI, Pasqual G, Rothenberger S, Martínez-Sobrido L, de la Torre JC, Kunz S (2012) Arenavirus nucleoprotein targets interferon regulatory factor-activating kinase IKKε. J Virol 86(15):7728–7738. https://doi.org/10.1128/JVI.00187-12

Raftery N, Stevenson NJ (2017) Advances in anti-viral immune defence: revealing the importance of the IFN JAK/STAT pathway. Cell Mol Life Sci 74(14):2525–2535. https://doi.org/10.1007/s00018-017-2520-2

Reguera J, Weber F, Cusack S (2010) Bunyaviridae RNA polymerases (L-protein) have an N-terminal, influenza-like endonuclease domain, essential for viral cap-dependent transcription. PLoS Pathog 6(9):e1001101. https://doi.org/10.1371/journal.ppat.1001101

Rizza P, Capone I, Urbani F, Montefiore E, Rapicetta M, Chionne P et al (2008) Evaluation of the effects of human leukocyte IFN-alpha on the immune response to the HBV vaccine in healthy unvaccinated individuals. Vaccine 26(8):1038–1049. https://doi.org/10.1016/j.vaccine.2007.12.044

Rodriguez JJ, Parisien J-P, Horvath CM (2002) Nipah virus V protein evades alpha and gamma interferons by preventing STAT1 and STAT2 activation and nuclear accumulation. J Virol 76(22):11476–11483. http://www.ncbi.nlm.nih.gov/pubmed/12388709

Saito T, Hirai R, Loo Y-M, Owen D, Johnson CL, Sinha SC et al (2007) Regulation of innate antiviral defenses through a shared repressor domain in RIG-I and LGP2. Proc Natl Acad Sci U S A 104(2):582–587. https://doi.org/10.1073/pnas.0606699104

Samuel CE (2001) Antiviral actions of interferons. Clin Microbiol Rev 14(4):778–809. https://doi.org/10.1128/CMR.14.4.778-809.2001

Sattler M, Salgia R (1998) Role of the adapter protein CRKL in signal transduction of normal hematopoietic and BCR/ABL-transformed cells. Leukemia 12(5):637–644. http://www.ncbi.nlm.nih.gov/pubmed/9593259

Schmitt JM, Stork PJS (2001) Cyclic AMP-mediated inhibition of cell growth requires the small G protein Rap1. Mol Cell Biol 21(11):3671–3683. https://doi.org/10.1128/MCB.21.11.3671-3683.2001

Schneider WM, Chevillotte MD, Rice CM (2014) Interferon-stimulated genes: a complex web of host defenses. Annu Rev Immunol 32(1):513–545. https://doi.org/10.1146/annurev-immunol-032713-120231

Schoggins JW, Rice CM (2011) Interferon-stimulated genes and their antiviral effector functions. Curr Opin Virol 1(6):519–525. https://doi.org/10.1016/j.coviro.2011.10.008

Shaw ML, García-Sastre A, Palese P, Basler CF (2004) Nipah virus V and W proteins have a common STAT1-binding domain yet inhibit STAT1 activation from the cytoplasmic and nuclear compartments, respectively. J Virol 78(11):5633–5641. https://doi.org/10.1128/JVI.78.11.5633-5641.2004

Sheahan T, Imanaka N, Marukian S, Dorner M, Liu P, Ploss A, Rice CM (2014) Interferon lambda alleles predict innate antiviral immune responses and hepatitis C virus permissiveness. Cell Host Microbe 15(2):190–202. https://doi.org/10.1016/j.chom.2014.01.007

Sjogren MH, Sjogren R, Lyons MF, Ryan M, Santoro J, Smith C et al (2007) Antiviral response of HCV genotype 1 to consensus interferon and ribavirin versus pegylated interferon and ribavirin. Dig Dis Sci 52(6):1540–1547. https://doi.org/10.1007/s10620-007-9757-9

Sommereyns C, Paul S, Staeheli P, Michiels T (2008) IFN-lambda (IFN-lambda) is expressed in a tissue-dependent fashion and primarily acts on epithelial cells in vivo. PLoS Pathog 4(3):e1000017. https://doi.org/10.1371/journal.ppat.1000017

Stark GR, Kerr IM, Williams BRG, Silverman RH, Schreiber RD (1998) How cells respond to interferons. Annu Rev Biochem 67(1):227–264. https://doi.org/10.1146/annurev.biochem.67.1.227

Stuart JH, Sumner RP, Lu Y, Snowden JS, Smith GL (2016) Vaccinia virus protein C6 inhibits type I IFN signalling in the nucleus and binds to the transactivation domain of STAT2. PLoS Pathog 12(12):e1005955. https://doi.org/10.1371/journal.ppat.1005955

Symons JA, Alcamí A, Smith GL (1995) Vaccinia virus encodes a soluble type I interferon receptor of novel structure and broad species specificity. Cell 81(4):551–560. http://www.ncbi.nlm.nih.gov/pubmed/7758109

Thompson MR, Kaminski JJ, Kurt-Jones EA, Fitzgerald KA (2011) Pattern recognition receptors and the innate immune response to viral infection. Viruses 3(6):920–940. https://doi.org/10.3390/v3060920

Uddin S, Yenush L, Sun XJ, Sweet ME, White MF, Platanias LC (1995) Interferon-alpha engages the insulin receptor substrate-1 to associate with the phosphatidylinositol 3′-kinase. J Biol Chem 270(27):15938–15941. http://www.ncbi.nlm.nih.gov/pubmed/7608146

Valmas C, Grosch MN, Schümann M, Olejnik J, Martinez O, Best SM et al (2010) Marburg virus evades interferon responses by a mechanism distinct from Ebola virus. PLoS Pathog 6(1):e1000721. https://doi.org/10.1371/journal.ppat.1000721

Varga ZT, Grant A, Manicassamy B, Palese P (2012) Influenza virus protein PB1-F2 inhibits the induction of type I interferon by binding to MAVS and decreasing mitochondrial membrane potential. J Virol 86(16):8359–8366. https://doi.org/10.1128/JVI.01122-12

Vidy A, Chelbi-Alix M, Blondel D (2005) Rabies virus P protein interacts with STAT1 and inhibits interferon signal transduction pathways. J Virol 79(22):14411–14420. https://doi.org/10.1128/JVI.79.22.14411-14420.2005

von Kobbe C, van Deursen JM, Rodrigues JP, Sitterlin D, Bachi A, Wu X et al (2000) Vesicular stomatitis virus matrix protein inhibits host cell gene expression by targeting the nucleoporin Nup98. Mol Cell 6(5):1243–1252. http://www.ncbi.nlm.nih.gov/pubmed/11106761

Weber M, Gawanbacht A, Habjan M, Rang A, Borner C, Schmidt AM et al (2013) Incoming RNA virus nucleocapsids containing a 5′-triphosphorylated genome activate RIG-I and antiviral signaling. Cell Host Microbe 13(3):336–346. https://doi.org/10.1016/j.chom.2013.01.012

Weber M, Sediri H, Felgenhauer U, Binzen I, Bänfer S, Jacob R, Brunotte L, García-Sastre A, Schmid-Burgk JL, Schmidt T, Hornung V, Kochs G, Schwemmle M, Klenk H-D, Weber F (2015) Influenza virus adaptation PB2-627K modulates nucleocapsid inhibition by the pathogen sensor RIG-I. Cell Host Microbe 17(3):309–319

Wen Z, Darnell JE (1997) Mapping of Stat3 serine phosphorylation to a single residue (727) and evidence that serine phosphorylation has no influence on DNA binding of Stat1 and Stat3. Nucl Acids Res 25(11):2062–2067. http://www.ncbi.nlm.nih.gov/pubmed/9153303

Wen Z, Zhong Z, Darnell JE (1995) Maximal activation of transcription by Stat1 and Stat3 requires both tyrosine and serine phosphorylation. Cell 82(2):241–250. http://www.ncbi.nlm.nih.gov/pubmed/7543024

Wu J, Chen ZJ (2014) Innate immune sensing and signaling of cytosolic nucleic acids. Annu Rev Immunol 32(1):461–488. https://doi.org/10.1146/annurev-immunol-032713-120156

Yoneyama M, Onomoto K, Jogi M, Akaboshi T, Fujita T (2015) Viral RNA detection by RIG-I-like receptors. Curr Opin Immunol 32:48–53. https://doi.org/10.1016/j.coi.2014.12.012

Yu C-Y, Chang T-H, Liang J-J, Chiang R-L, Lee Y-L, Liao C-L, Lin Y-L (2012) Dengue virus targets the adaptor protein MITA to subvert host innate immunity. PLoS Pathog 8(6):e1002780. https://doi.org/10.1371/journal.ppat.1002780

Zhang S-Y, Jouanguy E, Ugolini S, Smahi A, Elain G, Romero P et al (2007) TLR3 deficiency in patients with herpes simplex encephalitis. Science 317(5844):1522–1527. https://doi.org/10.1126/science.1139522

Zhou X, Michal JJ, Zhang L, Ding B, Lunney JK, Liu B, Jiang Z (2013) Interferon induced IFIT family genes in host antiviral defense. Int J Biol Sci 9(2):200–208. https://doi.org/10.7150/ijbs.5613

Züst R, Cervantes-Barragan L, Habjan M, Maier R, Neuman BW, Ziebuhr J et al (2011) Ribose 2'-O-methylation provides a molecular signature for the distinction of self and non-self mRNA dependent on the RNA sensor Mda5. Nat Immunol 12(2):137–143. https://doi.org/10.1038/ni.1979

Role of Cytokines in Infectious Viral Disease

5

Pavani Sanapala and Sudhakar Pola

Abstract

Cytokines are cell signaling polypeptides. They regulate and promote immune response, mostly related to autoimmunity. Role of cytokines in viral disease as cytokine biology in recent studies shows multifaceted interactions that are directly involved in the advancement of disease pathogenesis. Viruses induce the expression of numerous cytokine protein genes that are activated from the dormant state, and can also stimulate the expression or replication of extremely identical cytokines. Cytokines are also involved in treating viral diseases, especially human interferons. The ongoing developmental swift in the area of cytokines has led to the eventual growth of new antiviral therapeutic approaches rooted in mimicking the cytokine network. Some cytokines, especially IL-12 in combination with interferon γ, decrease the pathogenicity of viruses. Likewise, cytokines induce or increase the viral pathogenicity, for instance, Leishmania major, where IL-4 has a deleterious effect. This chapter details the salient features of cytokines, classification, inter-relation of cytokines- viral expression and signaling pathways of viruses inducing cytokines production.

Keywords

Cytokines · Chemokines · Viruses · NF-κβ activity · Signal transduction · viral infections

P. Sanapala · S. Pola (✉)
Department of Biotechnology, Andhra University, Visakhapatnam, Andhra Pradesh, India
e-mail: sudhakar@andhrauniversity.edu.in

P. V. Bramhachari (ed.), *Dynamics of immune activation in viral diseases*,
https://doi.org/10.1007/978-981-15-1045-8_5

5.1 Introduction

Cytokines, also called immunocytokines, are a broad group of polypeptides produced by both immune and non-immune cells (Ferreira et al. 2018). Cytokines modulate a wide range of biological function together with hematopoiesis, inflammation, proliferation, repair, differentiation, and innate and adaptive immunity even at a lower concentration (Chung 2009; Balkwill and Burke 1989; Whicher and Evans 1990). Cytokines act as a de facto host defense and amalgamate host metabolic function. They play a crucial role in guarding against viral and bacterial infections. Indeed, cytokines are also concerned with developing disease symptoms and pathogenesis (Balkwill and Burke 1989).

Viruses as small obligate parasites cannot replicate by themselves. For their replication and continued existence, they depend upon the cellular mechanism used during growth and differentiation of the susceptible cell they invade (Ramshaw et al. 1997). Viral infection characterizes an insightful challenge to host endurance, where the ability of the virus to duplicate or persevere in the host is intended in opposition to host antiviral defense mechanisms, i.e., both immune and non-immune responses. Certainly, the point of contact of a virus with its host is complex and active (Ramshaw et al. 1997; Campbell 1991), which rather seems not startling of the function played by cytokines in viral infections. Cytokines uphold viral replication and dodge host defense in the course of regulating humoral and cellular immune response (Campbell 1991).

The direct expression of multiple cytokines distinguishes many viral diseases. Both hemopoietic and non-hemopoietic together constitute a direct change in cytokine gene product expression by viruses. These approaches suggest that the law of complete and local antiviral responses may possibly be mediated with cell exterior to the hemopoietic cubicle. In this chapter, we review in brief general features of cytokines and its importance in biological functions and mainly focus on its role in infectious and viral diseases.

5.2 Cytokines

Cytokines are low molecular mass proteins habitually below 30 kDa in size. These polypeptides are active on signaling molecules and cells invigorating them toward sites of inflammation, infections, and traumas (Ferreira et al. 2018). Cytokines are emanated by various cells at confined higher concentrations, and entailed in the cell to cell communications either at the site they originated (autocrine effect) or at the adjacent cells (paracrine fashion), or on distant cells (endocrine effect/systemic effect) (Stadnyk 1994). As a result, the frontier between hormones and cytokines is quite unclear. Indeed, standard hormones, for instance, prolactin (PRL), growth hormone (GH), leptin, and erythropoietin (EPO), are all evidently cytokines confirmed by their receptor structure and their manner of signaling. Perhaps it is just simplest to accept that cell–cell communication and host defense went hand in hand during evolution, and so functional and structural resemblance exists among families of molecules that act on the immune, hematopoietic, endocrine, and nervous systems (O'Shea et al. 2019).

Earlier many attempts for a suitable nomenclature were made, but Cohen et al. emphasized that no single cell as a source is responsible for origination, as cytokines are made by one or more cell types (Cohen et al. 1974). The first defined cytokines were lymphokines a soluble product originated from lymphocytes in stimuli to the polyclonal antigen. Other such previous attempts were monokines for monocytes products. However, this nomenclature or classification did not seem apt since few cytokines, namely, interleukin 6 and tumor necrosis factor α, were made up of both cell types (Feldmann 1998). Cytokine may act interactively or antipathetically (Zhang and An 2007).

5.3 Classification and Source of Cytokines

Cytokines are recognized by extensive pleiotropism and factor of redundancy, with each cytokine comprising numerous overlie functions, and each role prospectively transmitted by more than one cytokine. This, by reasons made the classification of cytokines a difficult chore. Classifying cytokines in the best way was a challenge; hence they are grouped in different ways. One such categorization is based on the type of receptor that they bind. This includes the following: type I and type II receptors, i.e., hematopoietin and interferon family, tumor necrosis factor family receptors, interleukin-1 receptor, interleukin 17 receptor, toll-like receptors, tyrosine kinase receptors, serine kinases, a receptor of transforming growth factor-β family, and chemokine that binds to transmembrane domain receptors (Vilcek 2003). Other categorizations were by numerical cataloging based on discovery (IL-1 to IL-35), by their functional activity, by their kinetic role in the inflammatory response, by the origin of the primary cell, and as a recently added feature by structural homologies common with interrelated molecules (McInnes 2017) (Table 5.1).

Many cells are a source for the production of cytokines. As illustrated above, the most basic exemplary for cytokine action was resultant of T-lymphocytes that stimulated the production of immature T cells (Hamblin 1988). By analogy with cytokines derived from lymphocytes, two types of Helper T cells were known to secrete and generate cytokine products with varied functional properties (Barrett 1996). Later in line, a second cellular family of monocytes or macrophages has been an important source for cytokine production (Nathan 1987; Nicod 1993). Another immune and non-immune cell synthesizing cytokine within and by secondary nerve tissues at the time of physiological and pathological practices is by tissue-resident macrophages, recruited macrophages, mast cells, Schwann cells, and endothelial cells. Likely, herniated nucleus pulposus can produce cytokines in the spinal cord (DeLeo et al. 1996) and swollen dermis (Heijmans-Antonissen et al. 2006).

Table 5.1 Different classification of cytokines and cytokine receptor

Classification based on receptor binding	
Receptor family	*Cytokines*
Type I (hematopoetin)	GH, Prl, Epo, Leptin, IL-6, IL-11, Tpo, IL-27, IL-31, CNTF, LIF, Osm, CT-1, GM-CSF, IL-3, IL-5, IL-2, IL-4, IL-7, IL-9, IL-15, IL-21, IL-13, 12, 23, IL-35, TSLP
Type II (interferon)	IFN-α/β, IFN-γ, IL-10, IL-19, 20, 22, 24, 26, 28, 29, 30
Interleukin-1 family	IL-1 α/β, IL-1F, IL-18, 33
Interleukin-17 family	Il-17A, B, C, D, E, F
TGF-β receptor serine kinase	TGF-β 1,2,3
Receptor tyrosine kinase	Stem cell factor, CSF-1, Flt-3 ligand, IL-32, 16, 34
Other classifications	
Type of class	*Cytokines*
Pro-inflammatory cytokines	IL-1 α/β, TNF-α/β, IL-6, 11, 18, IFN-γ
Anti-inflammatory cytokines	IL-10, TGF-β, IL-1
Neutrophils	CXCL8/IL-8, IL-1 α/β, TNF-α/β, G-CSF, IL-17A, IL-17F
Eosinophil	IL-2, 3, 4, 5, GM-CSF, CCL5, CCL11, CCL7, CCL13
T-helper (Th) cells	*Th1:* IL-2, IFN-γ, IL-12, GM-CSF, TNF-α, TNF-β, Il-18 *Th2:* IL-4, 5, 6, 9, 10, 13, 25
T-regulatory (Tregs) cells	Il-10 and TGF-β
T cells	IL-16, CCL5, CCL3, CCL4, TSLP, CCL17, CCL22
Growth hormones	PDGF, VEGF, TGF-β, FGF, EGF, SCF

IL interleukin, *IFN* interferon, *TGF* transforming growth factor, *CSF* colony-stimulating factor, *G-CSF* granulocyte CSF, *CCL* chemokine, *TNF* tumor necrosis factor, *PDGF* platelet-derived growth factor, *VEGF* vascular endothelial growth factor, *FGF* fibroblast growth factor, *SCF* stem cell factor, *EGF* epidermal growth factor, *GM-CSF* granulocyte macrophage colony-stimulating factor

5.4 Viral Infection and Cytokines

Nearly all varied cytokines have a key role in many viral infections. Different approaches such as viruses modulating cytokine expression, cytokines promoting viral expression and replication, and a virus with cytokines or their receptors mimicking host immune response result in the cure of viral diseases.

5.4.1 Expression of Cytokine in Viral Disease

In general, cytokines were identified by its bioactivity and quantification by bioassay during its initial discovery. However, nowadays, tools such as sequence homology in gene database and homologous receptor binding sequence are in use. At present, knowledge on expression of cytokines in a diversity of viral infections is gathered as a result of tissues and cells analysis through in situ and ex vivo techniques.

The resistant reaction to viral diseases is a mix of numerous actions: viral antigen introduction to the immune framework, multiplication of particular lymphocytes, and cytokine mediation (Cavallo et al. 1994). A sign of viral contamination is an intense response by the affected cell. This incorporates enactment of a previous antiviral resistant mechanism, obligation headed for programmed cell death and generation of explicit cytokines. These actions add toward the decrease of viral duplication and the confinement of viral spread. Cytokines and chemokines have a vital role in the host reaction to viral contagion just like in the immunopathology related to numerous viral ailments. Numerous viral glycoproteins in association with specific cell receptors mediate the direct synthesis of cytokines. Furthermore, viral RNA and viral proteins endorse viral replication and pro-inflammatory protein expression when meddling with cell signaling factor activity such as transduction and transcription (Mogensen and Paludan 2001).

5.5 Signal Transduction in Viruses-Induced Cytokine Production

Subsequent to viral infection, viral surface proteins in many viruses principally act on the cellular surface protein initiating a cell response where in many cases directs the primary signal of cytokine production. As well, some viruses in the absence of viral proteins during initial virus infection can generate cytokines as proteins are produced at some stage of the viral life cycle. However, the increase of viral RNA and excess synthesis of the cellular macromolecule induce signs that activate early host response to the infection.

5.5.1 Transcriptional Factors

Signal transduction pathways stimulated by viruses comprise many transcriptional factors concerning cytokine induction. Recent studies revealed two factors, interferon regulatory factor 3 and interferon regulatory factor 7 (IRF-3 and IRF-7, respectively) play a significant role in IFN-α/β activation (Marié et al. 1998). Activation of transcription factors is elicited by serine/threonine (S/T) phosphorylation. The MAP (mitogen-activated protein) kinases p38 and JNK (Jun N-terminal Kinase) are stimulated in reaction to viruses. Downstream targets particularly activating transcription factor 2 (ATF-2) and Jun are phosphorylated by S/T kinases after activation of MAP kinases p38 (Arora and Houde 1992). The protein Jun forms homodimers with ATF-2 and heterodimers with Fos. The dimer ATF-2/Jun joins to cyclic AMP response element. Likewise, the Jun and Jun/Fos homodimer and heterodimer spot the TPA responsive element (Whitmarsh and Davis 1996). An additional transcriptional factor, nuclear factor of activated T cell (NF-AT) present in a latent form in the cytoplasm, is activated by viruses. The factor NF-AT is triggered by calmodulin-dependent phosphatase calcineurin by dephosphorylation process on increasing the cytoplasmic levels (Crabtree 1999). NF-κβ protein is generally originated in the cytoplasm in composite with Iκβ, an inhibitory protein.

5.5.2 Events in Signaling

Initiation of signal transduction is generated by activation of MAP 3 kinases that uphold the stimulation of kinase complex capable of phorphorylating Iκβ at two sites of N-terminal serine residues. The cofactors liable for Iκβ phosphorylation are Iκβ kinase α (IKKα) and Iκβ kinase β (IKKβ). Phosphorylated Iκβ is degraded by ubiquitin-dependent 26S proteasome pathway. Breakdown of Iκβ uncovers the target nuclear localization signal NF-κβ where it wanders toward the central cell nucleus and thereby initiates the transcription process. NF-κβ has a significant role in virus-independent cytokine expression and pathogenesis (Fig. 5.1). Viruses inducing the production of a range of cytokines and chemokines profile are exemplified below.

5.5.2.1 Cytokines in Influenza Virus Infection

The influenza virus is a part of the *Orthomyxoviridae* family. It is an encased virus with negative single-stranded RNA as a genome. Namely, four types of

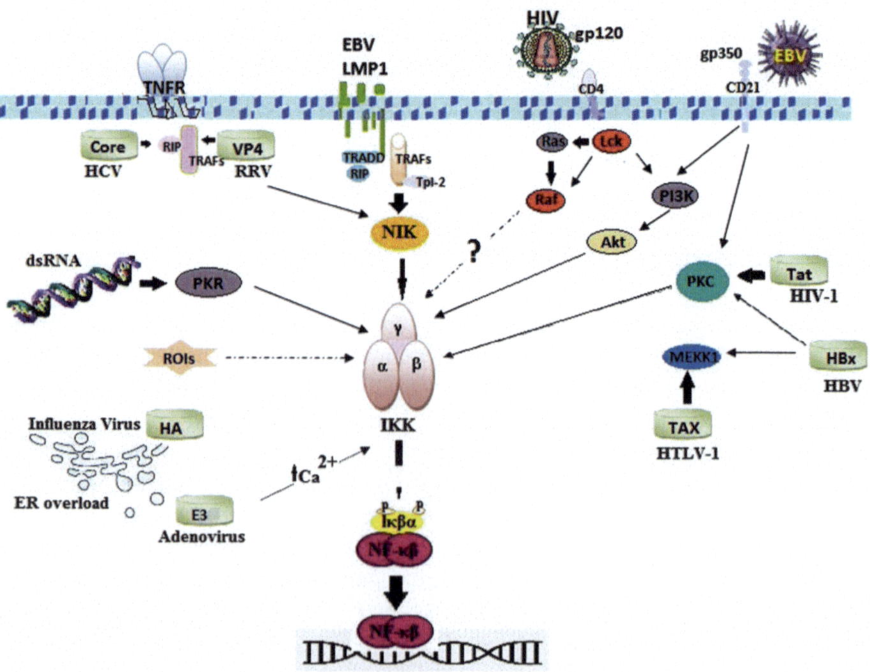

Fig. 5.1 Activation of NF-κβ by viruses. The picture depicts different viruses, their proteins, and pathways stimulating cellular response through cell membrane receptors

genera are distinguished: types A, B, C and Thogotovirus. The genera influenza A and B are the only two that are clinically related to a human being. The virus holds a pair of macromolecules: hemagglutinin (HA) and neuraminidase (NA) dependable for viral cell attachment via exchanges with sialic acids on top of the cell shell and breakdown of sialic acid that prop up viral disease respectively (Scholtissek 1991). The main source for transmission of the virus is through aerosol, i.e., via talking, coughing, or sneezing where the large droplets enter the mucosa (Blut 2009).

In influenza viral infection, cytokines come into sight in acute phase within 1–3 days of contagion. A wide array of cytokines and chemokines were stimulated in an influenza virus infection in various cell types. These comprise IL-1, 2, 6, 8, 10, 15, 18, TGF-β, TNF-α, IFN-α/β, IFN-γ, GM-CSF, and RANTES (Fawaz et al. 1999; Hayden et al. 1998; Hennet et al. 1992; Hofmann et al. 1997; Kallas et al. 1999). Studies showed a difference in the ability to influence virus inducing the cytokine INF-α/β (Chomik 1981). Of all the cytokines, interferon (IFN) plays a vital role in defense against influenza infection. IFN-α synthesized in virally infected lymphocytes, IFN-β processed in virally infected fibroblast, and IFN-γ produced in Th-1 cells are the three types of interferons studied for influenza virus. In a study to detect the role of interferons protecting from the influenza virus, the three subtypes were administrated as anti-interferons into mice model. When a mouse infected with the influenza virus is treated with anti-IFN-α and β, all the treated mice died within 7 days of postinfection. Similarly, treatment with anti-IFN-γ showed 60% death while controls in both the cases survived to indicate that IFN-α/β was active only in initial days of host-virus infection (Hoshino et al. 1983) and IFN-γ is important in guarding against influenza virus infection. Likewise, IL-6 which is produced in infected mice mast cells showed effective defense against this infection by preventing uncontrolled amassing of neutrophils to swollen tissues and also restricting rigorous pathogenesis of the lung (Graham et al. 2015; Hurst et al. 2001; Lauder et al. 2013; McLoughlin et al. 2004; Modur et al. 1997; Skoner et al. 1999). Houde and Arora illustrated that the protein NA promoted the production of IL-1 and TNF-α in murine peritoneal macrophages (Houde and Arora 1989, 1990).

Cell signaling in influenza virus infections is instigated by the activation of MAP kinases p38 and JNK and also by the downstream factors NF-κβ, AP-1, and CRE-binding protein (Fig. 5.1) (Hofmann et al. 1997; Bussfeld et al. 1997; Kujime et al. 2000). Studies have shown cellular response on exposure to HA triggering NF-κβ in excess in the endoplasmic reticulum (ER) (Baumann et al. 1998; Flory et al. 2000; Pahl and Baeuerle 1995). Two additional proteins of influenza virus (matrix protein and nucleoprotein) are stated to induce NF-κβ activation. For both the mechanical systems, the activation pathway is mediated by IKKβ (Pahl and Baeuerle 1997). Also, the double-stranded RNA of the virus is sensed to alter cell signal by dsRNA-activated protein kinase (PKR) through two mechanisms: one with PKR inhibitor p58 (Melville et al. 1999) and the other with nonstructural protein (Hatada et al. 1999).

5.5.2.2 Cytokine in Viral Myocarditis

Myocarditis is an inflammatory heart muscle disease, the main cause of dilated cardiomyopathy globally (Schultz et al. 2009). Cytokines IL-1a, IL-10, TNF-α, and G-CSF prominent in the serum of acute viral myocarditis probably play a role in myocardial injury (Matsumori et al. 1994).

In an earlier study to examine the effect of cytokines within viral myocarditis, a clinical trial in mice induced with virus encephalomyocarditis (EMCV) in the heart was carried out by RT-PCR, ELISA, and bioassay methods. The results confirmed that the expression of IL-15, 6, 10, TNF-α, IFN-α/β mRNA was enhanced in the heart for the period of acute stage 2–3 days after EMCV infection. The cytokines IL-6, TNF-α, IFN-γ, and IFN-β concentration were shown to be high in infected mice in comparison with uninfected mice. Later anti-cytokine antibody in vivo effects were observed, where anti-IL-6 and anti-IFN-γ resulted in considerably decreased survival and raised myocardial damage in mice. An anti-TNF-α antibody in EMCV-injected mice showed an improved survival rate and lesions. The study signifies that IL-6 and TNF-α have a key role in protecting and execrating of acute myocarditis correspondingly (Imanishi 2000).

5.5.2.3 Cytokines in the Hepatitis Virus

The hepatitis B virus (HBV) is a small DNA enveloped virus. It is a model virus of the *Hepadnaviridae* family. It is a common cause liver cancer and liver disease (Liang and Hepatitis 2009). The HBV genome is integrated into host DNA, leaving the infection in the dormant stage. Later, accrual of HBV surface antigen (HBsAg) leads to ground-glass hepatocyte pathology a feature of HBV infection (Chisari and Ferrari 1995). Studies reported showed production of a large number of pro-inflammatory cytokines and chemokines such as IL-1β, IL-2, 6, 8, TNF-α, and IFN-γ (Al-Wabel et al. 1993; Geneva-Popova and Murdjeva 1999; Mahe et al. 1991) and also anti-inflammatory cytokine IL-10 (Oquendo et al. 1997; Hsu et al. 1990; Vingerhoets et al. 1998). In another study, Penna et al. (1997) reported T helper 1 cytokine in acute self-stimulating HBV infection.

The viral component HBV x protein (HBx) is responsible for inducing cytokines and chemokines (IL-6, IL-8, and TNF-α) production. HBx functions as a transcriptional activator and is essential for viral infection. It is competent in upregulation of cellular and viral genes. Besides HBx, HBsAg and core antigens (HBc Ag and HBe Ag) trigger the production of cytokine (Hsu et al. 1990; Vingerhoets et al. 1998).

HBx protein stimulates the pro-inflammatory signaling cascade on activation of MAP kinase network via MEKK1, GTPase Ras, and S/T kinases Raf (Benn and Schneider 1994; Benn et al. 1996). This results in downstreaming of JNK and ERKs which stimulate transcriptional factors AP-1, ATF-2, and NF-κβ. AP-1 triggers Ras stimulation and NF-κβ (Cross et al. 1993; Natoli et al. 1994). Activation of NF-κβ is stimulated by two diverge cytoplasmic pathways. The first is phosphorylation followed by degradation of Iκβα and the latter is a decrease of p105 cytoplasmic levels (Haviv et al. 1996). Another such means entailed is the interaction of Hbx and transcriptional factor IIB (TFIIB), TFIIIB, RNA polymerase II (Su and Schneider 1996; Lin et al. 1997), and raised levels of TATA-binding proteins (Wang et al. 1995a).

5.5.2.4 Cytokines in Herpes Virus

Herpes simplex virus (HSV) a herpes virus is a member of *Alphaherpesvirinae*. HSV is reported to infect humans with disease like cold sores and eye and genital infections. HSV infection is categorized into primary and secondary, where the initial bearing more severity than later. Primary HSV infection in host produces a series of cytokines of interleukin family (IL-1β, IL-2, IL-6, IL-10, IL-12, and IL-13), TNF-α, interferon α/β and γ, and GM-CSF (Ellermann-Eriksen 1993; Ghanekar et al. 1996; He et al. 1999; Paludan et al. 2000). However, IL-8, macrophages, and inflammatory protein 1α and monocyte chemoattractant protein 1 and RANTES are produced additionally (Rösler et al. 1998; Thomas et al. 1998).

Another related herpes virus cytomegalovirus (CMV), a member of Betaherpesvirinae subgroup, is reported to induce the production of pro-inflammatory cytokines IL-1β, IL-6, 12, TNF-α, IFN α/β, and γ (Carlquist et al. 1999; Orange and Biron 1996; Peterson et al. 1992). CMV communicates as a disease in many cell types and institutes latency in leukocytes. Glycoprotein gB is the main protein of the CMV virion. The anti-gBcan blocks IL-1β in monocytes of the infected host. Likely, the ability to retain the induction of IL-6 is proliferated by UV-inactivated virus in human CMV (Carlquist et al. 1999; Baumforth et al. 1999).

Also, Epstein-Barr virus (EBV), a lymphotropic herpes virus, belongs to the *Gammaherpesvirinae* family. During its primary contagion, the virus ensures lytic duplication in epithelial and B lymphocyte cells, where it begins latency. EBV causes Burkitt's lymphoma, Hodgkin's disease, and nasopharyngeal malignancy (Hsu and Glaser 2000; D'addario et al. 1999). The initial gesture bearing cytokines and chemokines during EBV disease comprises IL-1β, IL-1 receptor antagonist, IL-6, 8, 18, TNF-α, IFN α/β and γ, GM-CSF, and inducible protein-10 (Lay et al. 1997; Setsuda et al. 1999; Tanner et al. 1999; D'Addario et al. 2000). Similar to the above-discussed herpesviruses, EBV surface protein activates early features of the infection. For instance, EBV gp350 a glycoprotein induces the products IL-1β and TNF-α (D'addario et al. 1999; McColl et al. 1997). Similar to CMV, EMV treated with UV induced IL-6, GM-CSF, MIP-1α, IL-8 (Roberge et al. 1996; Eliopoulos et al. 1999; Nakagomi et al. 1994). Moreover, a latent membrane protein 1 of EBV is proficient at producing cytokines interleukins 6, 8, and 10 (Roberge et al. 1996).

Cellular signaling response is almost similar in all the herpes virus infections. HSV infection is triggered by either UV insensible and reliant on gD or UV susceptible and dependent on an efficient viral genome. The actually gD-dependent signaling is not completely demonstrated, but HVS-2 inflammation of murine macrophages provokes improved DNA binding activity of NF-κβ and ATF-2/c-Jun an hour after the disease onset. The virus HSV uses Hve A as a receptor for signaling (Marsters et al. 1997). For CMV, signals are activated for cytokine induction (IL-1β and IL-6) by NF-κβ (Yurochko and Huang 1999). Part of signaling for EMV is similar to HSV and CMV, while the unique property of the pathway is the use of gp350 protein via cellular receptor complement receptor 2 (CD21) which acts as a signaling receptor. These receptors induce tyrosine phosphorylation, PKC, activation, NF-κβ, and phosphatidylinositol 3-kinase (Sugano et al. 1997). On the other side, proteins BZLF-1 and BRLF-1 trigger signaling cellular trauma (Adamson et al. 2000).

5.5.2.5 Cytokines in Human Immunodeficiency Virus (HIV)

HIV belongs to the family *Retroviridae* of subfamily *Orthoretroviridae*, grouped to the genus Lentivirus (Luciw 1996). Based on genetic features and variation in viral genes the virus is differentiated into type 1 and type 2 (HIV-1 and HIV-2). The virus infects CD4 T-lymphocytes and macrophages via CD 4 receptors and co-receptors CXCR4 and CCR5 (Oberlin et al. 1996; Virelizier 1999). The severe host reaction to initial HIV infection is differentiated by a T-helper cytokine profile of both pro -and anti-inflammatory cytokines. At an advanced stage, the cytokine production is triggered partially by Th-2 (Paludan et al. 1998; Rinaldo et al. 1990; Toso et al. 1990).

The virus glycoprotein gp120 links with CD 4 and the co-receptors to induce the production of IL-1α, IL-1β, IL 6, and 8, TNF-α, and interferon α/β/γ (Ameglio et al. 1994; Ankel et al. 1994). Cytokines IL-4 and -13 in basophils and IL-10 in mononuclear cells are also produced by gp120 signifying the role of gp120 with cell receptors. Also, the proteins transactivator (Trf), negative regulating factor (Nef), and virus protein r (Vpr) are likely able to trigger the secretion of cytokines (Chen et al. 1997).

Cellular activity of HIV infection in signal transduction is detailed profoundly. The protein gp120 induces signaling when in contact with CD_4 and CXCR4 and CCR5 chemokines. The binding of CD_4 with gp120 activates tyrosine kinase Lck and Serine/threonine kinase Raf-1 receptors (Popik and Pitha 1996). Subsequently, MAP kinases, ERK1, ERK2, and JNK, have also been activated which in turn activate NF-κβ, AP-1, and C/EBPβ (Lannuzel et al. 1997). The signaling mechanism for chemokines was by phosphorylation of protein tyrosine kinase. Tat, a protein inducer of many pro-inflammatory cytokines, activates viral replication by regulating transcriptional factors NF-κβ, NF-AT, Sp1, and C/EBPβ families (Ruocco et al. 1996; Vlach et al. 1995). Activation of NF-κβ by Tat protein is a prerequisite for PKR and the process occurs via IKK is active in the cells infected by HIV. Tat protein initially aims Ikβ-α for degradation (Demarchi et al. 1996). This mechanism of synthesis and degradation is sited at the promoter region in the presence of NF-κβ. However, JNK and ATF-2 are also controlled by Tat in HIV infections. Another feature regulated by Tat is the Sp1, a major source meant for basal and inducible expressions of many genes. On phosphorylation, Sp1 enhances the promoter activity. No such study is reported regarding Nef protein and its signal transduction. Vpr adapts the function of numeral DNA-binding proteins to stimulate HIV long-term repeats via direct protein–protein interaction. This is duly regulated by physical binding of Vpr and Sp1 factors (Wang et al. 1995b). Succeeding studies have revealed that Vpr acts together with p53 antagonizing Vpr/Sp-1-driven transcription (Schafer et al. 1996).

5.5.2.6 Cytokines in Human T-Lymphocyte Virus Type-1

The retrovirus HTLV-1 belongs to *Oncovirinae* subfamily. The viral disease is habitually asymptomatic. In its latency period, it causes mature T-cell lymphocytic leukemia (ATLL) or non-cancerous HTLV-1. HTLV 1 virus modulated the expression of many cellular genes, even for cytokines through posttranscriptional and transcriptional pathways. During the infectious period, the production of cytokines IL-1α, IL-1β, IL-2, IL-3, IL-4, IL-5, IL-6, IL-9, IL-15, TGF-β, TNF-α, interferon α/β/γ and GM-CSF is triggered by the host (Azimi et al. 1998). The proteins Tax and Rex enhance the secretion of cytokine, chemokines, and the receptors (Baba et al. 1996). However, Tax protein intercedes transactivation of TNF-β and TGF-β genes, where cellular transcriptional factor NF-κβ (Kim et al. 1990) and AP-1 (Paul et al. 1990) are essential. Posttranscriptional regulation through protein Rex adds to the stimulation of cellular genes cytokines. The product assists the transfer and expression of nucleus and cytoplasm of HTLV-1 gaga and env mRNAs and post-transcriptionally alleviates the IL-2 receptor protein mRNA, which considerably accumulates mRNA followed by the viral infection (Kanamori et al. 1990).

The protein Tax is certainly the major immune stimulator, not just regulating duplication but as well taking part in the expression of cell signaling. Signaling via Tax is a complex mechanism. Initially, Tax stimulates NF-κβ via IKK pathway that results in the release of active NF-κβ (Geleziunas et al. 1998). IKK-γ and MEKK-1 are involved in stimulating IKK by Tax protein. This process of specific interaction linking Tax and IKK-γ has a requirement of leucine zipper domains (Harhaj and Sun 1999; Xiao et al. 2000). Several discrete NF-κβ pathways are activated by Tax protein. One such pathway is the use of PKC isoforms (α, δ, and η) with phosphorylation and autophosphorylation of Tax and PKC correspondingly. A second factor ATF-2 in group with CREB transcription factor family activates viral transcription along CRE-like sites in HTLV-1 LTR (Franklin et al. 1993; Gachon et al. 2000). Next, Tax activates NF-AT that binds to CD28 responsive element in the promoter region of the genes triggering cell expression.

5.5.2.7 Cytokines in Filovirus Infection

Filovirus is nearly 80 nm in diameter recognized by its filamentous shape categorized under the family *Filoviridae*. The virus is divided into three genera: Ebolavirus, Marburgvirus, and Cuevavirus (Kuhn et al. 2010). The genome of the virus constitutes 19 kb single linear non-segmented negative-sense RNA. The virus codes for the proteins: nucleoprotein, polymerase factor VP35, the matrix proteins VP40 and VP24, transcriptional protein VP30, glycoprotein and the RNA-dependent RNA polymerase (Bixler and Goff 2015). This viral infection is differentiated by elevated levels of pro-inflammatory cytokines and chemokines that are produced by contaminated macrophages and monocytes. The cyto- and chemokines include IL-1β, IL-8, IL-15, IL-18, MIP-1α, MCP-1, IP-10, Gro-α, and eotaxin. However, at lower levels, IL-2 was observed in association with T cell (Geisbert et al. 2007).

5.6 Cytokines and Viral Replication

Cytokines besides their role as mediators in local and systematic host antiviral response are able to endorse viral replication and multiply some of viruses, especially by IFNs and TNF. The well-known suitable example explained is HIV. Activation of HIV expression from inactive state of replication is reliant partly on the stimulation of the host cell (Rosenberg and Fauci 1990). Activation of CD4$^+$ lymphocytes and monocytes is linked with the production of a number of cytokines that induce the expression of HIV from chronic infected T lymphocyte and monocyte cell lines. Specifically, the cytokines TNF-α and IL-6 upbeat transcriptional and posttranscriptional expression levels (Poli et al. 1990).

Acknowledgements The authors gratefully acknowledge Andhra University, Visakhapatnam, for the support extended.

Conflict of Interest The authors declare that they have no competing interests.

References

Adamson AL, Darr D, Holley-Guthrie E, Johnson RA, Mauser A, Swenson J, Kenney S (2000) Epstein-Barr virus immediate-early proteins BZLF1 and BRLF1 activate the ATF2 transcription factor by increasing the levels of phosphorylated p38 and c-Jun N-terminal kinases. J Virol 74(3):1224–1233

Al-Wabel A, Al-Janadi M, Raziuddin S (1993) Cytokine profile of viral and autoimmune chronic active hepatitis. J Allergy Clin Immunol 92(6):902–908

Ameglio F, Capobianchi MR, Castilletti C, Fei PC, Fais S, Trento E, Dianzani F (1994) Recombinant gpl 20 induces IL-10 in resting peripheral blood mononuclear cells; correlation with the induction of other cytokines. Clin Exp Immunol 95(3):455–458

Ankel H, Capobianchi MR, Castilletti C, Dianzani F (1994) Interferon induction by HIV glycoprotein 120: role of the V3 loop. Virology 205(1):34–43

Arora DJ, Houde M (1992) Modulation of murine macrophage responses stimulated with influenza glycoproteins. Can J Microbiol 38(3):188–192

Azimi N, Brown K, Bamford RN, Tagaya Y, Siebenlist U, Waldmann TA (1998) Human T cell lymphotropic virus type I Tax protein trans-activates interleukin 15 gene transcription through an NF-κB site. Proc Natl Acad Sci 95(5):2452–2457

Baba M, Imai T, Yoshida T, Yoshie O (1996) Constitutive expression of various chemokine genes in human T-cell lines infected with human T-cell leukemia virus type 1: role of the viral transactivator Tax. Int J Cancer 66(1):124–129

Balkwill FR, Burke F (1989) The cytokine network. Immunol Today 10(9):299–304

Barrett KE (1996) Cytokines: sources, receptors and signaling. Baillieres Clin Gastroenterol 10(1):1–5

Baumann B, Kistler B, Kirillov A, Bergman Y, Wirth T (1998) The mutant plasmacytoma cell line S107 allows the identification of distinct pathways leading to NF-κB activation. J Biol Chem 273(19):11448–11455

Baumforth KR, Young LS, Flavell KJ, Constandinou C, Murray PG (1999) The Epstein-Barr virus and its association with human cancers. Mol Pathol 52(6):307

Benn J, Schneider RJ (1994) Hepatitis B virus HBx protein activates Ras-GTP complex formation and establishes a Ras, Raf, MAP kinase signaling cascade. Proc Natl Acad Sci 91(22):10350–10354

Benn J, Su F, Doria M, Schneider RJ (1996) Hepatitis B virus HBx protein induces transcription factor AP-1 by activation of extracellular signal-regulated and c-Jun N-terminal mitogen-activated protein kinases. J Virol 70(8):4978–4985

Bixler S, Goff A (2015) The role of cytokines and chemokines in filovirus infection. Viruses 7(10):5489–5507

Blut A (2009) Influenza virus. Transfus Med Hemother 36(1):32

Bussfeld D, Bacher M, Moritz A, Gemsa D, Sprenger H (1997) Expression of transcription factor genes after influenza A virus infection. Immunobiology 198(1–3):291–298

Campbell IL (1991) Cytokines in viral diseases. Curr Opin Immunol 3(4):486–491

Carlquist JF, Edelman L, Bennion DW, Anderson JL (1999) Cytomegalovirus induction of interleukin-6 in lung fibroblasts occurs independently of active infection and involves a G protein and the transcription factor, NF-κB. J Infect Dis 179(5):1094–1100

Cavallo R, Cavallo F, Cavallo G (1994) The role of cytokines in viral infections. G Batteriol Virol Immunol 86(1–12):65–78

Chen P, Mayne M, Power C, Nath A (1997) The Tat protein of HIV-1 induces tumor necrosis factor-α production. Implications for HIV-1-associated neurological diseases. J Biol Chem 272(36):22385–22388

Chisari FV, Ferrari C (1995) Hepatitis B virus immunopathology. Springer Semin Immunopathol 17(2–3):261–281

Chomik M (1981) Interferon induction by influenza virus: significance of neuraminidase. Arch Immunol Ther Exp 29(1):109–114

Chung KF (2009) Cytokines. Asthma and COPD. Academic Press, Oxford, pp 327–341

Cohen S, Bigazzi PE, Yoshida T (1974) Similarities of T cell function in cell-mediated immunity and antibody production. Cell Immunol 12(1):150–159

Crabtree GR (1999) Generic signals and specific outcomes: signaling through Ca2+, calcineurin, and NF-AT. Cell 96(5):611–614

Cross JC, Wen P, Rutter WJ (1993) Transactivation by hepatitis B virus X protein is promiscuous and dependent on mitogen-activated cellular serine/threonine kinases. Proc Natl Acad Sci 90(17):8078–8082

D'addario M, Ahmad A, Xu JW, Menezes J (1999) Epstein-Barr virus envelope glycoprotein gp350 induces NF-κB activation and IL-1β synthesis in human monocytes-macrophages involving PKC and PI3-K. FASEB J 13(15):2203–2213

D'Addario M, Ahmad A, Morgan A, Menezes J (2000) Binding of the Epstein-Barr virus major envelope glycoprotein gp350 results in the upregulation of the TNF-α gene expression in monocytic cells via NF-κB involving PKC, PI3-K and tyrosine kinases. J Mol Biol 298(5):765–778

DeLeo JA, Colburn RW, Nichols M, Malhotra AM (1996) Interleukin-6-mediated hyperalgesia/allodynia and increased spinal IL-6 expression in a rat mononeuropathy model. J Interf Cytokine Res 16(9):695–700

Demarchi F, di Fagagna FD, Falaschi A, Giacca M (1996) Activation of transcription factor NF-kappaB by the Tat protein of human immunodeficiency virus type 1. J Virol 70(7):4427–4437

Eliopoulos AG, Gallagher NJ, Blake SM, Dawson CW, Young LS (1999) Activation of the p38 mitogen-activated protein kinase pathway by Epstein-Barr virus-encoded latent membrane protein 1 coregulates interleukin-6 and interleukin-8 production. J Biol Chem 274(3):16085–16096

Ellermann-Eriksen S (1993) Autocrine secretion of interferon-α/β and tumour necrosis factor-α synergistically activates mouse macrophages after infection with herpes simplex virus type 2. J Gen Virol 74(10):2191–2199

Fawaz LM, Sharif-Askari E, Menezes J (1999) Up-regulation of NK cytotoxic activity via IL-15 induction by different viruses: a comparative study. J Immunol 163(8):4473–4480

Feldmann M (1998) Cytokines. In: Encyclopedia of immunology. Elsevier, Oxford, pp 719–722

Ferreira VL, Borba HH, Bonetti AD, Leonart L, Pontarolo R (2018) Cytokines and interferons: types and functions. In: Autoantibodies and cytokines. IntechOpen, London, pp 65–87

Flory E, Kunz M, Scheller C, Jassoy C, Stauber R, Rapp UR, Ludwig S (2000) Influenza virus-induced NF-κB-dependent gene expression is mediated by overexpression of viral proteins and involves oxidative radicals and activation of IκB kinase. J Biol Chem 275(12):8307–8314

Franklin AA, Kubik MF, Uittenbogaard MN, Brauweiler A, Utaisincharoen PO, Matthews MA, Dynan WS, Hoeffler JP, Nyborg JK (1993) Transactivation by the human T-cell leukemia virus Tax protein is mediated through enhanced binding of activating transcription factor-2 (ATF-2) ATF-2 response and cAMP element-binding protein (CREB). J Biol Chem 268(28):21225–21231

Gachon F, Thebault S, Peleraux A, Devaux C, Mesnard JM (2000) Molecular interactions involved in the transactivation of the human T-cell leukemia virus type 1 promoter mediated by Tax and CREB-2 (ATF-4). Mol Cell Biol 20(10):3470–3481

Geisbert TW, Daddario-DiCaprio KM, Geisbert JB, Young HA, Formenty P, Fritz EA, Larsen T, Hensley LE (2007) Marburg virus Angola infection of rhesus macaques: pathogenesis and treatment with recombinant nematode anticoagulant protein c2. J Infect Dis 196(S2):S372–S381

Geleziunas R, Ferrell S, Lin X, Mu Y, Cunningham ET, Grant M, Connelly MA, Hambor JE, Marcu KB, Greene WC (1998) Human T-cell leukemia virus type 1 Tax induction of NF-κB involves activation of the IκB kinase α (IKKα) and IKKβ cellular kinases. Mol Cell Biol 18(9):5157–5165

Geneva-Popova M, Murdjeva M (1999) Study on proinflammatory cytokines (IL-1 β, IL-6, TNF-α) and IL-2 in patients with acute hepatitis B. Folia Med 41(1):78–81

Ghanekar S, Zheng L, Logar A, Navratil J, Borowski L, Gupta P, Rinaldo C (1996) Cytokine expression by human peripheral blood dendritic cells stimulated in vitro with HIV-1 and herpes simplex virus. J Immunol 157(9):4028–4036

Graham AC, Temple RM, Obar JJ (2015) Mast cells and influenza a virus: association with allergic responses and beyond. Front Immunol 6:238

Hamblin AS (1988) Lymphokines in the activation of T cells, B cells and macrophages. In: Lymphokines. IRL Press, Oxford, pp 35–37

Harhaj EW, Sun SC (1999) IKKγ serves as a docking subunit of the IκB kinase (IKK) and mediates interaction of IKK with the human T-cell leukemia virus Tax protein. J Biol Chem 274(33):22911–22914

Hatada E, Saito S, Fukuda R (1999) Mutant influenza viruses with a defective NS1 protein cannot block the activation of PKR in infected cells. J Virol 73(3):2425–2433

Haviv I, Vaizel D, Shaul Y (1996) pX, the HBV-encoded coactivator, interacts with components of the transcription machinery and stimulates transcription in a TAF-independent manner. EMBO J 15(13):3413–3420

Hayden FG, Fritz R, Lobo MC, Alvord W, Strober W, Straus SE (1998) Local and systemic cytokine responses during experimental human influenza A virus infection. Relation to symptom formation and host defense. J Clin Invest 101(3):643–649

He J, Ichimura H, Iida T, Minami M, Kobayashi K, Kita M, Sotozono C, Tagawa YI, Iwakura Y, Imanishi J (1999) Kinetics of cytokine production in the cornea and trigeminal ganglion of C57BL/6 mice after corneal HSV-1 infection. J Interf Cytokine Res 19(6):609–615

Heijmans-Antonissen C, Wesseldijk F, Munnikes RJ, Huygen FJ, van der Meijden P, Hop WC, Hooijkaas H, Zijlstra FJ (2006) Multiplex bead array assay for detection of 25 soluble cytokines in blister fluid of patients with complex regional pain syndrome type 1. Mediat Inflamm 2006:1

Hennet T, Ziltener HJ, Frei K, Peterhans E (1992) A kinetic study of immune mediators in the lungs of mice infected with influenza A virus. J Immunol 149(3):932–939

Hofmann P, Sprenger H, Kaufmann A, Bender A, Hasse C, Nain M, Gemsa D (1997) Susceptibility of mononuclear phagocytes to influenza A virus infection and possible role in the antiviral response. J Leukoc Biol 61(4):408–414

Hoshino A, Takenaka H, Mizukoshi O, Imanishi J, Kishida T, Tovey MG (1983) Effect of anti-interferon serum of influenza virus infection in mice. Antivir Res 3(1):59–65

Houde M, Arora DJ (1989) Human influenza virus neuraminidase, but not hemagglutinin, induces murine macrophage interleukin 1 in vivo and in vitro. Immunol Lett 22(1):41–46

Houde M, Arora DJ (1990) Stimulation of tumor necrosis factor secretion by purified influenza virus neuraminidase. Cell Immunol 129(1):104–111

Hsu JL, Glaser SL (2000) Epstein–Barr virus-associated malignancies: epidemiologic patterns and etiologic implications. Crit Rev Oncol Hematol 34(1):27–53

Hsu HY, Chang MH, Ni YH, Lee PI (1990) Cytokine release of peripheral blood mononuclear cells in children with chronic hepatitis B virus infection. J Pediatr Gastroenterol Nutr 29(5):540–545

Hurst SM, Wilkinson TS, McLoughlin RM, Jones S, Horiuchi S, Yamamoto N, Rose-John S, Fuller GM, Topley N, Jones SA (2001) Il-6 and its soluble receptor orchestrate a temporal switch in the pattern of leukocyte recruitment seen during acute inflammation. Immunity 14(6):705–714

Imanishi J (2000) Expression of cytokines in bacterial and viral infections and their biochemical aspects. J Biochem 127(4):525–530

Kallas EG, Reynolds K, Andrews J, Treanor JJ, Evans TG (1999) Production of influenza-stimulated tumor necrosis factor-alpha by monocytes following acute influenza infection in humans. J Interf Cytokine Res 19(7):751–755

Kanamori H, Suzuki N, Siomi H, Nosaka T, Sato A, Sabe H, Hatanaka M, Honjo T (1990) HTLV-1 p27rex stabilizes human interleukin-2 receptor alpha chain mRNA. EMBO J 9(12): 4161. -4166, 1990

Kim S, Kehrl JH, Burton J, Tendler CL, Jeang KT, Danielpour D, Thevenin C, Kim KY, Sporn MB, Roberts AB (1990) Transactivation of the transforming growth factor beta 1 (TGF-beta 1) gene by human T lymphotropic virus type 1 tax: a potential mechanism for the increased production of TGF-beta 1 in adult T cell leukemia. J Exp Med 172(1):121–129

Kuhn JH, Becker S, Ebihara H, Geisbert TW, Johnson KM, Kawaoka Y, Lipkin WI, Negredo AI, Netesov SV, Nichol ST, Palacios G (2010) Proposal for a revised taxonomy of the family Filoviridae: classification, names of taxa and viruses, and virus abbreviations. Arch Virol 155(12):2083–2103

Kujime K, Hashimoto S, Gon Y, Shimizu K, Horie T (2000) p38 mitogen-activated protein kinase and c-jun-NH2-terminal kinase regulate RANTES production by influenza virus-infected human bronchial epithelial cells. J Immunol 164(6):3222–3228

Lannuzel A, Tardieu M, Hery C, Barnier JV, Vincent JD, Guibert B, Van Tan H, Gray F (1997) Human immunodeficiency virus type 1 and its coat protein gp 120 induce apoptosis and activate JNK and ERK mitogen-activated protein kinases in human neurons. Ann Neurol 42(6):847–856

Lauder SN, Jones E, Smart K, Bloom A, Williams AS, Hindley JP, Ondondo B, Taylor PR, Clement M, Fielding C, Godkin AJ (2013) Interleukin-6 limits influenza-induced inflammation and protects against fatal lung pathology. Eur J Immunol 43(10):2613–2625

Lay JD, Tsao CJ, Chen JY, Kadin ME, Su IJ (1997) Upregulation of tumor necrosis factor-alpha gene by Epstein-Barr virus and activation of macrophages in Epstein-Barr virus-infected T cells in the pathogenesis of hemophagocytic syndrome. J Clin Invest 100(8):1969–1979

Liang TJ, Hepatitis B (2009) The virus and disease. Hepatology 49(5 Suppl):S13–S21

Lin Y, Nomura T, Cheong J, Dorjsuren D, Iida K, Murakami S (1997) Hepatitis B virus X protein is a transcriptional modulator that communicates with transcription factor IIB and the RNA polymerase II subunit 5. J Biol Chem 272(11):7132–7139

Luciw PA (1996) Human immunodeficiency viruses and their replication. In: Fields BN (ed) Virology, vol 2, 3rd edn. Lippincott-Raven, Philadelphia, pp 1881–1952

Mahe Y, Mukaida N, Kuno K, Akiyama M, Ikeda N, Matsushima K, Murakami S (1991) Hepatitis B virus X protein transactivates human interleukin-8 gene through acting on nuclear factor κB and CCAAT/enhancer-binding protein-like cis-elements. J Biol Chem 266(21):13759–13763

Marié I, Durbin JE, Levy DE (1998) Differential viral induction of distinct interferon-α genes by positive feedback through interferon regulatory factor-7. EMBO J 17(22):6660–6669

Marsters SA, Ayres TM, Skubatch M, Gray CL, Rothe M, Ashkenazi A (1997) Herpesvirus entry mediator, a member of the tumor necrosis factor receptor (TNFR) family, interacts with members of the TNFR-associated factor family and activates the transcription factors NF-κB and AP-1. J Biol Chem 272(22):14029–14032

Matsumori A, Yamada T, Suzuki H, Matoba Y, Sasa YS (1994) Increased circulating cytokines in patients with myocarditis and myocardiopathy. Br Heart J 72:561–566

McColl SR, Roberge CJ, Larochelle B, Gosselin J (1997) EBV induces the production and release of IL-8 and macrophage inflammatory protein-1 alpha in human neutrophils. J Immunol 159(12):6164–6168

McInnes IB (2017) Cytokines. In: Kelley and Firestein's textbook of rheumatology. Elsevier, Philadelphia, pp 396–407

McLoughlin RM, Hurst SM, Nowell MA, Harris DA, Horiuchi S, Morgan LW, Wilkinson TS, Yamamoto N, Topley N, Jones SA (2004) Differential regulation of neutrophil-activating chemokines by IL-6 and its soluble receptor isoforms. J Immunol 172(9):5676–5683

Melville MW, Tan SL, Wambach M, Song J, Morimoto R, Katze MG (1999) The cellular inhibitor of the PKR protein kinase, P58(IPK), is an influenza virus-activated co-chaperone that modulates heat shock protein 70 activity. J Biol Chem 274(6):3797–3803

Modur V, Li Y, Zimmerman GA, Prescott SM, McIntyre TM (1997) Retrograde inflammatory signaling from neutrophils to endothelial cells by soluble interleukin-6 receptor alpha. J Clin Invest 100(11):2752–2756

Mogensen TH, Paludan SR (2001) Molecular pathways in virus-induced cytokine production. Microbiol Mol Biol Rev 65(1):131–150

Nakagomi H, Dolcetti R, Bejarano MT, Pisa P, Kiessling R, Masucci MG (1994) The Epstein-Barr virus latent membrane protein-1 (LMP1) induces interleukin-10 production in Burkitt lymphoma lines. Int J Cancer 57(2):240–244

Nathan CF (1987) Secretory products of macrophages. J Clin Invest 79(2):319–326

Natoli G, Avantaggiati ML, Chirillo P, Puri P, Ianni A, Balsano C, Levrero M (1994) Ras-and Raf-dependent activation of c-jun transcriptional activity by the hepatitis B virus transactivatorp X. Oncogene 9(10):2837–2843

Nicod LP (1993) Cytokines overview. Thorax 48(6):660–667

O'Shea JJ, Gadina M, Siegel RM (2019) Cytokines and cytokine receptors. In: Clinical Immunology. Elsevier, London, pp 127–155

Oberlin E, Amara A, Bessia C, Virelizier JL, Arenzana-Seisdedos F, Schwartz O, Heard JM, Clark-Lewis I, Legler DF, Loetscher M, Baggiolini M (1996) The CXC chemokine SDF-1 is the ligand for LESTR/fusin and prevents infection by T-cell-line-adapted HIV-1. Nature 382(6594):833–835

Oquendo J, Karray S, Galanaud P, Petit MA (1997) Effect of hepatitis B virus on tumour necrosis factor (TNF α) gene expression in human THP-1 monocytic and Namalwa B cell lines. Res Immunol 148(6):399–409

Orange JS, Biron CA (1996) An absolute and restricted requirement for IL-12 in natural killer cell IFN-gamma production and antiviral defense. Studies of natural killer and T cell responses in contrasting viral infections. J Immunol 156(3):1138–1142

Pahl HL, Baeuerle PA (1995) Expression of influenza virus hemagglutinin activates transcription factor NF-κB. J Virol 69(3):1480–1484

Pahl HL, Baeuerle PA (1997) The ER overload response: activation of NF-κB. Trends Biochem Sci 22(2):63–67

Paludan SR, Ellermann-Eriksen S, Mogensen SC (1998) NF-kappaB activation is responsible for the synergistic effect of herpes simplex virus type 2 infection on interferon-gamma-induced nitric oxide production in macrophages. J Gen Virol 79(11):2785–2793

Paludan SR, Ellermann-Eriksen S, Malmgaard L, Mogensen SC (2000) Inhibition of NO production in macrophages by IL-13 is counteracted by HSV infection through TNF-α-induced activation of NF-κB. Eur Cytokine Netw 11(2):275–282

Paul NL, Lenardo MJ, Novak KD, Sarr T, Tang WL, Ruddle NH (1990) Lymphotoxin activation by human T-cell leukemia virus type I-infected cell lines: role for NF-kappaB. J Virol 64(11): 5412–5419

Penna A, Del Prete G, Cavalli A, Bertoletti A, D'Elios MM, Sorrentino R, D'Amato M, Boni C, Pilli M, Fiaccadori F, Ferrari C (1997) Predominant T-helper 1 cytokine profile of hepatitis B virus nucleocapsid-specific T cells in acute self-limited hepatitis B. Hepatology 25(4):1022–1027

Peterson PK, Gekker G, Chao C, Hu SX, Edelman C, Balfour HH, Verhoef J (1992) Human cytomegalovirus-stimulated peripheral blood mononuclear cells induce HIV-1 replication via a tumor necrosis factor-alpha-mediated mechanism. J Clin Invest 89(2):574–580

Poli G, Bressler P, Kinter A, Duh E, Timmer WC, Rabson A, Justement JS, Stanley S, Fauci AS (1990) Interleukin 6 induces human immunodeficiency virus expression in infected monocytic cells alone and in synergy with tumor necrosis factor alpha by transcriptional and post-transcriptional mechanisms. J Exp Med 172(1):151–158

Popik W, Pitha PM (1996) Binding of human immunodeficiency virus type 1 to CD4 induces association of Lck and Raf-1 and activates Raf-1 by a Ras-independent pathway. Mol Cell Biol 16(11):6532–6541
Ramshaw IA, Ramsay AJ, Karupiah G, Rolph MS, Mahalingam S, Ruby JC (1997) Cytokines and immunity to viral infections. Immunol Rev 159(1):119–135
Rinaldo JC, Armstrong JA, Kingsley LA, Zhou S, Ho M (1990) Relation of alpha and gamma interferon levels to development of AIDS in homosexual men. J Exp Pathol 5(3):127–132
Roberge CJ, Poubelle PE, Beaulieu AD, Heitz D, Gosselin J (1996) The IL-1 and IL-1 receptor antagonist (IL-1Ra) response of human neutrophils to EBV stimulation. Preponderance of IL-Ra detection. J Immunol 156(12):4884–4891
Rosenberg ZF, Fauci AS (1990) Immunopathogenic mechanisms of HIV infection: cytokine induction of HIV expression. Immunol Today 11:176–180
Rösler A, Pohl M, Braune HJ, Oertel WH, Gemsa D, Sprenger H (1998) Time course of chemokines in the cerebrospinal fluid and serum during herpes simplex type 1 encephalitis. J Neurol Sci 157(1):82–89
Ruocco MR, Chen X, Ambrosino C, Dragonetti E, Liu W, Mallardo M, De Falco G, Palmieri C, Franzoso G, Quinto I, Venuta S (1996) Regulation of HIV-1 long terminal repeats by interaction of C/EBP (NF-IL6) and NF-κB/Rel transcription factors. J Biol Chem 271(37):22479–22486
Schafer SL, Vlach J, Pitha PM (1996) Cooperation between herpes simplex virus type 1-encoded ICP0 and Tat to support transcription of human immunodeficiency virus type 1 long terminal repeat in vivo can occur in the absence of the TAR binding site. J Virol 70(10):6937–6946
Scholtissek C (1991) Synthesis and function of influenza A virus glycoproteins. Behring Inst Mitt 89:46–53
Schultz JC, Hilliard AA, Cooper LT Jr, Rihal CS (2009) Diagnosis and treatment of viral myocarditis. Mayo Clin Proc 84(11):1001–1009
Setsuda J, Teruya-Feldstein J, Harris NL, Ferry JA, Sorbara L, Gupta G, Jaffe ES, Tosato G (1999) Interleukin-18, interferon-γ, IP-10, and Mig expression in Epstein-Barr virus-induced infectious mononucleosis and posttransplant lymphoproliferative disease. Am J Pathol 155(1):257–265
Skoner DP, Gentile DA, Patel A, Doyle WJ (1999) Evidence for cytokine mediation of disease expression in adults experimentally infected with influenza A virus. J Infect Dis 180(1):10–14
Stadnyk AW (1994) Cytokine production by epithelial cells. FASEB J 8(13):1041–1047
Su F, Schneider RJ (1996) Hepatitis B virus HBx protein activates transcription factor NF-kappa B by acting on multiple cytoplasmic inhibitors of rel-related proteins. J Virol 70(7):4558–4566
Sugano N, Chen W, Roberts ML, Cooper NR (1997) Epstein-Barr virus binding to CD21 activates the initial viral promoter via NF-κB induction. J Exp Med 186(5):731–737
Tanner JE, Alfieri C, Chatila TA, Diaz-Mitoma F (1999) Induction of interleukin-6 after stimulation of human B-cell CD21 by Epstein-Barr virus glycoproteins gp350 and gp220. J Virol 70(1):570–575
Thomas J, Kanangat S, Rouse BT (1998) Herpes simplex virus replication-induced expression of chemokines and proinflammatory cytokines in the eye: implications in herpetic stromal keratitis. J Interf Cytokine Res 18(9):681–690
Toso JF, Ferbas JJ, Rappocciolo G, Armstrong JA, Ho M, Rinaldo JC (1990) Stimulation of IFN-alpha production in HLA-DR+, light density peripheral blood mononuclear cells by human immunodeficiency virus. J Exp Pathol 5(3):123–125
Vilcek J (2003) Thecytokines: anoverview. In: Lotte M, Thompson AW (eds) The cytokine handbook, vol 1, 4th edn. Academic Press, Amsterdam, pp 3–18
Vingerhoets J, Michielsen P, Vanham G, Bosmans E, Paulij W, Ramon A, Pelckmans P, Kestens L, Leroux-Roels G (1998) HBV-specific lymphoproliferative and cytokine responses in patients with chronic hepatitis B. J Hepatol 28(1):8–16
Virelizier JL (1999) Blocking HIV co-receptors by chemokines. Dev Biol Stand 97:105–109
Vlach J, Garcia A, Jacqué JM, Rodriguez MS, Michelson S, Virelizier JL (1995) Induction of Sp1 phosphorylation and NF-κB-independent HIV promoter domain activity in T lymphocytes stimulated by okadaic acid. Virology 208(2):753–761

Wang HD, Yuh CH, Dang CV, Johnson DL (1995a) The hepatitis B virus X protein increases the cellular level of TATA-binding protein, which mediates transactivation of RNA polymerase III genes. Mol Cell Biol 15(12):6720–6728

Wang L, Mukherjee S, Jia F, Narayan O, Zhao LJ (1995b) Interaction of virion protein Vpr of human immunodeficiency virus type 1 with cellular transcription factor Sp1 and trans-activation of viral long terminal repeat. J Biol Chem 270(43):25564–25569

Whicher JT, Evans SW (1990) Cytokines in disease. Clin Chem 36(7):1269–1281

Whitmarsh AJ, Davis RJ (1996) Transcription factor AP-1 regulation by mitogen-activated protein kinase signal transduction pathways. J Mol Med 74(10):589–607

Xiao G, Harhaj EW, Sun SC (2000) Domain-specific interaction with the IκB kinase (IKK) regulatory subunit IKKγ is an essential step in Tax-mediated activation of IKK. J Biol Chem 275(44):34060–34067

Yurochko AD, Huang ES (1999) Human cytomegalovirus binding to human monocytes induces immunoregulatory gene expression. J Immunol 162(8):4806–4816

Zhang JM, An J (2007) Cytokines, inflammation and pain. Int Anesthesiol Clin 45(2):27

B Cells and Their Role in Combating Viral Diseases

6

Devanabanda Mallaiah and Pallaval Veera Bramhachari

Abstract

Humoral immunity mediated by B cells plays an important role in combating different types of viral diseases. Two types of B cells (B-1 and B-2) originate in the primary lymphoid organ called bone marrow from hematopoietic stem cells (HSCs). Both B-1and B-2 cells develop separately through sequential steps (pro-B cells—pre-B cells—immature B cells) in bone marrow from distinct common lymphoid progenitors before their release as immature B cells into circulation. In the serous cavities, the immature B-1 cells differentiate into mature B-1 cells through transitional B cells. The immature B-2 cells differentiate into transitional B cells, which mature finally into marginal zone (MZ) and follicular (FO) B cells in the secondary lymphoid organs. The B-1 cells produce poly-specific natural antibodies (antibodies produced before infection), which provide first line of defense. The B-1 cells mainly defend mucosal and blood-borne pathogens in a T-cell independent manner. The MZ B cells produce immune response against blood-borne pathogens and undergo both T-independent and T-dependent activation. In addition, both B1 and MZ B cells behave like innate immune cells by expressing toll-like receptors (TLR) and produce immune response without or with their membrane-bound poly-reactive B-cell receptors (BCR). Finally, FO B cells are the conventional B cells of adaptive immunity and primarily responsible for T-dependent immune response by their membrane-bound mono-reactive BCR. The antigen-activated B-2 cells differentiate into antibodies secreting plasma cells and memory B cells. Along with natural antibodies, the non-neutralizing, neutralizing, and broadly neutralizing specific

D. Mallaiah (✉)
Department of Biotechnology and Bioinformatics, Yogi Vemana University, Kadapa, Andhra Pradesh, India

P. V. Bramhachari
Department of Biotechnology, Krishna University, Machilipatnam, Andhra Pradesh, India

P. V. Bramhachari (ed.), *Dynamics of immune activation in viral diseases*,
https://doi.org/10.1007/978-981-15-1045-8_6

antibodies are important in combating various viral diseases. Additionally, B cells modulate T-cell functions by presenting antigens, providing co-stimulation, and secreting cytokines. This chapter describes different types of B cells, antibodies, and their role in combating viral diseases (HIV, influenza and hepatitis).

Keywords
B cells · Humoral immunity · Antibodies · Viral diseases

6.1 Introduction

Serum therapy by Von Behring and Kitasato provided the first evidence that humoral response played an important role in the host resistance against infectious agents such as diphtheria and tetanus toxins. They explained that serum therapy gives disease resistance due to the presence of antibodies in the serum. Paul Ehrlich demonstrated the basis for serum therapy and established the field of humoral immunity. Today, the importance of antibodies as immunotherapeutic is well characterized and is used to treat a wide range of infectious diseases. Antibodies are main components in many effective vaccines and have been used to prevent many human diseases caused by viruses (Yamada 2011; Burton 2002).

The success of immune system in combating different types of viruses depends on both non-specific innate immune system and specific-adaptive immune system. Humoral immunity is a branch of adaptive immunity mediated by antibodies secreted from B cells, which plays an important role in combating viruses along with cell-mediated immunity (Bonilla and Oettgen 2010). Humoral immune response to antigens is categorized as primary and secondary response. In the primary response, T helper cells are of primary importance whereas in the secondary immune response, the specific-neutralizing antibodies play a primary role against reinfection and T cells are of secondary importance. The primary immune response can take 1–2 weeks and IgM is the predominant type of antibody elicited. The secondary immune response is more rapid than primary immune response and predominant type of antibody formed is IgG (Lefevre et al. 2009).

The T helper cells, typically follicular T helper cells (T_{FH} cells), were activated by the same antigen and cooperate with antigen-activated mature B cells to differentiate into plasma cells and memory B cells. The activated B cells remain in the margins of T-cell zone and differentiate into short-lived plasma cells. The activated B cells which initiate germinal centers differentiate into long-lived plasma cells and memory B cells. The high-affinity and antigen-specific antibodies are secreted and maintained by both short-lived and long-lived plasma cells. The long-lived plasma cells migrate to survival niches in the bone marrow and persist for several years (Gourley et al. 2004). The protective natural antibodies are produced before the foreign antigen exposure or pathogens by B-1 cells and marginal zone B cells in a T-cell independent manner. About 80% of all circulating natural antibodies are IgM type and remaining are IgG and IgA (Palma et al. 2018). B cells provide several lines of protection against viruses. In the first line, natural antibodies provide defense against initial viral infections. The long-lived plasma cells generate high amounts of neutralizing IgG and provide second line of defense, and reactive long-lived memory B cells provide third line of defense against viruses (Dörner and Radbruch 2007).

Natural antibodies are encoded by germ line variable genes and not shaped by post-recombination processes such as somatic hypermutation and class switching. Natural antibodies recruit antigens into the spleen to prevent infection of vital organs and also to induce early neutralizing antibodies without T-cell help. Natural antibodies activate complement to enhance B- and T-cell-specific immune responses. Therefore, natural antibodies are considered an important link between innate and adaptive immunity (Matter and Ochsenbein 2008). The natural antibodies are poly-reactive that they can bind to different structurally unrelated antigens and also auto-reactive that they can bind to self-antigens. Along with natural antibodies, poly-reactive antigen-binding B cells provide protection against multiple pathogens. In addition to poly-reactive antibodies, the specific neutralizing and cross-reactive antibodies also provide effective antiviral immune response (Warter et al. 2012).

The highly specific antibodies protect from viral infection by both neutralizing and Fc-mediated effector mechanisms. Neutralizing antibodies inhibit the viruses by binding through their antigen-binding sites or Fab region and also through their constant or Fc region. The non-neutralizing antibodies bind to many epitopes including neutralizing antibody targets and induce the destruction of microbe or infected host cells by innate immune system (Hua and Ackerman 2017). Recently, novel broadly neutralizing antibodies were characterized, which are able to neutralize diverse isolates of viruses (Sok and Burton 2018). This chapter describes about different types of B cells, antibodies, and their role in combating different types of viruses (HIV, influenza, and hepatitis).

6.2 B Cells

The B lymphocytes are subset of lymphocytes, which express diverse specific surface immunoglobulins and constitute 5–15% of total circulating lymphoid cells. The antigen-independent development of B cells starts from the fetal liver and bone marrow of an adult, where the hematopoietic stem cells differentiate into pro-B cells, which in turn differentiate pre-B cells to immature B cells. Along with the development of phenotypically distinct precursor cells, rearrangements of VDJ genes occur parallelly, which contribute to formation of diverse specific immature B cells with membrane-bound IgM. Immature B cells differentiate into mature B cells with membrane-bound IgM and IgD as its BCR in the secondary lymphoid organs. Two types of B lymphocytes (B-1 and B-2) have been identified. Mature B lymphocytes are activated by different types of antigens. There are three types of antigens activating the B lymphocytes. (1) T-independent antigen type 1, for example, LPS (lipopolysaccharide), which is polyclonal activator of both mature and immature B cells. LPS activates the B cells through TLR. (2) T-independent antigen type 2, for example, bacterial polysaccharides, which have repeated epitopes, require help from cytokines secreted by T helper cells but not their direct contact. (3) T-dependent antigens, for example, soluble protein antigens, require T helper cells' direct contact and cytokines secreted by them to activate B lymphocytes (Maddaly et al. 2010).

The origin and development of B-1 and B-2 cells are described by two competing models. The selection model proposes that the response to a particular antigen decides the differentiation of B cell to become either B-1 or B-2. The other layered immune system hypothesis of Herzenberg proposes that they emerge from separate distinct progenitors. The B-1 cells present in serous cavities (peritoneal and pleural cavities) further differentiates into B-1a and B-1b. The B-2 cells in secondary lymphoid organs also differentiate into two types of subsets such as follicular (FO) and marginal zone (MZ) B cells after passing through T1, T2, and T3 transitional stages (Montecino-Rodriguez and Dorshkind 2012).

The B-1a cells produce poly-specific IgM natural antibodies in the serum of newborn or germ-free animals. The B-1b cells secrete antibodies after stimulation with thymus-independent antigens (Matter and Ochsenbein 2008). In addition to IgM, the B-1 cells also produce poly-reactive IgA antibodies for mucosal immunity (Suzuki et al. 2010). The MZ B cells express poly-reactive BCR and produce poly-specific IgM antibodies against blood-borne microorganisms. Both B-1 and MZ B cells express TLR and produce response to pathogen-associated or endogenous TLR ligands with or without recognition through BCR. Unlike B-1 cells, the MZ B cells generate response to T-dependent proteins and produce high-affinity isotype-switched antibodies. Although the FO B cells respond to T-independent antigens, they are primarily responsible for high-affinity IgG antibodies production against T-dependent antigens. The T-dependent antigens activate the FO B cells in the presence of T helper cells and their secreted cytokines. Antigen-activated FO B cells involves in the formation of germinal center (GC) reaction. The clonal expansion, isotype-switching, somatic hypermutation, and affinity maturation are characteristics of GC reaction. The activated FO B cells undergo proliferation and differentiation into antibody secreting long-lived plasma cells and memory B cells (Fig 6.1) (Hoffman et al. 2016). The external antigens are captured through BCR of B cells and they are degraded inside the B cells and the corresponding antigen fragments presented by B cells on MHC-II molecule to helper T cells are an important step in adaptive cellular immunity (Yuseff et al. 2013).

B cell surface not only consists of membrane-bound Ig, but also other complement component receptors and Fc receptors are expressed. In addition to that with the advancement of monoclonal antibody technology, many B-cell-specific surface molecules were identified. To bring common nomenclature, monoclonal antibodies were designated as clusters of differentiation (CD). The CD designation is embraced as a label for the target molecule rather than grouping of monoclonal antibodies with common reactivity. The target molecules or CD are involved in B-cell development, function and communication with extracellular environment. These CD molecules also provide cellular context which interprets the BCR signals; for example, in BCR signaling and B-cell development, CD79a (Igα) and CD79b (Igβ) are

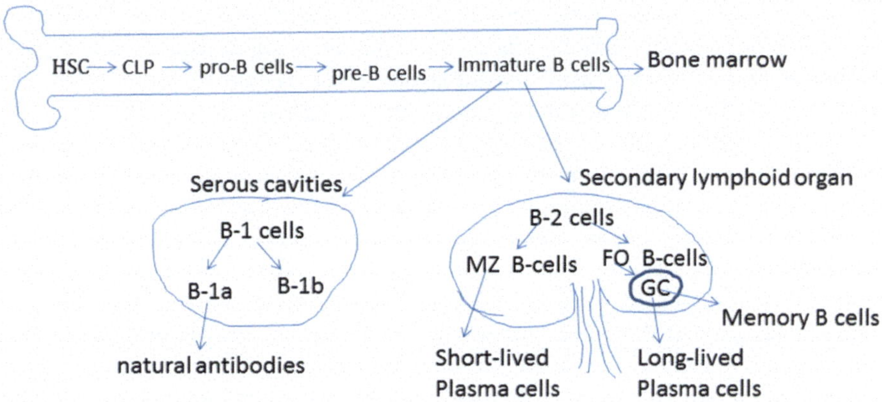

Fig. 6.1 Schematic diagram of B-cell development and different humoral effector immune cells

non-covalently attached with BCR and their cytoplasmic domains certainly contains conserved motifs for tyrosine phosphorylation and Src family kinase docking respectively (LeBien and Tedder 2008). In addition to production of antibodies, B cells also actively participate in cellular immune response mediated by T cells. The B cells directly modulate effector, memory, and regulatory T-cell functions via antigen-specific but antibody-independent mechanisms. B cells modulate T cells by antigen presentation, by providing co-stimulation, and by secreting cytokines. B cells are functionally subdivided into two types: (1) Be-1 cells and (2) Be-2 cells. In the presence of TH1-type cytokines, the Be-1 cells do not secrete significant amounts of IL-4, IL-13, or IL-2 but secrete IFN-γ and IL-12 including IL-10, TNF-α, and IL-6. In the presence of TH2-type cytokines, Be-2 cells do not secrete significant amounts of IFN-γ and IL-12 but secrete IL-4, IL-13, or IL-2 including IL-10, TNF-α, and IL-6. The regulatory B cells also called B-10 cells produce IL-10 cytokines that suppress the $CD4^+$ T-cell responses. In addition to suppression, cytokines producing B cells enhance the T-cell-mediated immune responses (Lund and Randall 2010).

6.3 Antibodies

The B lymphocytes differentiate finally into plasma cells, which are the source for antibodies. The two competing natural selection and clonal selection theories for antibody formation were proposed by Jerne and Burnet, respectively, but finally, clonal selection theory was supported by experimental evidence (Burnet 1976).

Antibodies are chemically glycoprotein molecules. They are also called immunoglobulins and are present on the surface of B cells as BCR or free molecules in blood, plasma, and extracellular fluids. The fluids formerly were called humors and are part of humoral immune response. They have two principal functions in humoral defense. The first function is that it recognizes and binds to antigenic determinants or epitopes on the surface of foreign molecules, for example, envelope spikes on the virus surface. The second function is that it triggers elimination mechanisms, e.g., complement activation and phagocytosis by neutrophils and macrophages. The first function requires huge diversified specific antibodies and the second function requires common features in all different types of antibodies. The Y-shaped antibodies are made up of four proteins chains: the upper and lower parts of heavy and light chains are called variable and constant regions, respectively. The two heavy and two light chains are covalently linked by disulfide bonds at constant section. The two heavy chains again linked at their constant section contain hinge region, which provide flexibility to the antibodies (Fig. 6.2). The antibodies consist of three units, of which two units present at N-terminus of the chains are identical and responsible for antigen binding—the Fab (fragment antigen binding) arms. The variable domains of both heavy and light chains form the antigen-binding site or paratope. The variable domain again contains special hypervariable segments called complementary-determining regions, which are specific to antibodies. The third unit is called Fc (fragment crystalline) arm involved in effector functions (Hey 2015; Chan et al. 2009).

There are five classes or isotypes (IgG, IgM, IgA, IgD, and IgE), which are different in their C-terminus region of H chains termed γ, μ, α, δ, and ε, respectively. There are four subclasses of IgG (IgG1, IgG2, IgG3, and IgG4. The effector mechanisms are mainly dependent on the type of antibodies; for example, IgM and IgG 3 are complement activators, whereas IgG1 and IgGE activate macrophages and mast cells, respectively (Hoffman et al. 2016).

Antibodies act on viruses directly by neutralization or indirectly via interaction with complement or Fc receptors on innate immune cells (Fig. 6.3). The multiple effector functions of antibodies include (1) neutralization; (2) antibody-dependent cellular toxicity (ADCC) or antibody-dependent cellular phagocytosis (ADCP) of both viruses and infected cells through interaction with Fc receptors on innate

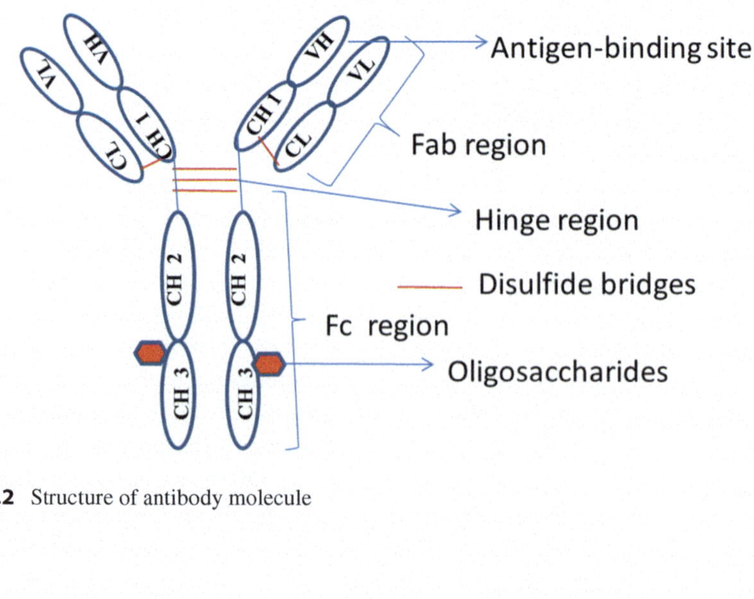

Fig. 6.2 Structure of antibody molecule

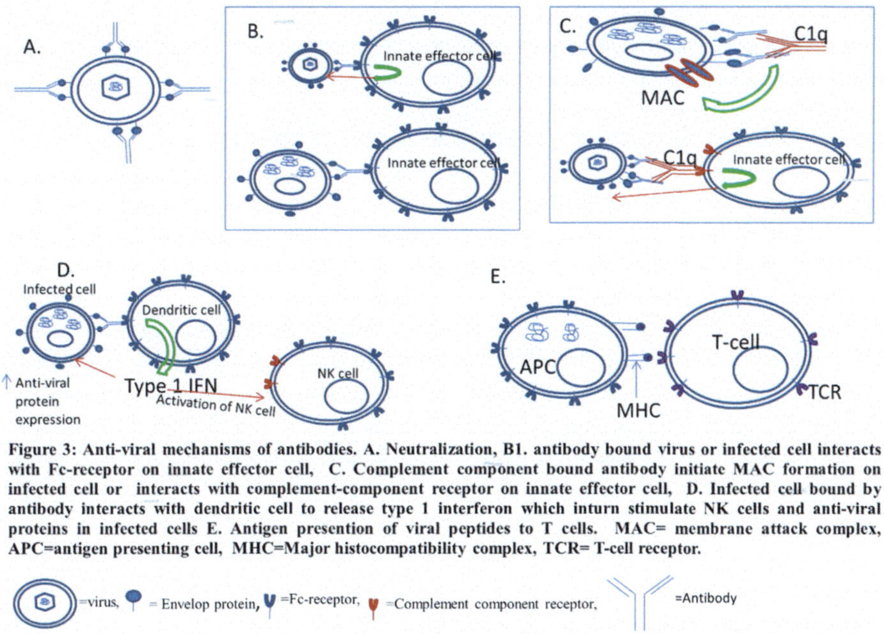

Fig. 6.3 Antiviral mechanism of antibodies. (**a**) Neutralization, (**b**) antibody-bound virus or infected cell interacts with Fc receptor on innate effector cell, (**c**) complement component-bound antibody initiates MAC formation on infected cell or interacts with complement-component receptor on innate effector cell, (**d**) infected cell bound by antibody interacts with dendritic cell to release type 1 interferon which in turn stimulates NK cells and antiviral proteins in infected cells, (**e**) antigen presentation of viral peptides to T cells. *MAC* membrane attack complex, *APC* antigen-presenting cell, *MHC* major histocompatibility complex, *TCR* T-cell receptor

immune cells; (3) complement-dependent cytotoxicity (CDC), where complement component C1q mediates formation of membrane attack complex on infected cells or opsonization-based phagocytosis mediated through complement receptors on innate immune effector cells; (4) antibodies bound to infected cells interact with dendritic cells to release type 1 interferons, which stimulate antiviral activity of NK cells and antiviral functions in infected cells, or antigen-presenting cells phagocytize the antibody–virus immune complex and present specific viral peptides to T cells for cellular immunity (Forthal 2014; Hua and Ackerman 2017).

Antibodies neutralize the pathogen by multiple mechanisms such as (1) pre-attachment neutralization, where antibodies directly bind with pathogen to agglutinate. For example, the neutralizing IgG antibodies aggregate the poliovirus to decrease their infectivity. (2) Interference with pathogen attachment—antibodies bind to pathogen ligands which are essential for pathogen attachment. For example, antibodies against HIV-1 gp120 interfere with their attachment to $CD4^+$ T cells. (3) Post-attachment neutralization-inhibition of fusion or entry. For example, a monoclonal antibody 2F5 inhibits the fusion of viral and cellular membranes. (4) Inhibition of various steps in pathogen life cycle—once the pathogen internalized, in order to neutralize pathogen, antibodies must be internalized and interfere with the replication, genetic material expression, and release of pathogen. (5) Inhibition at later steps—antibodies inhibit the liberation of virus or budding (Forthal 2014).

6.4 Humoral Immune Response Against to Viruses

Viruses are intracellular pathogens and need host cells machinery for their survival. After synthesizing many viral copies, viruses burst out and reinfect healthy new host cells. These viruses directly kill the host cells and are called acutely cytopathic viruses (polioviruses, rabies, and small pox viruses). Some viruses do not kill the host cells and they are called non-cytopathic viruses (hepatitis B and C, herpes virus, herpes simplex virus 1 and 2, cytomegalovirus, and Epstein-Barr viruses). The role of immune system in antiviral defense is both protective and pathogenic, where it eliminates the viruses by effective immune response and in addition also causes more damage to host cells than viruses themselves. With regard to effective immunity, the viruses have developed an array of strategies to escape elimination by immune system through co-evolution (Dörner and Radbruch 2007).

Viruses are chemically nucleoprotein particles with a size of ~20–200 nm. Due to their small size, viruses move freely in the lymphatic system and interact with B cells in secondary lymphoid organs. The highly repetitive structures of viral particles allow the cross-linking of BCRs, which is an initial step in the activation of B cells. Further, the repetitive viral structures also bind with natural antibodies, fix complement, and also carry TLR ligands (DNA or RNA) for TLR 7/8 or 9 to activate B cells directly (Zabel et al. 2013).

6.4.1 Human Immunodeficiency Virus

AIDS (acquired immune deficiency syndrome) is caused by HIV (human immunodeficiency virus). HIV is genetically highly diverse virus and can be found in two types: HIV-1 and HIV-2. HIV-1 is the most common and falls into three groups: M, N, and O. Group M is the most common and divided into subtypes or clades (A–D, F–H, J, and K), of which B is predominant subtype in the Western world and C in India, China, and Africa. The $CD4^+$ T cells, macrophages, and different subsets of dendritic cells are the major targets of HIV-1 infection. HIV-1 belongs to Retroviridae family and lentivirus genus. HIV-1 is an enveloped virus containing two positive sense RNA strands. The enveloped glycoproteins of HIV-1 present as trimers of gp120/gp41 heterodimers on the surface of virus (Phogat et al. 2007).

After few days of following HIV-1 postinfection, anti-gp41 antibodies and anti-gp120 antibodies are produced, but these antibodies are unable to neutralize the infecting virus strain or autologous virus. The autologous neutralizing antibodies are produced several months postinfection, which are not able to neutralize the heterologous virus. The cross-reacting antibodies, which can neutralize many heterologous viruses, develop 2–4 years after seroconversion, but in some patients develop earlier. Some HIV patients like "Elite neutralizers" develop highly cross-reacting antibodies or broadly neutralizing antibodies (bNAbs) 20-months postinfection. Diversity of HIV strains and molecular peculiarities of HIV envelope protein are some obstacles in recognition and neutralization of HIV strains. However, some patients produce bNAbs that neutralize many HIV strains. The discovery and well characterization of bNAbs has given new hope in the treatment of HIV, so that vaccines should be prepared that induce broadly neutralizing antibodies and antibodies with particular functional activity. Not only systemic immune response but also mucosal immune response is produced against HIV virus. Even though, very little is known about mucosal immune response, some studies have shown that both sero-negative and positive patients develop HIV-specific mucosal IgA, which inhibits HIV transcytosis and replication in epithelial cells (Mouquet 2014; Baum 2010).

Non-neutralizing antibody plays an important role in the prevention and control of HIV in humans. In addition, limited protection was observed with RV144 vaccine in the absence of neutralizing antibodies, which indicates the role of non-neutralizing functions by antibodies (Zolla-Pazner 2016; Mayr et al. 2017). Some studies reported that they found anti-HIV antibodies are part of innate immune system. They defend HIV by binding to the HIV protein Tat. Natural antibodies for host receptor CCR5 prevent HIV infection effectively at major sites of virus entry such as mucosal tissues (Ward 2001; Lopalco 2010).

6.4.2 Influenza Virus

Influenza or flu viruses cause respiratory diseases and continuously threaten human health. Genome high mutational rates of influenza viruses cause emergences of new strains by genetic drifts. Influenza affects 5–30% of global population and causes hospitalization and death of many people. They are enveloped viruses and contain single-stranded, segmented, and negative sense 7–8 RNA strands. Influenza viruses are categorized into four types: A, B, C, and D. Influenza A virus is again classified based on the antigenic properties of two surface viral glycol proteins such as hemagglutinin (HA) and neuraminidase (NA) with 18 (H1-H18) and 11 (N1-N11) antigenic subtypes, respectively. Influenza A (H1N1 and H3N2) virus causes seasonal influenza epidemics in humans along with other members of influenza viruses such as influenza B virus (Sautto et al. 2018).

Two important hypotheses such as original antigenic sin (OAS) and immune imprinting or antigenic seniority explain how humoral immunity gets affected by influenza viruses. Both are dependent on humoral immune memory response rather than on de novo humoral immune response to drifted or altered influenza viruses (Guthmiller and Wilson 2018). The original antigenic sin concept refers that the first exposure of influenza variant in early life dictates lifelong immunity to all variants of influenza viruses in subsequent encounters. The immune memory produced by the first influenza variant influences the immune response to subsequently expose influenza distinct variants, but how this sequential exposure shapes the immune response remains obscure. The antigenic seniority concept better explains the hierarchical nature of immune response to previously exposed variants of influenza virus. According to the antigenic seniority concept, the first exposed influenza variant in childhood takes the senior antigenic position in immune repertoire and subsequent exposed strains take the junior positions. The immune response to the first exposure is larger than the responses to subsequent exposures. Understanding of how previous exposure shapes the antibody responses to vaccination and infection is critical for the development of universal influenza vaccine (Henry et al. 2018).

Neutralizing antibodies produced against HA head region of influenza are strain-specific and bind to highly variable regions. Antibodies neutralize more strains of influenza and are called broadly neutralizing or cross-reactive antibodies. The broadly neutralizing antibodies produced against influenza virus mainly target the conserved regions of HA stem domain. The existence of non-neutralizing antibodies also clears the influenza virus infection by exploiting non-neutralizing effector mechanisms (Sicca et al. 2018).

The natural antibodies against influenza virus are produced by B-1 cells and therefore reported to suggest that innate and acquired humoral immunity comes from separate effector arms of immune system (Baumgarth et al. 1999).

6.4.3 Hepatitis C Virus

Hepatitis C (HCV) is non-cytopathic virus which infects millions of people around the world, and it is transmitted mainly by unsafe injections and transfusions. During acute infections few people (20%) only clear the virus spontaneously and remaining 70–80% people suffer from chronic infection. The acute hepatitis C infections are asymptomatic. The chronic viral hepatitis by HCV leads to cirrhosis, hepatocellular carcinoma, and end-stage liver disease (Webster et al. 2015).

Hepatitis C virus is small enveloped virus and has positive single-stranded RNA genome encoding a single polyprotein which undergoes posttranslational modification by cellular and viral-encoded proteases into structural (core proteins and envelope proteins) and nonstructural proteins. The HCV RNA genome interacts with core proteins to form the nucleocapsid that is surrounded by a lipid membrane, called the viral envelope, in which envelope glycoproteins are anchored (Irshad et al. 2008).

The clearance of HCV is associated with induction of cellular immune response. In addition, evidence is accumulating that neutralizing antibodies also contribute to HCV clearance. Acutely HCV-infected individuals produce antibodies against epitopes present on the structural and nonstructural proteins of HCV virus. A small fraction of antibodies called neutralizing antibodies are able to inhibit virus binding, entry, and uncoating. The glycoproteins E1 and E2 are major targets for the neutralizing antibodies. Neutralizing antibodies are generated against HVR1 region of envelope glycoprotein E2. The generation of neutralizing antibodies in the early phase of infection correlates with the resolution of HCV infection in some acutely infected people. In contrast, chronic infected people showed delayed induction of neutralizing antibodies (Lapa et al. 2019).

The HVR1 antibodies are type-specific but some evidence shows that anti-HVR1 antibodies are cross-reactive. Some neutralizing antibodies against E1 and E2 envelope proteins show cross-neutralizing potential (Drummer 2014). However, there is strong evidence on the production of broadly neutralizing antibodies against HCV infections (Kinchen et al. 2018). The HCV virus also has immunologic regions for virus escape or non-neutralizing antibodies in the envelope glycoproteins (Fuerst et al. 2018). Similar to HIV virus, the humoral immune response also evolves with time in HCV. The evolution of Ab specificity (non-neutralizing—autologous neutralizing—heterologous neutralizing—broadly neutralizing) during the course of infection produces increased breadth in the neutralizing activity (Murira et al. 2016). There are also reports on natural antibodies which recognize the linear epitopes on hepatitis C envelope glycoprotein E2 to confer additional neutralization (Tarr et al. 2012).

6.5 Future Perspectives and Conclusions

B cells are the main players of humoral immunity against viruses. B cells also act as antigen-presenting cells, contributing to cell-mediated immunity. Both B-1 and B-2 cells secrete natural and highly specific antibodies for the prevention and control of viral infections, respectively. B-2 cells capture the viral antigens through BCR and differentiate into memory B cells and plasma cells, which secrete different types of specific adaptive antibodies. Recently, many studies reported the generation of broadly neutralizing antibodies, which can neutralize different strains of viruses. Understanding the production of antibodies and their antiviral mechanisms will provide new ways to develop novel, efficient prophylactic and therapeutic vaccines.

Acknowledgements The authors are grateful to YVU Kadapa and Krishna University, Machilipatnam, for the support extended.

Conflict of Interest The authors declare that they have no competing interests.

References

Baum LL (2010) Role of humoral immunity in host defense against HIV. Curr HIV/AIDS Rep 7(1):11–18

Baumgarth N, Herman OC, Jager GC, Brown L, Herzenberg LA, Herzenberg LA (1999) Innate and acquired humoral immunities to influenza virus are mediated by distinct arms of the immune system. Proc Natl Acad Sci U S A 96(5):2250–2255

Bonilla FA, Oettgen HC (2010) Adaptive immunity. J Allergy Clin Immunol 125(2 Suppl 2): S33–S40

Burnet FM (1976) A modification of Jerne's theory of antibody production using the concept of clonal selection. CA Cancer J Clin 26(2):119–121

Burton DR (2002) Antibodies, viruses and vaccines. Nat Rev Immunol 2(9):706–713

Chan CE, Chan AH, Hanson BJ, Ooi EE (2009) The use of antibodies in the treatment of infectious diseases. Singap Med J 50(7):663–672; quiz 673
Dörner T, Radbruch A (2007) Antibodies and B cell memory in viral immunity. Immunity 27(3):384–392
Drummer HE (2014) Challenges to the development of vaccines to hepatitis C virus that elicit neutralizing antibodies. Front Microbiol 5:329
Forthal DN (2014) Functions of antibodies. Microbiol Spectr 2(4):1–17
Fuerst TR, Pierce BG, Keck ZY, Foung SKH (2018) Designing a B cell-based vaccine against a highly variable hepatitis C virus. Front Microbiol 8:2692
Gourley TS, Wherry EJ, Masopust D, Ahmed R (2004) Generation and maintenance of immunological memory. Semin Immunol 16(5):323–333
Guthmiller JJ, Wilson PC (2018) Harnessing immune history to combat influenza viruses. Curr Opin Immunol 53:187–195
Henry C, Palm AE, Krammer F, Wilson PC (2018) From original antigenic sin to the universal influenza virus vaccine. Trends Immunol 39(1):70–79
Hey A (2015) History and practice: antibodies in infectious diseases. Microbiol Spectr 3(2):AID-0026-2014
Hoffman W, Lakkis FG, Chalasani G (2016) B cells, antibodies, and more. Clin J Am Soc Nephrol 11(1):137–154
Hua CK, Ackerman ME (2017) Increasing the clinical potential and applications of anti-HIV antibodies. Front Immunol 8:1655
Irshad M, Khushboo I, Singh S, Singh S (2008) Hepatitis C virus (HCV): a review of immunological aspects. Int Rev Immunol 27(6):497–517
Kinchen VJ, Zahid MN, Flyak AI, Soliman MG, Learn GH, Wang S, Davidson E, Doranz BJ, Ray SC, Cox AL, Crowe JE Jr, Bjorkman PJ, Shaw GM, Bailey JR (2018) Broadly neutralizing antibody mediated clearance of human Hepatitis C virus infection. Cell Host Microbe 24(5):717–730.e5
Lapa D, Garbuglia AR, Capobianchi MR, Del Porto P, Hepatitis C (2019) Virus genetic variability, human immune response, and genome polymorphisms: which is the interplay? Cells 8(4):E305
LeBien TW, Tedder TF (2008) B lymphocytes: how they develop and function. Blood 112(5):1570–1580
Lefevre EA, Carr BV, Prentice H, Charleston B (2009) A quantitative assessment of primary and secondary immune responses in cattle using a B cell ELISPOT assay. Vet Res 40(1):3
Lopalco L (2010) Natural anti-CCR5 antibodies in HIV-infection and exposure. J Transl Med 9(Suppl 1):S4
Lund FE, Randall TD (2010) Effector and regulatory B cells: modulators of CD4+ T cell immunity. Nat Rev Immunol 10(4):236–247
Maddaly R, Pai G, Balaji S, Sivaramakrishnan P, Srinivasan L, Sunder SS, Paul SF (2010) Receptors and signaling mechanisms for B-lymphocyte activation, proliferation and differentiation--insights from both in vivo and in vitro approaches. FEBS Lett 584(24):4883–4894
Matter MS, Ochsenbein AF (2008) Natural antibodies target virus-antibody complexes to organized lymphoid tissue. Autoimmun Rev 7(6):480–486
Mayr LM, Su B, Moog C (2017) Non-neutralizing antibodies directed against HIV and their functions. Front Immunol 8:1590
Montecino-Rodriguez E, Dorshkind K (2012) B-1 B cell development in the fetus and adult. Immunity 36(1):13–21
Mouquet H (2014) Antibody B cell responses in HIV-1 infection. Trends Immunol 35(11):549–561
Murira A, Lapierre P, Lamarre A (2016) Evolution of the humoral response during HCV infection: theories on the origin of broadly neutralizing antibodies and implications for vaccine design. Adv Immunol 129:55–107
Palma J, Tokarz-Deptuła B, Deptuła J, Deptuła W (2018) Natural antibodies – facts known and unknown. Cent Eur J Immunol 43(4):466–475
Phogat S, Wyatt RT, Karlsson Hedestam GB (2007) Inhibition of HIV-1 entry by antibodies: potential viral and cellular targets. J Intern Med 262(1):26–43

Sautto GA, Kirchenbaum GA, Ross TM (2018) Towards a universal influenza vaccine: different approaches for one goal. Virol J 15(1):17

Sicca F, Neppelenbroek S, Huckriede A (2018) Effector mechanisms of influenza-specific antibodies: neutralization and beyond. Expert Rev Vaccines 17(9):785–795

Sok D, Burton DR (2018) Recent progress in broadly neutralizing antibodies to HIV. Nat Immunol 19(11):1179–1188

Suzuki K, Maruya M, Kawamoto S, Fagarasan S (2010) Roles of B-1 and B-2 cells in innate and acquired IgA-mediated immunity. Immunol Rev 237(1):180–190

Tarr AW, Urbanowicz RA, Jayaraj D, Brown RJ, McKeating JA, Irving WL, Ball JK (2012) Naturally occurring antibodies that recognize linear epitopes in the amino terminus of the hepatitis C virus E2 protein confer noninterfering, additive neutralization. J Virol 86(5):2739–2749

Ward S (2001) Natural anti-HIV antibodies. Trends Immunol 22(10):544

Warter L, Appanna R, Fink K (2012) Human poly- and cross-reactive anti-viral antibodies and their impact on protection and pathology. Immunol Res 53(1–3):148–161

Webster DP, Klenerman P, Dusheiko GM (2015) Hepatitis C. Lancet 385(9973):1124–1135

Yamada T (2011) Therapeutic monoclonal antibodies. Keio J Med 60(2):37–46

Yuseff MI, Pierobon P, Reversat A, Lennon-Duménil AM (2013) How B cells capture, process and present antigens: a crucial role for cell polarity. Nat Rev Immunol 13(7):475–486

Zabel F, Kündig TM, Bachmann MF (2013) Virus-induced humoral immunity: on how B cell responses are initiated. Curr Opin Virol 3(3):357–362

Zolla-Pazner S (2016) Non-neutralizing antibody functions for protection and control HIV in humans and SIV and SHIV in non-human primates. AIDS 30(16):2551–2553

Part II

Role of Adaptive Immunity in Combating Viral Diseases

Significant Role of Cellular Activation in Viral Infections

7

Bojjibabu Chidipi, Samuel Ignatious Bolleddu, Ganugula Mohana Sheela, and Alavala Matta Reddy

Abstract

Immune response during viral infection is the result of pathogen recognition and antibody production by the memory cells residing in the host target tissue. These cells by origin are differentiated from the cytotoxic $CD8^+$ T cells and play a major role in developing adaptive immunity. Early activation of these T cells by ribosomal binding proteins during various cell death pathways can induce a critical autoimmune response suitable to develop enhanced strategic immune therapies for many chronic viral infections. In addition, there are various genes which get expressed in good amount in the process of T-cell activation during infections like Epstein-Barr virus. Focusing on these changes can help in determining the pathogenic process and orientation of the viral infection. Since we can only control but not cure most of the viral infections using present medications, following these novel strategies can help in focusing on the host cell death pathways, thereby developing the drugs that can give complete immunity against virus invasion by facilitated viral cell elimination from the humoral environment.

Keywords

Cytotoxic $CD8^+$ T cells · Epstein-Barr virus · Humoral environment · Cell death pathways · Memory cells

B. Chidipi · S. I. Bolleddu
Molecular Pharmacology and Physiology, University of South Florida, Tampa, FL, USA

G. M. Sheela
Department of Biotechnology, Krishna University, Machilipatnam, Andhra Pradesh, India

A. Matta Reddy (✉)
School of Life and Health Sciences, Adikavi Nannaya University, Rajamahendravaram, Andhra Pradesh, India

P. V. Bramhachari (ed.), *Dynamics of immune activation in viral diseases*,
https://doi.org/10.1007/978-981-15-1045-8_7

7.1 Introduction

The main role of the immune system is to identify the presence of any foreign protein complexes or antigens and initiate an immune response to eliminate them without harming the metabolism of the host by implementing the pathogen-specific defense response. Immune response during viral infection is the result of pathogen recognition and antibody production by the memory cells residing in the host target tissue. B lymphocytes produce the antibodies during the humoral immune response while T cells initiate cell-mediated immunity (Finotti et al. 2018).

During viral infections, the cellular activation of various lymphocytes leads to degranulation of various inflammatory mediators like histamines, TNFα, and various amino acids within the first few seconds. The host inborn resistant reaction is a profoundly strong boundary of barrier containing numerous intrinsic safe cells, for example, dendritic cells (DCs), macrophages, and common executioner (NK) cells. These cells are fit for perceiving a broad scope of viral items through a complex and covering atomic system of pattern recognition receptors (PRRs), including toll-like receptors (TLRs) just as cytosolic sensors, for example, the helicases, retinoic acid-inducible gene 1 (RIG-I) and MDA5. Actuation of intrinsic cells through these PRRs advances fast popular detecting and the arrival of critical antiviral cytokines with the possibility of quickly controlling viral replication and driving further natural and versatile invulnerable reactions. Specifically, the sort I interferon (IFN-I) is a pleiotropic cytokine family comprising 13 IFNα isoforms and one IFNβ isoform that move through a common pervasively communicated receptor (IFNαβR).

Eicosanoids like leukotrienes and prostaglandins are released as a secondary mediator after few minutes. In a prolonged immune response, cytokines and chemokines along with growth factors are released (Bull and Plummer 2014; Frossi et al. 2018; Yu et al. 2016; Brown et al. 2018; Gieseck et al. 2018).

The viral reproduction process in the host cells after the viral invasion starts with the synthesis of new viral proteins and nucleic acids intracellularly. The new viral particles or virions are then self-assembled by the aggregation of the macromolecules. These virions are seen in the blood circulation invading the healthy cells after the rupture of the infected host cells. The envelope or capsids of the virus particles offer protection and platform for inducing an immune response. Motioning through IFNαβR prompts both autocrine and paracrine actuation, bringing about phosphorylation of flag transducer and activator of interpretation 1 and 2 (STAT1 and STAT2). This actuates several IFN-I-animated qualities (ISGs) that advance an antiviral state and initiate various safe cells (Brown et al. 2018). Despite this complex inborn reaction, numerous infections, including those that accomplish interminable disease, can stifle and additionally avoid these early natural host reactions to support their replication and transmission. In this segment, we centered on IFN-I to show the dynamic changes that the natural resistant framework experiences and the critical and expansive results that such changes can have amid perpetual viral diseases (Ning et al. 2018).

This new wave of viral invasion directly stimulates glycoprotein inducing cells for the production of the antiviral proteins called interferons. Vital cells of the immune system like leucocytes, fibroblasts, and natural killer cells play a vital role in the production of interferons and lysosomes that lyse a wide variety of virus particles, thereby recognizing the antigen and developing a suitable antibody.

7.1.1 Type I Interferon Induction by Persisting Viruses

Within days after HIV discovery in plasma from tainted patients, and inside hours after contamination with LCMV Cl13 in mice, raised dimensions of fundamental IFN-I are distinguished. Moreover, upgraded ISG articulation is visible in tissues as well as T cells after LCMV Cl13, HCV, and SIV diseases (Fig. 7.1). Interestingly, qualities with intrinsic insusceptible capacity are imperceptible right on time after in vivo HBV contamination in chimpanzees (Haas and Trumpp 2018).

7.2 Role in Adaptive Immunity

Early activation of these T cells by ribosomal binding proteins can induce a critical immune response suitable to develop enhanced strategic immune therapies for many chronic infections. Several studies confirm T lymphocytes functional modifications as the key pathological change responsible for adaptive immunity. A few non-hematopoietic and hematopoietic cells, including macrophages, ordinary DCs (cDCs), and plasmacytoid DCs (pDCs), can add to the first vigorous IFNI reaction

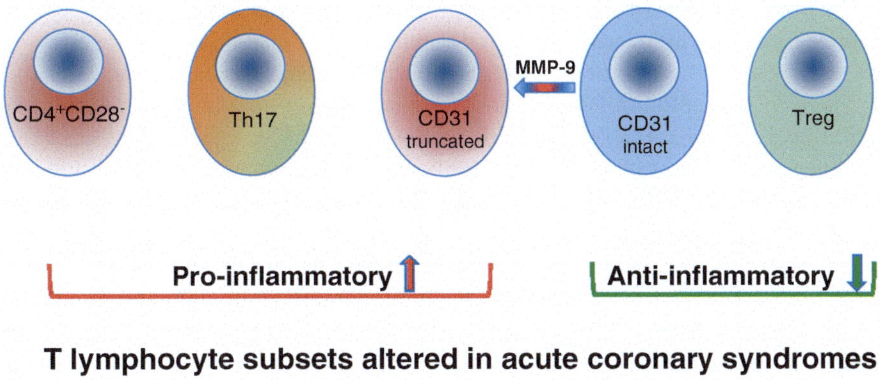

Fig. 7.1 depicts several subtypes of T lymphocytes implicated in dysregulated immune responses during acute coronary syndromes. Regulatory T cells counterbalance the pro-inflammatory properties of $CD4^{+}$ CD 28^{-} and Th17 T cells. IL-17A produced by Th17 cells may not only promote inflammation but also favor fibrosis in atheromata. The rainbow hue of the Th17 cell portrays its potential dual roles. T lymphocytes that express intact CD31 on their surface interfere with adaptive immune responses. Proteolytic cleavage of CD31 by the enzyme matrix metalloproteinase-9 removes T cells from inhibition and licenses them to contribute to the stimulation of adaptive immunity. As patients with acute coronary syndromes have higher circulating concentrations of matrix metalloproteinase-9, this mechanism for boosting T-cell responses may operate particularly during acute ischemic episodes (Libby and Hansson 2018)

by perceiving and reacting to viral disease through PRR-intervened flagging pathways. For instance, in LCMV Cl13 disease, cDCs and macrophages produce IFNI through MDA5/MAVS-subordinate pathways. Even though the phone source and the kind of IFN (e.g., IFN-I as well as IFN-III) that prompts raised ISGs in the liver upon HCV disease are as yet hazy, demonstrating that hepatocytes are fit for creating IFN-I through RIG-I motioning in light of HCV contamination (Libby and Hansson 2018; Shuai et al. 2016).

For instance, the CD4 type of cells in the lymphocyte population plays a key role in the process of adaptive immunity during cellular response by initiating the effective antibodies against the foreign pathogens. This is usually done by secreting the cytokines that are capable of the effector mechanisms that combat with antigen interactions with the host system. At long last, monocyte determined DCs can detect HIV-2 (and HIV-1 when the host catalyst SAMHD1 does not limit its replication) using cyclic guanosine monophosphate–adenosine monophosphate synthase (cGAS) pathways, initiating expert incendiary cytokines and IFN-I creation. Regardless of the way that practically all phone types can create IFN-I upon viral experience, pDCs are very particular for discharging these cytokines, delivering multiple times more IFN-I than some other platelet type, and integrating a more extensive scope of IFN-I isoforms upon TLR incitement. pDCs can straightforwardly perceive few steady infections, incorporating HIV and HCV in people and LCMV in mice, through TLR7. pDCs were determined to be basic for T-cell reactions and viral control amid ceaseless LCMV disease, although another examination utilizing an alternate LCMV variation and an inducible pDC cancelation framework showed just insignificant impacts (Gao et al. 2018).

The microbes present in the host body also play a vital role in the process of adaptive immunity by interacting with the antigen–host responses. The macromolecules present in the microbes interact with the immune mediators and alter the specificity of the autoimmune response, thereby leading to adaptive immunity in natural process. Remarkably, rather than IFN-I creation in most cell types, PDC IFN-I generation does not require viral replication, and most of DC IFN-I originates from uninfected pDCs right on time after LCMV Cl13 contamination. Correspondingly, uninfected pDCs can detect neighboring tainted hepatocytes by perceiving HCV RNA-containing exosomes, and neither viral combination nor replication is required for PDC acknowledgment of HIV. All in all, these investigations demonstrate that most of industriously recreating infections can be detected by inborn cells using PRRs, bringing about a vigorous IFN-I reaction in the beginning of in vivo unending disease (Ost and Round 2018).

1. Antibodies play a vital role in the treatment of viral infections, but a new class of human monoclonal antibodies called super-antibodies are drawing attention in immunotherapy. These antibodies are rarely seen in human infections. They are majorly generated during screening using the B-cell approach. This strategy is mainly implemented to increase the half-life and immune functions in the prophylaxis and treatment of a wide range of viral infections. This strategy can be a promising approach in the future (Walker and Burton 2018).

2. The current antiviral therapies for HBV are effective but need to be administered for a long period without allowing the virus to rebound or replicate. There are some immunotherapies which focus on the response of the host to the virus and maintaining the viral load under control. One of the best strategies is to develop the antiviral drug based on the toxic outcome and therapeutic efficiency to reduce the viral levels by immunotherapy studies in the varied populations (Bertoletti and Le Bert 2018).
3. The new approach in the targeting of the immunodeficiency diseases like HIV, HBV, and HCV is to target the T regulatory cells which play a major role in protecting important organs like the liver and kidneys from injury due to immunosuppression. This approach will help protect a vast majority of the white blood cells that are destroyed by viral infections whose primary goal is to eliminate the large populations of the macrophages (Karkhah et al. 2018).
4. Another very successful strategy in the treatment of viral infections soon after transplantation of the stem cells of the blood is to utilize the T lymphocytes taken from the infected donors who are specific to virus. This can treat patients suffering from primary immunodeficiency disorders like HIV. This type of therapy is also known as adoptive immunotherapy (Keller et al. 2018).
5. In one approach DNA was collected from different patients with immunodeficiency disorders and transcriptional analysis was done on the genome to study the genetic sequences coding for viral replication by nuclear protein binding. The outcome is studied through mass spectroscopy and the allele or non-allele dependent messenger RNA transcription is done in possible ways (Lummus et al. 2018).
6. Both the lymphocytes aggregate and get compromised in the lung injuries like asthma induced by dust particles. To study the extent of the exposure and the injury, various fluids from the bronchiolar tissues and the alveoli are collected and antibodies are confirmed by ELISA or western blotting techniques to study autoimmunity associated with it. Even the lung tissues can be stained for the identification of particular proteins (Poole et al. 2018).

7.3 T-Cell Activation During Infections

There are various genes which get expressed in good amount during T-cell activation during infections like Epstein-Barr virus. The natural regulatory T cells are usually involved in the activation of the receptors on the cell surface like the CD25. The T lymphocytes develop in the thymus and are differentiated into regulatory and non-regulatory cells when they migrate to the peripheral region of the thymus gland.

These regulatory cells are further divided into natural regulatory and the inducible regulatory T cells and include various other subtypes. All these cells are classified based on their role in the process of eliciting the immune response according to the invading mechanism of the pathogen.

7.3.1 Quick Attenuation and Persistence of Type I Interferon

Underlying IFN-I reaction after contamination, the foundational IFN-I winds up imperceptible at later stages, regardless of kept up viral burdens. This quick attenuation is collated by a significant decrease in pDC IFN-I creation limit amid LCMV, HIV, and HCV diseases that frequently harmonize with changed natural reactions to auxiliary, disconnected pathogens. Notwithstanding, it is vital to take note of that in LCMV, HCV, HIV, and SIV contaminations; IFN-I or ISG transcripts are tenaciously recognized in all-out DCs, CD4+, and CD8+ T cells, or potentially tissues, but at lower sums than those distinguished amid the pinnacle IFN-I reaction. A few components may add to the weakening of IFN-I in the post-acute phases of interminable viral contaminations, including (an) immediate concealment of intrinsic pathways by viral items, and have immunomodulatory systems. Instances of direct systems of IFN-I restraint incorporate HCV NS3–4A protein hindrance of TRIF (the TLR3 connector), RIG-I, and MAVS viral detecting pathways upstream of IFN creation just as LCMV nucleoprotein (NP) barricade of IRF3 actuation (Gasteiger et al. 2017).

7.4 Viral Pathogenesis

The pathophysiology of viral infection is very complex unless we can identify the type of virus species more specifically under careful correlation with the clinical symptoms of the infection. This can help in determining the pathogenic process and orientation of the disease. All the cells in the immune system rely on the energy pathways to survive and produce the response related to their metabolism. Amino acids like arginine, tryptophan, and alanine play a major role in the creation of the immune response during pathological conditions and inhibition of the unwanted immune response during viral attack. Also, the HIV proteins Vpr and Vif initiate IRF3 debasement (31), and direct HIV gp120 communication with pDCs restrains TLR9 (however not TLR7) reactions, including PDC actuation and IFNα discharge. Additionally the HIV capsid connects with chosen have cofactors to avoid PRR detecting in macrophages. The previously mentioned hidden dimensions of natural qualities ahead of schedule after in vivo HBV contamination might be mostly clarified by HBV polymerase impedance with IRF3 phosphorylation or potentially by hindrance of MAVS motioning by the viral HBx protein. Also, by the diminished TLR2 articulation in hepatocytes, monocytes, and Kupffer cells (liver-inhabitant macrophages) from ceaseless HBV patients, HBV dissolvable antigen can decrease articulation of TLR2 in hepatic cell lines. Thus, HBV dissolvable antigen likewise stifles TLR9 (however not TLR7) motioning in pDCs, and TLR9 articulation is diminished in blood pDCs from incessantly contaminated patients. Notwithstanding the direct IFN-I inhibitory exercises of viral proteins, infections that build up unending diseases additionally advance host immunomodulatory reactions that may cause deviant working of natural cells (Coe et al. 2014).

In chronic immunosuppressing diseases like HIV, there is a huge loss of T-cell CD4 population along with the increased blood levels of the mediators like cytokines. Initially they compromise the host immune system alarmingly leading to the clinical symptoms of opportunistic infections in HIV patients. Combinational therapy with both antiretroviral therapy and antibiotics will help reduce the levels of immunosuppression (Moreno-Fernandez et al. 2012).

However, in hepatitis virus infection the number of regulatory T lymphocytes increases drastically in chronic conditions to induce a good suppression effect on the virus particles. This is mainly due to the hepatic-cellular invasion of the carcinogenic virus particles. This mechanism will also help in protecting the morphology of the liver in the severe inflammatory conditions (Li et al. 2016).

7.5 Host Cell Death Pathways

Since we can only control but not cure most of the viral infections, these strategies can help in focusing on the host cell death pathways, thereby developing drugs that can give complete immunity against the virus by facilitated viral cell elimination. Some of the virus will escape the host immune system and become persistent in the human body causing infection in the future immunocompromised pathological conditions.

7.5.1 Advantageous and Deleterious Effects of Type I Interferon on Antiviral Immunity

IFN-I antiviral impacts are interceded by acceptance of a few ISGs that have the capacity to impede viral replication through different systems [e.g., protein interpretation restraint, viral RNA debasement (investigated by Brown et al. 2018). IFN-I likewise coordinates both inborn and versatile resistant cells, for example, DCs, macrophages, NK cells, and T cells following viral diseases, giving enactment and survival signals. It reliably examines persevering LCMV contamination, in which the infection is cleared from the blood and most tissues by 2–3 months postinfection, showing that mice lacking IFNαβR build up long-lasting viremia. Besides, Sandler et al. exhibited that treatment of rhesus macaques with IFN αβR killing counteracting agent amid serious SIV contamination brought about improved CD4+ T-cell consumption, diminished antiviral quality articulation, decreased extents of cytotoxic NK cells, and expanded SIV repositories, all associated with quicker movement to simian AIDS. What is more, treatment with recombinant IFN-I can regularly support the host barrier and upgrade viral control (Cooney et al. 2018).

A plethora of examples of such infections are human immunodeficiency virus, Epstein-Barr virus, hepatitis B and C virus, and so on. These viruses can be prevalent irrespective of the advancement in the treatment and vaccination. For instance, amid incessant HCV disease, PDC IFN-I creation limit is disabled, in any event to a limited extent, by monocyte-determined tumor rot factor α (TNFα) and interleukin

(IL)-10. It ought to be noticed that not all parts of natural reactions are weakened amid unending viral diseases; in fact, select pathways are improved. For example, though LCMV Cl13-tainted mice react ineffectively to TLR9 ligand (CpG) challenge, they produce unusually high IFN-I levels upon TLR4 ligand (LPS) incitement, prompting quick passing. By these perceptions, unending HCV contamination likewise advances hyperresponsive macrophages, including Kupffer cells, adding to the constant liver irritation that is normal for endless HCV disease. At long last, macrophages treated with HIV and antigen-displaying cells detached from HIV-contaminated patients show hyperresponsiveness to random TLR ligands and commensal microbes, individually. Together, these investigations feature that although natural resistant cells are equipped for IFN-I reaction in the beginning of most constant viral contaminations, blocking of intrinsic pathways by viral proteins associated with host immunomodulatory atoms prompts a significant (yet inadequate) constriction of IFN-I at later phases of the disease (Alvarado-Mora and Pinho 2013).

There are various novel treatment strategies focusing on the host cell death pathways to cure these persistent diseases. One involves clearing the latent infected cells first which have the potential to cause infection; in the future, this can be done by screening and eluting these cells. But these latent cells are very hard to differentiate from the healthy cells in the absence of the infection. This strategy is very successful in targeting the HIV virus with antiretroviral therapy. The very high rate of mutations in the virus genome can also be a problem to cure the infection when they acquire resistance to the antiretroviral drugs administered after the diagnosis and treatment of HIV (Liang et al. 2015).

Wang and colleagues demonstrated that treatment with recombinant IFNα5 and IFNβ between days 2 and 5 after LCMV Cl13 contamination (maybe identical to deferring the IFN-I lessening mentioned above) brought about early popular regulation joined by improved infection explicit CD8+ T-cell reactions, despite the fact that a similar treatment in the unending period of disease had no impact.

Also, prophylactic treatment with pegylated IFNα2a (pIFNα) preceding SIV mucosal test prompted expanded protection from contamination interceded to some degree by an expansion in CD56+ NK cells; in any case, persistent pIFNα treatment brought about an IFN-I desensitization state with decreased ISG levels and expanded cell-related viral burdens. Outstandingly, pIFNα additionally improves articulation of a few ISGs known to be direct enemy of HIV movement. In any case, treatment with pIFNα in HIV-tainted patients has yielded mixed outcomes; only few investigations have appeared.

In some infections like hepatitis, there are few transcriptionally very active fragments which can code for the viral DNA or RNA and produce new viral fragments. They invade the host cells and remain attached to the host genome. These fragments may remain in the body for long periods of time and relapse during favorable conditions producing the antigens. These infective active fragments may get inserted in the host genome and remain inactive for a period of time which is known as the latent phase during which screening for the infection is very tiresome (Battivelli et al. 2018).

Acknowledgments The authors are grateful for the support extended from Adikavi Nannaya University, Osmania University, Krishna University, and the University of South Florida, USA.

Conflict of Interest The authors declare that they have no competing interests.

References

Alvarado-Mora MV, Pinho JR (2013) Epidemiological update of hepatitis B, C and delta in Latin America. Antivir Ther 18(3 Pt B):429–433

Battivelli E, Dahabieh MS, Abdel-Mohsen M, Svensson JP, Da Silva IT, Cohn LB, Verdin E (2018) Chromatin functional states correlate with HIV latency reversal in infected primary CD4+ T cells. bioRxiv:242958

Bertoletti A, Le Bert N (2018) Immunotherapy for chronic hepatitis B virus infection. Gut Liver 12:497

Brown GD, Willment JA, Whitehead L (2018) C-type lectins in immunity and homeostasis. Nat Rev Immunol 18(6):374–389

Bull MJ, Plummer NT (2014) Part 1: the human gut microbiome in health and disease. Integr Med 13(6):17–22

Coe DJ, Kishore M, Marelli-Berg F (2014) Metabolic regulation of regulatory T cell development and function. Front Immunol 5:590

Cooney J, Allison C, Preston S, Pellegrini M (2018) Therapeutic manipulation of host cell death pathways to facilitate clearance of persistent viral infections. J Leukoc Biol 103(2):287–293

Finotti G, Migliorati D, Costantini M (2018) Multisensory integration, body representation and hyperactivity of the immune system. Conscious Cogn 63:61–73

Frossi B, Mion F, Sibilano R, Danelli L, Pucillo CEM (2018) Is it time for a new classification of mast cells? What do we know about mast cell heterogeneity? Immunol Rev 282(1):35–46

Gao Y, Deason KL, Jain A, Dozmorov I, Wakeland EK, Pasare C (2018) Transcriptional profiling of pathogen-specific CD4 T cells reveals key events that regulate Th17 cell priming and differentiation. J Immunol 200(1):47

Gasteiger G, D'Osualdo A, Schubert DA, Weber A, Bruscia EM, Hartl D (2017) Cellular innate immunity: an old game with new players. J Innate Immun 9:111–125

Gieseck RL, Wilson MS, Wynn TA (2018) Type 2 immunity in tissue repair and fibrosis. Nat Rev Immunol 18(1):62–76

Haas S, Trumpp A (2018) An intrinsic interferon program protects stem cells from viral infection. Dev Cell 44(3):279–280

Karkhah A, Javanian M, Ebrahimpour S (2018) The role of regulatory T cells in immunopathogenesis and immunotherapy of viral infections. Infect Genet Evol 59:32–37

Keller MD, Hanley PJ, Hoover J, Roesch L, McCormack S, Lang H, Abraham A, Williams K, Davila B, Adams R, Leiding JW, Bollard C (2018) Third-party T cell immunotherapy for viral infections in primary immunodeficiency disorders. J Allergy Clin Immunol 141(2):AB198

Li W, Han J, Wu H (2016) Regulatory T-cells promote hepatitis B virus infection and hepatocellular carcinoma progression. Chronic Dis Transl Med 2(2):67–80

Liang S, Shen Z, Yan J, Liang F, Tang Z, Liu W, Xing H (2015) Low virologic failure and drug resistance among HIV-infected patients receiving hospital-based ART while care and outreach through community in Guangxi, China. Front Public Health 3:244

Libby P, Hansson GK (2018) Adaptive immunity in acute coronary syndromes: chicken or egg? Eur Heart J 39(13):1098–1099

Lummus ZL, Kesavalu B, Kaufman K, Yao J, Weirauch MT, Kottyan LC, Muñoz X (2018) Regulatory variants of ATF3, CDH17 and FAM71A are risk factors for diisocyanate induced occupational asthma (DA). J Allergy Clin Immunol 141(2):AB198

Moreno-Fernandez ME, Presicce P, Chougnet CA (2012) Homeostasis and function of regulatory T cells in HIV/SIV infection. J Virol 86(19):10262–10269

Ning X, Luckenbaugh L, Liu K, Bruss V, Sureau C, Hu J (2018) Common and distinct capsid and surface protein requirements for secretion of complete and genome-free hepatitis B virions. J Virol 92(14):e00272–e00218

Ost KS, Round JL (2018) Communication between the microbiota and mammalian immunity. Annu Rev Microbiol 72(21):399

Poole JA, Mikuls TR, Duryee MJ, Warren KJ, Wyatt TA, Nelson AJ, Romberger DJ, West WW, Thiele GM (2018) A role for B cells in organic dust induced lung injury. J Allergy Clin Immunol 141(2):AB198

Shuai Z, Leung MWY, He X, Zhang W, Yang G, Leung PSC, Eric Gershwin M (2016) Adaptive immunity in the liver. Cell Mol Immunol 13(3):354–368

Walker LM, Burton DR (2018) Passive immunotherapy of viral infections: 'super-antibodies' enter the fray. Nat Rev Immunol 18(5):297–308

Yu Y, Blokhuis BR, Garssen J, Redegeld FA (2016) Non-IgE mediated mast cell activation. Eur J Pharmacol 778(5):33–43

Macrophages/Microvesicles and Their Task in Viral Diseases

8

Bojjibabu Chidipi, Samuel Ignatius, Madhavi Maddala, Pallaval Veera Bramhachari, and Alavala Mattareddy

Abstract

Viral infectious diseases are mainly associated with immunological responses triggered by immunocytes present in the circulation like macrophages. A better understanding of these immunity mechanisms will help us to control the progression of the disease to chronic conditions. Microvesicles are an extracellular group of plasma membrane secretions, which play a significant role in the spread of infection or evasion of viral entities to surrounding healthy cells acting as a viral vehicle. Their involvement in the viral cycle can be utilized productively in hampering the period by developing the antibodies that shield the evasion. These extracellular vesicles will facilitate the communication between the cells in the body humoral environment leading to host and viral interactions in the immune system. Thereby attempting to study the role of macrophages in the immunity process will help us to eliminate the viral pathogens by either delaying or damaging their spread to healthy cells. Focusing on these mechanisms of the microvesicles and macrophages will help lead to the early cure of many life-threatening infectious diseases like HIV, Hepatitis, and Dengue by developing a single specific antibiotic.

Keywords

Macrophages · Viral cycle · HIV · Hepatitis · Dengue · Humoral environment · Extracellular vesicles

---

B. Chidipi (✉) · S. Ignatius
Molecular Pharmacology and Physiology, University of South Florida, Tampa, FL, USA

M. Maddala
Department of Zoology, Nizam College, Osmania University, Hyderabad, Telangana, India

P. V. Bramhachari
Department of Biotechnology, Krishna University, Machilipatnam, Andhra Pradesh, India

A. Mattareddy
School of Life and Health Science, Adikavi Nannaya University, Rajahmundry, India

P. V. Bramhachari (ed.), *Dynamics of immune activation in viral diseases*,
https://doi.org/10.1007/978-981-15-1045-8_8

8.1 Introduction

Viral infectious diseases are mainly associated with immunological responses triggered by immunocytes like Macrophages. A better understanding of these immunity mechanisms and the knowledge related to the prevalence of the infection will help us to control the progression of the disease to chronic conditions by actual estimation of the number of infected individuals by screening methods (Birdwell et al. 2018). The primary route of entry of the virus into the human body is through the epithelial lining, every time epithelial cells will be the first cells to get infected or encounter the virus particles. When infected, one of the surface of the epithelial cells will get polarized since both the inner and outer surfaces of the epithelial cells are morphologically distinct (Hoebe et al. 2018). With or without finishing the life cycle inside the epithelium these viruses may exit the cells to invade one or other surfaces (Baraldo et al. 2018).

The process by which a viral infection leads to disease is known as viral pathogenesis, the consequences and the period of the viral infection depend on various host and viral-related interplaying factors. They can develop as an acute infection but may last for months to years depending on the favorable viral factors. If the disease becomes chronic it may easily get transmitted to others and also recovery from such conditions is sporadic (Jaimes and Whittaker 2018). The cell damage created by the infection depends on the type of virus because not all viruses can damage cells even though they replicate and widely spread throughout the body. Some viruses belonging to the class known as retrovirus can cause persistent infection but cannot cause cell death as they are released from the cells without any lysis by a unique phenomenon called budding (Shaheen 2018).

Whereas picornavirus can cause lysis of cell and cause cell damage and rhinovirus cause mucus secretion and poliovirus paralysis the tissues causing tremendous cell damage. The progressive clinical symptoms are noticed with these infections until the viral population minimizes to noninfectious levels (Carrasco et al. 2018).

8.2 Viral Infection Cycle

An infection must utilize cell procedures to repeat. The viral replication cycle can create dramatic biochemical and fundamental changes in the host cell, which may cause cell harm. These changes, called cytopathic (causing cell harm) impacts, can change cell function. Some infected cells, for example, those contaminated by the regular cold infection known as rhinovirus, kick the bucket through lysis (bursting) or apoptosis (customized cell demise or "cell suicide"), discharging all offspring virions without a moment's delay. The side effects of viral maladies result from the insusceptible reaction to the infection, which endeavors to control and dispense with the virus from the body, and from cell harm brought about by the virus (McDougall et al. 2019).

Numerous viral infections, for example, HIV (human immunodeficiency infection), leave the contaminated cells of the safe framework by a procedure known as sprouting, where virions go to the cell separately. Amid the growing process, the cell does not experience lysis and is not promptly slaughtered. In any case, the harm to the cells that the infection taints may make it outlandish for the cells to work ordinarily, even though the cells stay alive for a timeframe. Most beneficial viral diseases pursue comparative strides in the infection replication cycle: connection, entrance, un-coating, replication, gathering, and discharge. In Fig. 8.1, the host cell is annihilated toward the finish of the replication cycle—recollect this does not generally occur: here and there the host cell lives on and keeps on reproducing the infection (Khalili and Malcolm 2019).

Connection An infection joins to a particular receptor site on the host cell film through connection proteins in the capsid or using glycoproteins inserted in the viral envelope. The explicitness of this communication decides the host—and the cells inside the host—that can be tainted by a specific infection. This can be outlined by thinking about a few keys and a few locks, where each key will fit just a single explicit lock (Zhao et al. 2019).

Entry Phase The nucleic acid of bacteriophages enters the host cell exposed, leaving the capsid outside the cell. Plant and creature infections can enter through endocytosis, in which the cell film encompasses and overwhelms the whole virus. Some wrapped infections come to the cell when the viral envelope melds legitimately with the cell layer. Once

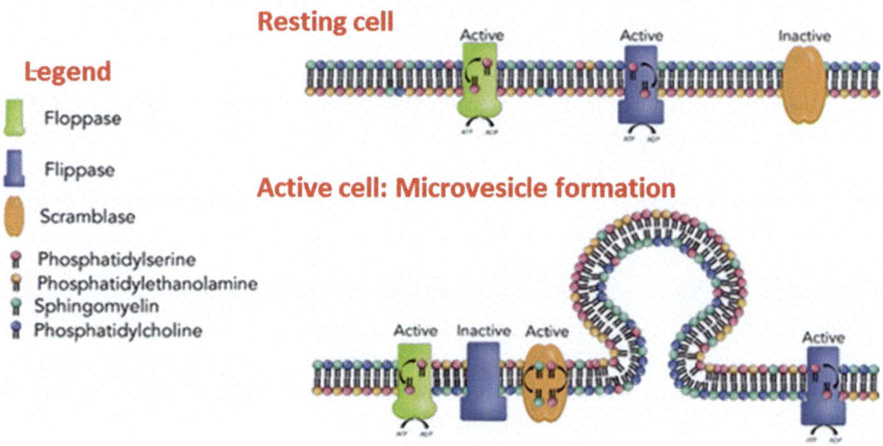

Fig. 8.1 Microvesicle formation mechanism. (Top panel) In resting cells, both phosphatidylserine and phosphatidylethanolamine are negatively charged and segregated in the inner side of the plasma membrane, on the other hand sphingomyelin and phosphatidylcholine, both are positively charged head group, are in the outer side of the plasma membrane (Seigneuret et al. 1984). (Lower panel) The "flippase" and "floppase are an ATP-dependent transmembrane enzyme. Flippase reacts with phosphatidylserine and help to inward folding, meanwhile floppase intermediate the both anionic and cationic phospholipids to outward fold (Daleke 2003; Laberge et al. 2018)

inside the cell, the viral capsid is debased, and the viral nucleic acid is discharged, which at that point ends up accessible for replication and translation (Appaiah et al. 2019).

Replication Phase The replication instrument relies upon the viral genome. The viral infections generally use cell proteins and compounds to make extra DNA that is translated to ambassador RNA (mRNA), which is then used to coordinate protein union. RNA infections, for the most part, utilize the RNA center as a layout for the union of viral genomic RNA and mRNA. The viral mRNA guides the host cell to orchestrate viral catalysts and capsid proteins, and gather new virions (Ilan 2019).

There are individual cases to this example. On the off chance that a host cell does not give the compounds essential to viral replication, viral qualities supply the data to coordinate the union of the missing proteins. Retroviruses, for example, HIV, have an RNA genome that must be turned around translated into DNA, which at that point is consolidated into the host cell genome (Solomon and Geretti 2019).

Assembly Phase They are inside gathering VI of the Baltimore characterization plot. To change over RNA into DNA, retroviruses must contain qualities that encode the infection explicit protein switch transcriptase that deciphers an RNA format to DNA. Invert translation never happens in uninfected host cells—the required compound switch transcriptase is just gotten from the outflow of viral qualities inside the infected host cells (Jin and Musier-Forsyth 2019).

The way that HIV creates its very own portion compounds not found in the host has enabled scientists to develop drugs that hinder these chemicals. These medications, including the turnaround transcriptase inhibitor AZT, inhibit HIV replication by diminishing the movement of the protein without influencing the host's digestion. This methodology has prompted the advancement of an assortment of medications used to treat HIV and has been successful at decreasing the number of irresistible virions (duplicates of viral RNA) in the blood to non-discernible dimensions in numerous HIV-contaminated people (Roder et al. 2019).

Budding Phase The last phase of viral replication is the arrival of the new virions delivered in the host living being, the place they can taint neighboring cells and rehash the replication cycle. As you have adapted, some infections are discharged when the host cell bites the dust, and different infections can leave contaminated cells by sprouting through the layer without legitimately slaughtering the phone (May 2019).

8.3 Microvesicles

Microvesicles are the extracellular group of plasma membrane secretions, which play a significant role in the spread or evasion of viral entities to surrounding healthy cells acting as a viral vehicle. Microvesicles formed from the lipid rafts also called cholesterol- and sphingolipid-rich membrane microdomains (Fig. 8.1). These microdomains involved in many cellular processes including microvesicles release.

These being the cell-derived vesicles encapsulated by a bilayer of lipids are present throughout the body in the amniotic fluids, breast milk, nasal secretions, cerebrospinal fluid, blood, and serum. Higher concentrations of these cysts will be noticed during the pregnancy and tumorous conditions in a diameter of about 30–100 nm (Chen et al. 2018) (Fig. 8.2).

The primary biological function of microvesicles in the human body is to eliminate unwanted molecules like proteins from the healthy cells during the process like blood coagulation. Their secondary purpose is intercellular communication between the cells during the exchange of materials between them. During the dangerous conditions, their role is to propagate the pathogens for the immune response by inhibitory or regulatory mechanisms and also in the presentation of antigens (Laberge et al. 2018).

The formation of microvesicles is by a unique process known as the membrane remodeling in which the integrity of the cell membrane is changed due to increased levels of calcium deposition leading to the formation of small vesicles on the surface and are released by the disruption of the cell membrane. These vesicles are released into the blood circulation and reach their target cells. These vesicles bind to the surface of the host cell and undergo ligand-specific interaction to enter the host cell or may undergo fusion or endocytosis to gain entry into the cell cytoplasm (Bello-Morales et al. 2018a).

The main constituents of the microvesicles are proteins, lipids, messenger RNA, or Micro RNA. These are spontaneously released or shed by many types of cells like microvesicles during cell activation or apoptosis. They help in communication between cells with their surface receptors and play a potential role in the exchange of genetic material (Yao et al. 2018).

8.4 Role of Microvesicles in the Viral Cycle

Microvesicles have many common characteristics with that of the enveloped viruses. During viral infections, many infected cells secrete a vast number of microvesicles that may or may not be similar to the biogenic and biophysical characteristics of viral counterparts. For instance, HIV virions will shed from the plasma membrane to but not from the cytoplasm in a similar budding mechanism like microvesicles spread from cell to cell. Moreover, it is tough to separate HIV particles from exosome vesicles during centrifugation as they both have identical biological properties (Bello-Morales et al. 2018a).

Microvesicles as discussed above will play a vital role in assisting the virally infected cells in promoting the intercellular microvesicle-mediated communication. These vesicles, when fused to the surface of the other cells, will eventually transfer the cellular components exerting a biological function to the recipient cells. During infection with herpes simplex virus, microvesicles promote the viral pathogenesis and infectivity by using the exosome components and endosome makers like tetraspanin.

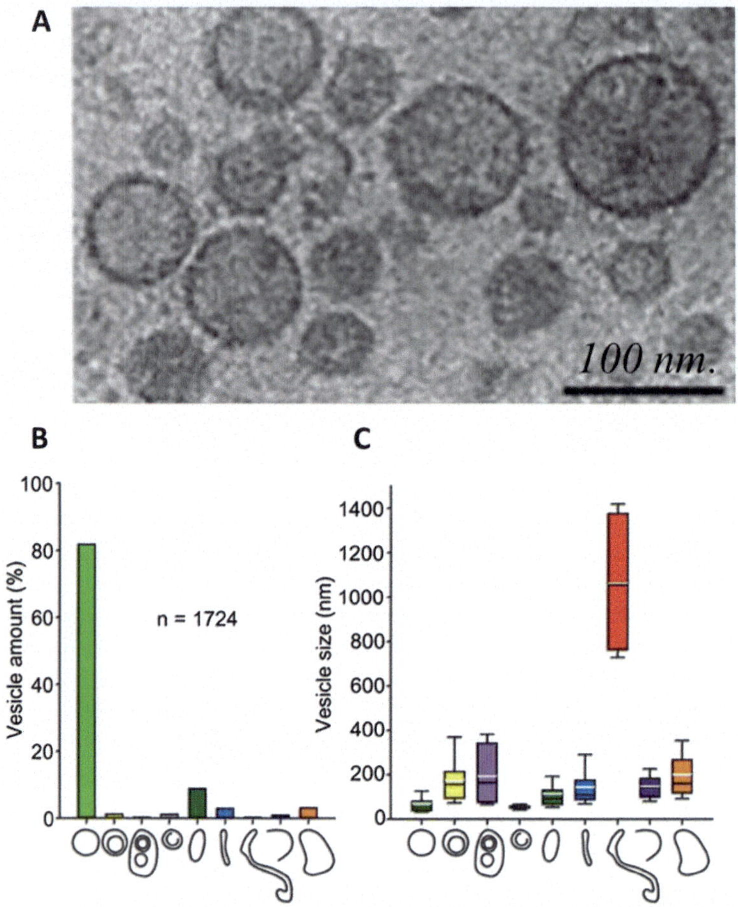

Fig. 8.2 Size and number of microvesicles. (**a**) Cryo-electron micrograph of extracellular vesicles secreted by MLP-29 cells (Conde-Vancells et al. 2008). The percentage of vesicles that belonged to each category and their size distribution within each category are shown in cultured HMC-1 cells (**b**, **c**) (Zabeo et al. 2017)

These particles are similar to the endosome vesicles in size making them use the exosome cellular pathways for their invasion and evasion (Sharma et al. 2018).

The primary functions of the microvesicles in the viral infection are to modulate the cellular processes like angiogenesis, cell proliferation, cell invasion, gene regulation, and immune regulation. These can also generate the essential ligands for receptor cell surfaces to initiate the signaling pathways and release their contents into the plasma membrane by endocytosis. Their involvement in the viral cycle can be utilized in hampering the period by developing antibodies that shield the evasion. Virally infected cells will sometimes secrete microvesicles to remove unwanted cellular proteins from the cytoplasm thereby regulating the signaling complexes and pathological process (Meckes and Raab-Traub 2011).

8.5 The Communication Between the Cells

Extracellular vesicles facilitate the connection between the cells in the body humoral environment leading to host and viral interactions in the immune system. These extracellular vesicles are usually the exosome or microvesicles containing RNA's, proteins, and lipids (Birdwell et al. 2018). These involve in the extracellular communication by a unique mechanism in which the cells selectively discharge membrane and elute the internal components. This mechanism is identified using the exosome secreted by antigen-presenting cells major histocompatibility complex 2 using the immunogold labels (Paolicelli et al. 2018).

Even in the tumor environment, extracellular vesicles (EVs) play a vital role in the mediation of communication between the tumor cells. EVs contribute to regulating the bystander effects in which the stressed tumor cells will cause DNA damage to the neighboring healthy cells (Birdwell et al. 2018). Evidence shows that exosomes help in the cross-communication between cells from which they were derived from, for example, between macrophages, dendrocytes, both T and B lymphocytes, and natural killer cells in the cancer environment. These exosomes will influence the functions of the cells by modifying their genotype due to mutations in the cancer cells. These are playing a vital role in the generation of the vaccines for various diseases involved with macrovesicle mediation (Raposo et al. 1996).

The significant contribution of these vesicles in their release and function by biogenesis will be the novel tool for the detection and treatment of the viral diseases. It has been revealed that they contribute to homeostasis in the central nervous system trauma conditions (Baraldo et al. 2018). During the therapeutic purpose, the same exosomes play a vital role in targeting and modifying specific tissue cells and combating the significant barriers in immunological diseases. They also help in the development of various vaccines against those infectious diseases that mimic the microvesicles for exploiting the host immune system (Samuel et al. 2017).

8.6 Macrophages in the Immunity Process

Macrophages play a vital role in the generation of the immune response in the human body (Baraldo et al. 2018). Thereby attempting to study the role of macrophages in the immunity process will help us to eliminate the viral pathogens by either delaying or damaging their spread to healthy cells. Macrophages also contribute a vital role in increasing the genome load for viral infections like infectious bronchitis virus as they accumulate in the respiratory tract (Shen and Ren 2018).

During viral infections, a large number of infected cells will undergo a programmed cell death process called apoptosis during which the macrophages eliminate the cells preventing further necrosis (Shaheen 2018). They also play a supportive role in the activation of the tissue repair process by activating the inflammatory cascade by recruiting inflammatory mediators thereby initiating the wound healing process (Pegtel et al. 2014).

8.7 Role of Macrophages in Infectious Diseases

In infections like hepatitis, the microvesicles budding from the infected cells will participate in the viral cycle along with the virus. This was the first time discovered under the electron microscope when observing the infected liver cells for the herpes simplex virus virions. In hepatitis infection, microvesicles harbor the virions as part of the viral cycle, and when these are ingested by the macrophages they are infecting the naïve cells as part of the productive infection. In hepatitis, virus particles are depending on the microvesicles to promote the disease effectively in the host, which can be used therapeutically by shielding the microvesicles with antibodies (Huang-Doran et al. 2017).

In dengue infection, the virus uses the arthropod vesicles to promote the disease from the mosquito to humans. These extracellular vesicles contain the viral RNA and proteins that are infectious. In the treatment of the dengue infection effectively scientists are using the glycoprotein present in the tetraspanin domain of the microvesicle as a marker to identify the pathophysiology and blocking the spread of disease by targeting the microvesicle-mediated transmission by DNA silencing and gene isolation studies. These changes are confirmed by immunoprecipitation studies (Vora et al. 2018).

In HIV infection microvesicles play a vital role in the remodeling of the pulmonary vasculature. The actual reason for the cardiac regeneration in the humans is anonymous for recent times and later discovered to be due to the extracellular vesicles released from the human monocyte-derived macrophages. The main constituent of these vesicles is miRNA-130a that plays a significant role in the decrease in the expression of the phosphatase and tensin molecules that are responsible for the perivascular inflammation in the HIV patients (Sharma et al. 2018).

With respect to entry and departure of virus, four noteworthy stages have been proposed to portray these procedures: capsid get together and DNA bundling in the core; essential envelopment and de-envelopment at the atomic envelope; tegumentation and optional envelopment in the cytoplasm; and, at long last, exocytosis of viral particles at the plasma layer and additionally cell-to-cell transmission at cell intersections (Lannes et al. 2019). A unique method of viral communication in human tissues is cell-to-cell spread, that is, the immediate entry of descendant's infection from a contaminated cell to a nearby one (Bracq et al. 2018) (Fig. 8.3).

It is broadly acknowledged that this system of dissemination speaks to a resistant avoidance technique since it shields the infection from insusceptible observation. Be that as it may, as referenced above, HSV-1 may utilize a few methods of spread to go from tainted to uninfected cells (You et al. 2018). Numerous viewpoints concerning the procedure of viral spread, for example, the instruments of viral departure from epithelial cells and spread to neurons and the other way around, are not seen yet (Bayliss and Piguet 2018).

Elucidating the devices of viral scattering and the ensuing passage into neighboring cells remains a vital advance to comprehend the viral cycle in the host. In this unique situation, discharged vesicles have risen as another object of consideration given their capacity to take part in the intercellular correspondence process amid

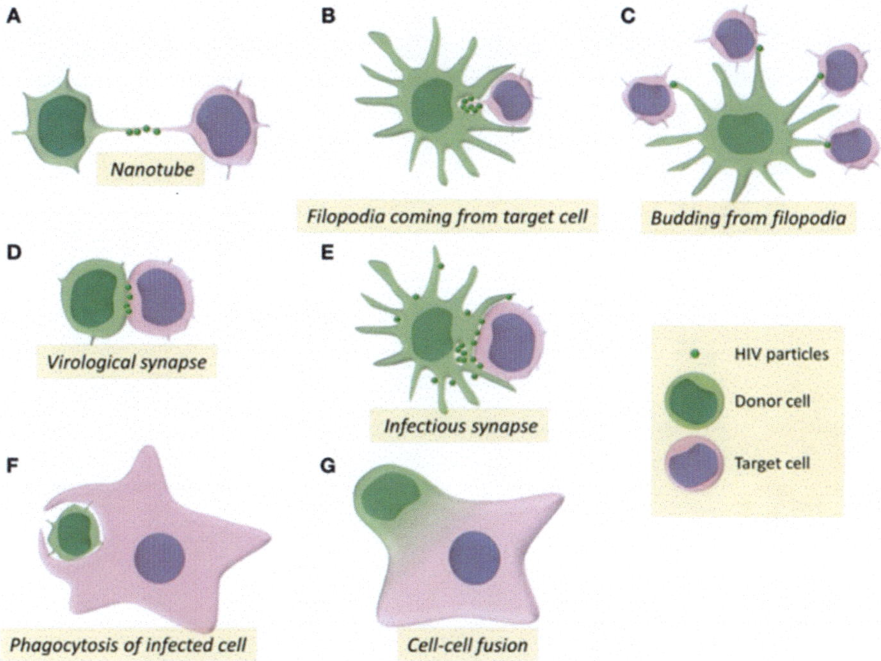

Fig. 8.3 Intercellular structures and processes involved in cell-to-cell transmission of HIV-1. (**a**–**g**) Schemes represent the different pathways for HIV-1 cell-to-cell transfer between donor cells (in green) and target cells (in pink) (Bracq et al. 2018)

viral diseases. Extracellular vehicles (EVs) are a very different gathering of emitted layer vesicles, which have been separated from most cell types and natural liquids. Three noteworthy subgroups of EVs have been recognized: apoptotic bodies; microvesicles (MVs), which get from the shedding of the plasma film; and exosomes, which are intraluminal vesicles discharged to the endless extracellular supply of multivesicular bodies (MVBs) with the plasma layer (Fig. 8.4) (Roth et al. 2019).

Exosome usually is 30–100 nm in the distance across, while MVs have a different size running from 100 nm to 1 m in breadth. EVs have been appeared to be associated with various physiological and neurotic procedures, for example, aggravation and the insusceptible reaction, cell bond, coagulation, squander the board, tumor movement, and viral spread. Oligodendrocytes (OLs) are the myelin-shaping cells of the focal sensory system (CNS). The myelin sheath is an electrically protecting layer that encompasses axons in both the focal and fringe sensory systems, permitting salutatory conduction of activity potential (Domingues et al. 2018).

All neural cell types discharge EVs, which have a focal job in procedures, for example, myelination or guideline of synaptic movement and may be associated with the pathogenesis of a few neurodegenerative maladies or, despite what might be expected, in neuroprotection. To be sure, OLs discharge exosomes conveying myelin proteins, for example, proteolipid protein (PLP), the real myelin protein in the CNS;

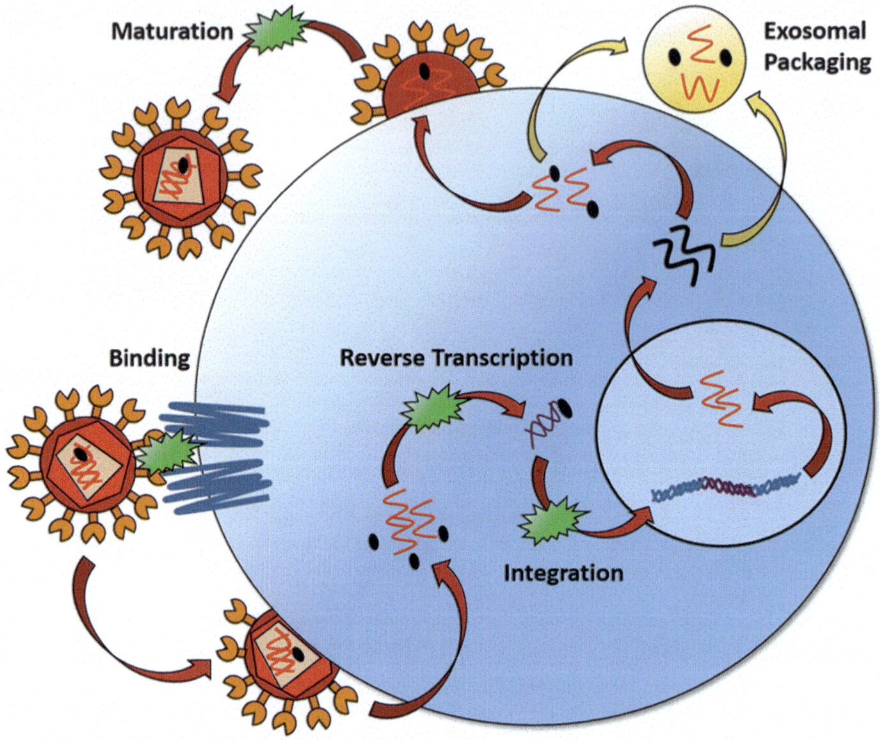

Fig. 8.4 Multiple stages of the retroviral life cycle, however, no current antiretroviral drug blocks the expression of HIV proteins and RNA from integrated viral DNA, or their subsequent sorting into exosomes and secretion from infected cells (Patters and Kumar 2018)

2=3=-cyclic nucleotide phosphodiesterase (CNPase); myelin essential protein (MBP); and myelin oligodendrocyte glycoprotein (MOG), just as various chemicals, for example, glycolytic or oxidative pressure easing compounds (Cinelli et al. 2019). Past works completed by our gathering have demonstrated that OLs are profoundly vulnerable to HSV-1 disease. Consequent actions uncovered that oligodendrocyte antecedent cells (OPCs) and cells of the human oligodendroglia HOG cell line refined under separation conditions become progressively vulnerable to HSV-1, a reality that was steady with the later finding that PLP, a myelin protein upregulated all through separation, is engaged with the viral passage (Ueda et al. 2018).

EVs got from OLs have likewise been associated with such procedures, and there is expanding proof for the job of EVs in myelination and neuron–glia correspondence. EVs have additionally been engaged with viral contamination as a method for infections to enter have cells, improve spread, or avoid the host resistant reaction (Charlton et al. 2018). Two noteworthy elements of EVs amid viral diseases are the exchange of viral genomes into target cells and the strengthening of contamination by changing the physiology of target cells. Generation of discharged vesicles by HSV-1-contaminated cells has been known for quite a while. The first to be found, named L-particles, are like the virions in appearance and do not have the viral nucleocapsid and the genome and along these lines are not irresistible (Bello-Morales et al. 2018a). In any case, L-particles have been appeared to encourage HSV-1 disease by conveying viral proteins and the cellular factors required for infection replication and invulnerable avoidance (Bello-Morales and López-Guerrero 2018).

Hence, it has been confirmed that the comparative capacities watched for vhs and TIF among virions and L-particles recommend that viral connection, combination, and the arrival of covering proteins are the equivalent for both. Likewise, L-particles share comparative gathering and departure pathways with virions, recommending that the covering and glycoproteins are adequate to provoke auxiliary envelopment. It has been exhibited that useful viral proteins can be exchanged to uninfected onlooker cells by means of L-particles, a procedure that may show a methodology for viral invulnerable break. Different particles, the previral DNA replication-encompassed particles (PREPs), are morphologically like L-particles; however, they vary in their relative protein arrangements (Sun et al. 1999). Nonetheless, to date, there is no proof of virions being bundled inside EVs. Here, we propose a unique job for MVs in spread of infection. Recent discoveries show out of the blue that virions might be exchanged from tainted to uninfected cells through MVs. By methods of transmission electron microscopy (TEM), we identified miniaturized scale vesicles containing HSV-1 virions. What is more, we found that the no lenient Chinese hamster ovary (CHO) cell line was powerless to HSV disease directly after immunization with infection containing MVs recently detached from supernatant of infected HOG cells. These outcomes recommend that MVs emitted by HOG cells tainted with HSV-1 may be associated with viral spread and may add to evading resistant reconnaissance (Bello-Morales et al. 2018b).

Acknowledgments The authors are grateful for the support extended from Adikavi Nannaya University, Krishna University, and University of South Florida, USA.

Conflict of Interest The authors declare that they have no competing interests.

References

Appaiah CB, Manur J, Sriram B (2019) Reducing conjugative plasmids in bacteria. Google Patents

Baraldo S, Contoli M, Bonato M et al (2018) Deficient immune response to viral infections in children predicts later asthma persistence. Am J Respir Crit Care Med 197(5):673–675

Bayliss RJ, Piguet V (2018) Masters of manipulation: viral modulation of the immunological synapse. Cell Microbiol 20(10):e12944

Bello-Morales R, López-Guerrero JA (2018) Extracellular vesicles in herpes viral spread and immune evasion. Front Microbiol 9:2572

Bello-Morales R, Praena B, de la Nuez C et al (2018a) Role of microvesicles in the spread of herpes simplex virus 1 in oligodendrocytic cells. J Virol 92(10):e00088

Bello-Morales R, Praena B, de la Nuez C et al (2018b) Role of microvesicles in the spread of herpes simplex virus 1 in oligodendrocytic cells. J Virol 92(10):e00088. https://doi.org/10.1128/jvi.00088-18

Birdwell CE, Prasai K, Dykes S et al (2018) Epstein-Barr virus stably confers an invasive phenotype to epithelial cells through reprogramming of the WNT pathway. Oncotarget 9(12):10417–10435. https://doi.org/10.18632/oncotarget.23824

Bracq L, Xie M, Benichou S et al (2018) Mechanisms for cell-to-cell transmission of HIV-1. Front Immunol 9:260

Carrasco L, Sanz MA, González-Almela E (2018) The regulation of translation in alphavirus-infected cells. Viruses 10(2):E70

Charlton CL, Babady E, Ginocchio CC et al (2018) Practical guidance for clinical microbiology laboratories: viruses causing acute respiratory tract infections. Clin Microbiol Rev 32(1):e00042–e00018

Chen Q, Rutten V, Cheng W-T et al (2018) Phagocytosis of extracellular vesicles extruded from the placenta by ovarian cancer cells inhibits growth of the cancer cells. Int J Gynecol Cancer 28(3):545–552

Cinelli I, Destrade M, McHugh P et al (2019) Head-to-nerve analysis of electromechanical impairments of diffuse axonal injury. Biomech Model Mechanobiol 18(2):361–374

Conde-Vancells J, Rodriguez-Suarez E, Embade N, Gil D, Matthiesen R, Valle M, Elortza F, Lu SC, Mato JM, Falcon-Perez JM et al (2008) Characterization and comprehensive proteome profiling of exosomes secreted by hepatocytes. J Proteome Res 7(12):5157–5166

Daleke DL et al (2003) Regulation of transbilayer plasma membrane phospholipid asymmetry. J Lipid Res 44(2):233–242

Domingues HS, Cruz A, Chan JR et al (2018) Mechanical plasticity during oligodendrocyte differentiation and myelination. Glia 66(1):5–14

Hoebe E, Wille C, Hagemeier S et al (2018) Epstein–Barr virus gene BARF1 expression is regulated by the epithelial differentiation factor ΔNp63α in undifferentiated nasopharyngeal carcinoma. Cancer 10(3):76

Huang-Doran I, Zhang C-Y, Vidal-Puig A (2017) Extracellular vesicles: novel mediators of cell communication in metabolic disease. Trends Endocrinol Metab 28(1):3–18

Ilan Y (2019) Microtubules: from understanding their dynamics to using them as potential therapeutic targets. J Cell Physiol 234(6):7923–7937

Jaimes JA, Whittaker GR (2018) Feline coronavirus: insights into viral pathogenesis based on the spike protein structure and function. Virology 517:108–121

Jin D, Musier-Forsyth K (2019) Role of host tRNAs and aminoacyl-tRNA synthetases in retroviral replication. J Biol Chem 294(14):5352–5364

Khalili K, Malcolm T (2019) Methods and compositions for RNA-guided treatment of HIV infection. Google Patents

Laberge A, Arif S, Moulin VJ (2018) Microvesicles: intercellular messengers in cutaneous wound healing. J Cell Physiol 233(8):5550–5563

Lannes N, Garcia-Nicolàs O, Démoulins T et al (2019) CX 3 CR1-CX 3 CL1-dependent cell-to-cell Japanese encephalitis virus transmission by human microglial cells. Sci Rep 9(1):4833

May MK (2019) Characterizing bacterial antibiotic resistance, prevalence, and persistence in the marine environment. Massachusetts Institute of Technology and Woods Hole Oceanographic Institution, Woods Hole

McDougall WM, Perreira JM, Hung H-F et al (2019) Viral infection or IFN-α alters mitotic spindle orientation by modulating pericentrin levels. iScience 12:270–279

Meckes DG, Raab-Traub N (2011) Microvesicles and viral infection. J Virol 85(24):12844–12854

Paolicelli RC, Bergamini G, Rajendran L (2018) Cell-to-cell communication by extracellular vesicles: focus on microglia. Neuroscience 405:148–157

Patters BJ, Kumar S (2018) The role of exosomal transport of viral agents in persistent HIV pathogenesis. Retrovirology 15:79

Pegtel D, Peferoen L, Amor S (2014) Extracellular vesicles as modulators of cell-to-cell communication in the healthy and diseased brain. Philos Trans R Soc Lond B Biol Sci 369(1652):20130516

Raposo G, Nijman HW, Stoorvogel W et al (1996) B lymphocytes secrete antigen-presenting vesicles. J Exp Med 183(3):1161–1172
Roder AE, Vazquez C, Horner SM (2019) The acidic domain of the hepatitis C virus NS4A protein is required for viral assembly and envelopment through interactions with the viral E1 glycoprotein. PLoS Pathog 15(2):e1007163
Roth R, Hillmer S, Funaya C et al (2019) Arbuscular cell invasion coincides with extracellular vesicles and membrane tubules. Nat Plants 5(2):204
Samuel P, Mulcahy LA, Furlong F et al (2017) Cisplatin induces the release of extracellular vesicles from ovarian cancer cells that can induce invasiveness and drug resistance in bystander cells. Philos Trans R Soc Lond B Biol Sci 373(1737):20170065
Seigneuret M, Zachowski A, Hermann A, Devaux PF et al (1984) Asymmetric lipid fluidity in human erythrocyte membrane: new spin-label evidence. Biochemistry 23(19):4271–4275
Shaheen A (2018) A study on types of skin diseases and its impact on quality of life of patients coming to private hospitals in Dhaka city: East West University
Sharma H, Chinnappan M, Agarwal S et al (2018) Macrophage-derived extracellular vesicles mediate smooth muscle hyperplasia: role of altered miRNA cargo in response to HIV infection and substance abuse. FASEB J 32(9):5174–5185
Shen M, Ren X (2018) New insights into the biological impacts of immune cell-derived exosomes within the tumor environment. Cancer Lett 431:115–122
Solomon T, Geretti AM (2019) Introduction and general principles. In: Hunter's tropical medicine and emerging infectious disease. Elsevier, London, p 228
Sun M, Zhang GR, Yang T et al (1999) Improved titers for helper virus-free herpes simplex virus type 1 plasmid vectors by optimization of the packaging protocol and addition of noninfectious herpes simplex virus-related particles (previral DNA replication enveloped particles) to the packaging procedure. Hum Gene Ther 10(12):2005–2011. https://doi.org/10.1089/10430349950017365
Ueda Y, Misumi S, Suzuki M et al (2018) Disorganization of oligodendrocyte development in the layer II/III of the sensorimotor cortex causes motor coordination dysfunction in a model of white matter injury in neonatal rats. Neurochem Res 43(1):127–137
Vora A, Zhou W, Londono-Renteria B et al (2018) Arthropod EVs mediate dengue virus transmission through interaction with a tetraspanin domain containing glycoprotein Tsp29Fb. Proc Natl Acad Sci 115(28):E6604–E6613
Yao Z-Y, Chen W-B, Shao S-S et al (2018) Role of exosome-associated microRNA in diagnostic and therapeutic applications to metabolic disorders. J Zhejiang Univ 19(3):183–198
You Y, Liu T, Wang M et al (2018) Duck plague virus Glycoprotein J is functional but slightly impaired in viral replication and cell-to-cell spread. Sci Rep 8(1):4069
Zabeo D, Cvjetkovic A, Lässer C, Schorb M, Lötvall J, Höög JL et al (2017) Exosomes purified from a single cell type have diverse morphology. J Extracell Vesicles 6(1):1329476
Zhao L, Kopylov M, Potter CS et al (2019) Engineering the PP7 virus capsid as a peptide display platform. ACS Nano 13(4):4443–4454

T Cells in Viral Infections: The Myriad Flavours of Antiviral Immunity

9

Achanta Jagadeesh, A. M. V. N. Prathyusha,
Ganugula Mohana Sheela, and Pallaval Veera Bramhachari

Abstract

Viral diseases are a major cause of morbidity and mortality and result in a significant public health burden. T lymphocytes first identified in the chordate lineage and constitute a highly sophisticated branch of adaptive immune system. Apart from B cells, it is the only cell type that exhibits antigenic specificities; achieved by gene rearrangement. T cells are unique with respect to diversity of their subsets, which have distinct effector specificities, proliferative abilities, memory generation, and life span. T cells are impactful in viral infections by virtue of their capability to combat intracellular pathogens. The effector functions of T cells are mediated through cytokines/chemokines and by direct cytotoxicity of infected cells. T cell response can be beneficial or detrimental to host; prognosis depending on qualitative and quantitative differences in the response. Persistent viral infections are associated with functionally suboptimal, exhausted T cell responses, which are unable to clear virus. Specific subsets such as regulatory T cells (Tregs) dampen antiviral responses; thereby favouring viral persistence. However, Tregs protect the host from immunopathology by limiting perpetual inflammation. Certain other subsets such as Th17 cells may contribute to autoimmune component of viral infections. The importance of T cells is highlighted by the fact that modern vaccination and therapeutic approaches focus on modulating T cell frequencies and effector functions. This chapter emphasises the understanding how T cells influence outcomes of viral infections, modern vaccination and therapeutic strategies with thrust on T cell biology.

Achanta Jagadeesh, A. M. V. N. Prathyusha, G. Mohana Sheela and Pallaval Veera Bramhachari contributed equally with all other contributors.

A. Jagadeesh
Tumor Microenvironment Global Core Research Center, College of Pharmacy, Seoul National University, Seoul, South Korea

A. M. V. N. Prathyusha · G. M. Sheela · P. V. Bramhachari (✉)
Department of Biotechnology, Krishna University, Machilipatnam, Andhra Pradesh, India

P. V. Bramhachari (ed.), *Dynamics of immune activation in viral diseases*,
https://doi.org/10.1007/978-981-15-1045-8_9

Keywords
Adaptive immune system · T cell responses · Viral infections · Therapeutic strategies

9.1 Introduction

Viruses are infectious agents consists of nucleic acids coated in a simple protein casing, infects, replicates in host cells and causes acute, chronic infections. Mammals established a refined immune system to cope with several viral and bacterial infections (García-Sastre and Biron 2006). Remarkably, adaptive arm of immune responses is significant in action against virus particles and infected cells. T Lymphocytes are the key players in adaptive, cell-mediated immune responses and also in elimination of foreign pathogens by activating various immune responses. There exist many viruses for which both $CD8^+$ and $CD4^+$ T cells, reported to play key role in viral control viz. measles virus (Nelson et al. 2017), cytomegalovirus (CMV) (Wehrens et al. 2016), hepatitis C virus (HCV) (Sheiko et al. 2016) and HIV (Jones and Walker 2016). Typically, T_H1 cells are assumed to be efficient in antiviral T cell response. Nonetheless, many viruses can inhibit T_H1 response by downregulating interferons release from infected cells, which greatly influence outcome of virus infection (Laidlaw et al. 2016). The humoral immune response involves antibodies specific for virus to block host–virus interactions, neutralize virus and recognise viral antigens on infected cells and activates antibody-dependent cytotoxic cells (ADCC) or complement-mediated lysis to kill infected cells. However, if these antibodies are ineffective, viruses are able to infect host cells where adaptive arm acts to control viral pathogenesis (Rosendahl Huber et al. 2014).

Virus infects host cells if humoral immunity fails, viruses use protein synthesis and replication machinery of host cells for their replication and synthesis of their own proteins. Some newly synthesized proteins may degrade into peptide fragments. If these peptides have sufficient binding affinity, to class I MHC molecules they appear on cell surface of an infected cell as class I MHC–peptide complex. This complex activates $CD8^+$ T cells and induces infected cell apoptosis by releasing cytotoxic granules and production of TNF-α and IFN-γ. Activation of $CD8^+$ T cells also occurs in draining lymph nodes, where antigen-presenting cells (APCs) encounter naïve T cells. Priming of naïve T cells will not only occur through classical pathway via infection of cell, but also through cross-presentation of viral peptides on MHC class I molecules, taken up from extracellular sources. Priming of T cells triggers a massive expansion of antigen-specific T cells. Their progeny usually accumulate in large numbers of armed effector T cells and these normally contribute to the eradication of viral pathogens.

T cells in chronic viral infections typically exhibit strong impairments in the production of cytokines (IFN-g, TNF and IL-2) and express high levels of inhibitory receptors viz. PD-1 (programmed cell death-1) and Lymphocyte-activation gene 3 (Lag-3). These phenotypic changes along with failure of immune system to clear pathogens in chronic infection exhaust functional T cell response. These primarily promote terminally differentiated T cells and inhibiting formation of $CD8^+$ T cell memory. Furthermore, antiviral activity and $CD8^+$ T cell response drastically increase when signalling through PD-1 is prevented. These observations signify that T cell immune response has ceased reversibly. However, molecular mechanisms that conserve terminally differentiated T cells in chronic infections and improvement in T cell response after checkpoint inhibition remain poorly understood of time, which still mediates certain level of virus control.

The pathogenesis of T cell also depends on processing pathogen components by Antigen Presentation Cells (APCs), and their presentation via Major Histocompatibility Complex (MHC). Antigenic diversity of peptides enhances viral pathogenesis, where diversity of MHC and TCR repertoire in viral pathogenesis is yet unexplored. The different methods are administered likewise in vivo and ex vivo to elucidate T cell responses and their mechanism in viral pathogenesis. Therefore, to explore the diversity of the viral proteins and its pathogenesis in effector T cell immune reaction and their mechanisms are important. As a result, there is a potential for designing new therapeutics to combat the viral pathogenicity. Additionally, viruses also exploit complement system for cellular entry as well as their spread.

9.2 T Cell Responses in SARS Virus

There are more than 8000 cases of respiratory diseases among which the Severe acute Respiratory Syndrome (SARS) is caused by novel coronavirus (SARS-CoV), and contributed to 10% of mortality in 2002–2003. Pro-inflammatory responses enhance disease progression. The mechanism of immune evasion is mainly characterised by poor antigen presentation by antigen presentating cells (Legge and Braciale 2003). The antigen presentation is key for activation of T cell and produces chemokines and cytokines that regulate disease progression (Seder et al. 2008). The immune evasion of novel coronavirus targets APC and suppresses T cell activation. Dendritic cell immunisation activates T cells. As a result, the production of IFN-gamma, IL-2 and TNF-alfa are released (Zhao et al. 2010). Therefore, the viral titre gets reduced as immunisation with Dendritic cells (DC) by T cell activation successfully suppressed viral pathogenesis. It can be characterised as a potent immunogen to activate immune response.

9.3 T Cell Responses in West Nile Virus

West Nile Virus is a positive sense single-stranded RNA belongs to Flaviviridae and transmitted by mosquito vectors and originated in the USA in 1999, it causes mosquito borne encephalitis. It is asymptomatic in majority of individuals and symptoms usually are arthralgia, myalgia and cephalea. The minor part of pathogenesis can be neurologic deficits and neuroinvasive in elderly people. Viral pathogenesis mainly favours by efferocytosis activator TIM-3, which inhibits $CD8^+$ T cell activation (Lanteri et al. 2014). The viral RNA plays major role in dampening the $CD8^+$ T cell activation and it activates $CD4^+$ T cell expresses Th1 and Th17 cytokines that favours neuroinvasion (James et al. 2016). viral RNA mainly responsible in regulating T cell pathogenesis and responsible for neuroinvasion. As per study, $CD8^+$ T cell activation controls viral replication, tissue tropism and infection (Aguilar-Valenzuela et al. 2018). Despite the fact that $CD8^+$ T cell activation reaches peak expansion in the periphery of west nile virus by 7-day post infection followed by chemokine activation of CXCR3. As a result, viral clearance activated by membrane-mediated apoptosis (Shrestha and Diamond 2007) was recently evidenced to combat viral replication, novel recombinant TCR can enhance immune response (Aguilar-Valenzuela et al. 2018).

9.4 T Cell Responses in Japanese Encephalitis Virus

Japanese encephalitis virus (JEV) is an arthropod borne member belongs to family Flaviviridae, which is endemic to rural parts of South and Southeast Asia. The virus mainly infects children and it causes death. The JEV possess a single-stranded, positive sensed RNA genome and its 10 kb open reading frame (ORF) encodes four structural proteins, envelope (E), premembrane (prM), core (C) and seven non-structural (NS). The JEV viral pathogenesis mainly inhibits infiltration of T cells due to production of low levels of IFN-gamma and regulates production of IL-2 (Kumar et al. 2004). As a result, JEV is successful in disease progression. The IFN-gamma plays important role in the activation of T cells. In recent study, modes of activation of TCR by various viral epitopes and maintaining levels of IFN-gamma and increasing infiltration of activated T cells (Turtle et al. 2016).

9.5 T Cell Responses in Acute Dengue Virus

The immune evasion strategy of viruses mainly by producing diversity of viral proteins and enhances viral pathogenesis. The HLA alleles on $CD8^+$ T cell and diversity of viral protein presentation are not uniform in viral pathogenic environment makes the strong presentation of viral pathogenesis (Weiskopf et al. 2013). As a result, T cell activation is suppressed due to less polymorphism of HLA. The T cell cross-reactivity plays an important role in disease pathogenesis. Likewise, disease serotypes fail to provide immune response against later stage secondary infection of dengue virus and unresponsiveness to T memory cells to secondary infection, which are produced in primary infection and it also produces Types 1 and 2 cytokines and responsible for other pathologies (Duangchinda et al. 2011). Therefore, cytotoxic memory T cell enhances viral replication and it is considered as original antigenic sin (Mongkolsapaya et al. 2003). Therefore inorder to combat the strategy was applied to activate $CD8^+$ T cell, as a result IFN-gamma enhances immune response against DENV peptides.

9.6 T Cell Responses in Viral Hemorrhage Fever

The viral hemorrhage fever (VHF) majorly regulates T cell priming, effector T cell response and T cell activation (Dahlke et al. 2017). Early during VHF, the antigen-presenting cells produce high levels of IFN-gamma and TNF-alfa and IL-6. As a result, activation of T cells increases viremia and overproduction of IFN-gamma and TNF-alfa this results in over activation of T cells (Perdomo-Celis and Salvato 2019) This condition is termed as cytokine storm. During VHF the proportions of regulatory T lymphocytes decreases and therefore, this enhances disease pathogenicity.

9.7 T Cell Responses in Chronic and Acute Viral Infections

The chronic and viral infections effect a plethora of T cell populations drastically. The $CD4^+$ T cells and $CD8^+$ T cells are major players in regulating viral pathogenesis. The other differentiated T cell called regulatory T cell, which regulates immune responses and inflammatory responses. During chronic and viral infections, T reg plays important role in regulating immune responses and various cytokines (Keynan et al. 2008). As a result, it regulates effector T cells population. In chronic viral infections, depletion of Treg cells, $CD8^+$ T cell proliferation (Boettler et al. 2005), IFN-gamma production and increase cytolytic activity are predominant (Haeryfar et al. 2005). In HIV infection, due to decrease in Treg population, hyper activation of $CD4^+$ and $CD8^+$ T cells is predominant (Oswald-Richter et al. 2004) As a result, it favours viral replication.

9.8 T Cell Responses in Human Papilloma Virus (HPV)

Human Papilloma Virus (HPV), a small DNA virus majorly infects birds, reptiles and mammals, there are 300 viral genotypes that have been discovered till now (Vande and Klingelhutz 2013). The HPV infects mucosal and/or cutaneous skin and causes benign or malignant tumours. HPV associates with cervical cancer, oral squamous carcinoma and Head and Neck cancers (Forman et al. 2012). The HPV oncoproteins, E6 and E7, majorly regulates the host immune responses and plays vital role in tumorigenesis (den Boon et al. 2015). The HPV16 viral particle involves in the host immune dysregulation by epigenetic mechanisms. The viral pathogenesis regulates the synthesis of chemokines that required for T cell activation, i.e. CXCL14 (Cicchini et al. 2016). The CXCL14 synthesis regulated epigenetically by HPV viral protein, i.e. E7. Likewise, the HPV viral protein interacts with the host DNMT1 and stimulates the methylase activation (Burgers et al. 2007) and as a result, CXCL14 is repressed. Therefore, evasion of immune responses against virus, by inhibiting the T cell activation. The HPV16 E7 suppresses the production of pro-inflammatory responses like IL-8, IL18, CCL2, CCL20 (Cho et al. 2001; Guess and McCance 2005; Huang and McCance 2002; Kleine-Lowinski et al. 2003)

Additionally, HPV16 E7 upregulates the immunosuppressive genes like IDO1 and it triggers the activation of Treg cells (Mittal et al. 2013). Strikingly, the HPA18 E6 and E7 proteins jointy dysregulate the T cell activation by binding directly to the proteins involved in non-receptor tyrosine kinase signalling pathway (Li et al. 1999). As result, it downregulates the IFN-alfa and other cytokines like IL-6 and IL12. The HPV proteins regulate the IFN signalling for its major mechanism in evading the host T cell responses. The MHC restriction is a typical immune escape used by the HPV protein E5 by downregulating the MHC-I complex (Ashrafi et al. 2005, 2006). Hence HPV hinders the activation of $CD8^+$ T cell and evasion of host responses are dysregulated by viral proteins and favours for the tumour progression.

9.9 T Cell Responses in Hepatitis B Virus

There are more than 350 million people infected with the Hepatitis B virus. The virus mainly infects liver and causes chronic and acute pathology. The infection carried commonly by mother to child during birth. The one in all among the viruses, which is a non-retroviruses that uses the host RNA polymerase for its transcription and as a result, the closed circular DNA gets transcribed and involved in the host disease pathogenesis. The adaptive immunopathogenesis plays important role in the disease progression. The liver residing APC plays an important role in the activation of naïve CD4 T and CD8 T cells via cross-priming and facilitates the persistence of the virus (Lan et al. 2016). The downregulation of the TLR-7 and TLR-9 in plasmacytoid dendritic cell and as a result, the IFN signalling is inactivated and cytokines are inhibited (Seeger and Mason 2000). The NK cells are also major cells providing immune to the HBV by expressing death ligand and inactivating the CD4 T cells Bertoletti and Ferrari 2016). The CD8 T cells exhaustion is the major episode in the disease pathology by expressing death receptors such as TIM-3, PD-1 and 2B4, poor proliferative signal, IL-2 and IFN-gamma (Raziorrouh et al. 2010). The CTL apoptosis of CTL by upregulation of various apoptotic genes likewise, Bim and TRAIL-R2. Due to downregulation of T-bet results in exhaustion of the T cells.

9.10 T Cell Responses in Zika Virus

Zika virus is mainly infected by mosquito borne flavivirus and it has unexpected links with microcephaly and Guillain–Barre syndrome. It is hypothesised as the infection resultant to the testis damage. The Zika viral infection results in the expression of TIM-4, which is one of the phagocytic markers and facilitates viral replication (Osuna et al. 2016; Zhang et al. 2018). The population of lymphocyte subsets are reduced in the period of infection. The virus getting successfully evading the host immune responses. The current study in combatting the viral immune evasion mainly on designing the RNA vaccines and T cell epitope tetramer for activation of T cells, recombinant vector-based vectors (Zang et al. 2018).

9.11 Future Perspectives and Conclusions

Acquired immunity plays an important role to eliminate the pathogen by activating APC, CD4 T and CD8 T cells. The APC presentation by MHC and the presented processed peptide activates the TCR and CD4 and CD8 associated with MHC stabilises the TCR activation. The TCR activation supported by the co-stimulatory molecules like CD28/B7 activates the T cell proliferation. As a result, the IL-2 and other cytokines and chemokines are released. The TLR's are the receptors which activates the innate immunity and as a result, the activation of chemokines and cytokines are upregulated and various proliferative signals are activated with the response to the PAMP's. The naïve T cells, as aresult differentiates into CD4 and further has different subsets like, Th1, Th2, Th17 and T reg. The other CD8 T and memory T cells are prominent cells in elimination of disease. The virus infection evades the acquired immunity and immune responses by various mechanisms. The viral antigen manipulates the host immune responses by inhibiting the proliferative signals, expressing exhausting receptors, MHC restriction and various anti-inflammatory cytokines. The viral proteins upregulate apoptotic genes and inhibits immune responses of T cells. Due to high diversity of viral proteins, the HLA polymorphism is restricted, hence the MHC restriction and MHC cross presentation makes more significant in the viral persistence. The TLR-7 and TLR-9 are majorly downregulated in the HBV and provide immune to the viral propagation. The differentiation of the CD4 Th cells into Regulatory T lymphocytes hinders the immune activation and provide immune-resistance to the virus. In few viral serotypes, the T memory cells fail to respond to the secondary infection. In VHF, the overexpression of cytokines like TNF-alfa, IFN-gamma and IL-6 are responsible for 'cytokine storm' and increase viremia and favour in disease progression. The NK cell-mediated immune evasion in cancer causing viruses is unique by causing exhaustion of T cell and therefore, inhibiting immune responses.

There are vast potential targets to design the therapeutic and diagnosing markers for the virus. Recently, immunotherapy is a novel application for treating various virus pathogenicity by PD-1/PD-L1, this infers the overcoming the exhaustion of T cells (Pauken and Wherry 2015). The active T cell induces immune responses against viral titre and resolving the immune evasion of viruses. The tcf1 can be a potential biomarker for the chronic viral infections. The acquired immunity is a potential target to combat against the viral infection. The RNA-based Next-generation sequencing play an imperative role in transplanting the γδ T cell antigen receptors from the human cohort (Utzschneider et al. 2016). Ex vivo application of IFN-gamma is also a potential therapeutic activation of viral immune responses (Turtle et al. 2016). Immunisation of acquired immune responses by activating the DC cells viral peptide and evade the anergic condition (Zhao et al. 2010). The maintenance of antiviral peptides from the healthy infected individuals is the potential strategy for combatting the immune evasion.

Acknowledgments The authors gratefully acknowledge Krishna University, Machilipatnam and Seoul National University, Seoul, South Korea for the support extended.

Conflict of Interest The authors declare that they have no competing interests.

References

Aguilar-Valenzuela R, Netland J, Seo Y-J, Bevan MJ, Grakoui A, Suthar MS (2018) Dynamics of tissue-specific CD8+ T cell responses during West Nile virus infection. J Virol 92:e00014–e00018

Ashrafi GH, Haghshenas MR, Marchetti B, O'Brien PM, Campo MS (2005) E5 protein of human papillomavirus type 16 selectively downregulates surface HLA class I. Int J Cancer 113:276–283

Ashrafi GH, Brown DR, Fife KH, Campo MS (2006) Down-regulation of MHC class I is a property common to papillomavirus E5 proteins. Virus Res 120(1–2):208–211

Bertoletti A, Ferrari C (2016) Adaptive immunity in HBV infection. J Hepatol 64:S71–S83

Boettler T, Spangenberg HC, Neumann-Haefelin C et al (2005) T cells with a CD4+CD25+ regulatory phenotype suppress in vitro proliferation of virus-specific CD8+ T cells during chronic hepatitis C virus infection. J Virol 79:7860–7867

Burgers WA, Blanchon L, Pradhan S, de Launoit Y, Kouzarides T, Fuks F (2007) Viral oncoproteins target the DNA methyltransferases. Oncogene 26:1650–1655

Cho YS, Kang JW, Cho M, Cho CW, Lee S, Choe YK, Kim Y, Choi I, Park SN, Kim S, Dinarello CA, Yoon DY (2001) Down modulation of IL-18 expression by human papillomavirus type 16 E6 oncogene via binding to IL-18. FEBS Lett 501:139–145

Cicchini L, Westrich JA, Xu T, Vermeer DW, Berger JN, Clambey ET, Lee D, Song JI, Lambert PF, Greer RO, Lee JH, Pyeon D (2016) Suppression of antitumor immune responses by human papillomavirus through epigenetic downregulation of CXCL14. MBio 7:e00270-16

Dahlke C, Lunemann S, Kasonta R, Kreuels B, Schmiedel S, Ly ML, Fehling SK, Strecker T, Becker S, Altfeld M et al (2017) Comprehensive characterization of cellular immune responses following ebola virus infection. J Infect Dis 215(2):287–292

Den Boon JA, Pyeon D, Wang SS, Horswill M, Schiffman M, Sherman M, Zuna RE, Wang Z, Hewitt SM, Pearson R, Schott M, Chung L, He Q, Lambert P, Walker J, Newton MA, Wentzensen N, Ahlquist P (2015) Molecular transitions from papillomavirus infection to cervical precancer and cancer: role of stromal estrogen receptor signaling. Proc Natl Acad Sci U S A 112:E3255–E3264

Duangchinda T, Dejnirattisai W, Vasanawathana S, Limpitikul W, Tangthawornchaikul N, Malasit P et al (2011) Immunodominant T-cell responses to dengue virus NS3 are associated with DHF. Proc Natl Acad Sci U S A 107(39):16922–16927

Forman D, de Martel C, Lacey CJ, Soerjomataram I, Lortet-Tieulent J, Bruni L, Vignat J, Ferlay J, Bray F, Plummer M, Franceschi S (2012) Global burden of human papillomavirus and related diseases. Vaccine 30(Suppl 5):F12–F23

García-Sastre A, Biron CA (2006) Type 1 interferons and the virus-host relationship: a lesson in detente. Science 312(5775):879–882

Guess JC, McCance DJ (2005) Decreased migration of Langerhans precursor-like cells in response to human keratinocytes expressing human papilloma virus type 16 E6/E7 is related to reduced macrophage inflammatory protein-3α production. J Virol 79:14852–14862

Haeryfar SM, DiPaolo RJ, Tscharke DC, Bennink JR, Yewdell JW (2005) Regulatory T cells suppress CD8+ T cell responses induced by direct priming and cross-priming and moderate immunodominance disparities. J Immunol 174:3344–3351

Huang SM, McCance DJ (2002) Down regulation of the interleukin-8 promoter by human papillomavirus type 16 E6 and E7 through effects on CREB binding protein/p300 and P/CAF. J Virol 76:8710–8721

James EA, Gates TJ, LaFond RE, Yamamoto S, Ni C, Mai D, Gersuk VH, O'Brien K, Nguyen QA, Zeitner B, Lanteri MC, Norris PJ, Chaussabel D, Malhotra U, Kwok WW (2016) Neuroinvasive West Nile infection elicits elevated and atypically polarized T cell responses that promote a pathogenic outcome. PLoS Pathog 12:e1005375

Jones RB, Walker BD (2016) HIV-specific CD8+ T cells and HIV eradication. J Clin Invest 126(2):455–463

Keynan Y, Card CM, McLaren PJ, Dawood MR, Kasper K, Fowke KR (2008) The role of regulatory T cells in chronic and acute viral infections. Clin Infect Dis 46(7):1046–1052

Kleine-Lowinski K, Rheinwald JG, Fichorova RN, Anderson DJ, Basile J, Munger K, Daly CM, Rosl F, Rollins BJ (2003) Selective suppression of monocyte chemoattractant protein-1 expression by human papillomavirus E6 and E7 oncoproteins in human cervical epithelial and epidermal cells. Int J Cancer 107:407–415

Kumar P, Sulochana P, Nirmala G, Chandrashekar R, Haridattatreya M, Satchidanandam V (2004) Impaired T helper 1 function of non-structural protein 3-specific T cells in Japanese patients with encephalitis with neurological sequelae. J Infect Dis 189:880–891

Laidlaw BJ, Craft JE, Kaech SM (2016) The multifaceted role of CD4+ T cells in CD8+ T cell memory. Nat Rev Immunol 16(2):102

Lan S, Wu L, Wang X, Wu J, Lin X, Wu W et al (2016) Impact of HBeAg on the maturation and function of dendritic cells. Int J Infect Dis 46:42e8

Lanteri MC, Diamond MS, Law JP, Chew GM, Wu S, Inglis HC, Wong D, Busch MP, Norris PJ, Ndhlovu LC (2014) Increased frequency of Tim-3 expressing T cells is associated with symptomatic West Nile virus infection. PLoS One 9:e92134

Legge KL, Braciale TJ (2003) Accelerated migration of respiratory dendritic cells to the regional lymph nodes is limited to the early phase of pulmonary infection. Immunity 18:265–277

Li S, Labrecque S, Gauzzi MC, Cuddihy AR, Wong AH, Pellegrini S, Matlashewski GJ, Koromilas AE (1999) The human papilloma virus (HPV)-18 E6 oncoprotein physically associates with Tyk2 and impairs Jak-STAT activation by interferon-α. Oncogene 18:5727–5737

Mittal D, Kassianos AJ, Tran LS, Bergot AS, Gosmann C, Hofmann J, Blumenthal A, Leggatt GR, Frazer IH (2013) Indoleamine 2,3-dioxygenase activity contributes to local immune suppression in the skin expressing human papillomavirus oncoprotein e7. J Invest Dermatol 133:2686–2694

Mongkolsapaya J, Dejnirattisai W, Xu XN, Vasanawathana S, Tangthawornchaikul N et al (2003) Original antigenic sin and apoptosis in the pathogenesis of dengue hemorrhagic fever. Nat Med 9:921–927

Nelson AN, Putnam N, Hauer D, Baxter VK, Adams RJ, Griffin DE (2017) Evolution of T cell responses during measles virus infection and RNA clearance. Sci Rep 7(1):11474

Osuna CE et al (2016) Zika viral dynamics and shedding in rhesus and cynomolgus macaques. Nat Med 22(12):1448–1455

Oswald-Richter K, Grill SM, Shariat N et al (2004) HIV infection of naturally occurring and genetically reprogrammed human regulatory T-cells. PLoS Biol 2:E198

Pauken KE, Wherry EJ (2015) Overcoming T cell exhaustion in infection and cancer. Trends Immunol 36(4):265–276

Perdomo-Celis F, Salvato MS (2019) T-cell response to viral hemorrhagic fevers. Vaccine 7:11

Raziorrouh B, Schraut W, Gerlach T, Nowack D, Gruner NH, Ulsenheimer A et al (2010) The immunoregulatory role of CD244 in chronic hepatitis B infection and its inhibitory potential on virus-specific CD8+ T-cell function. Hepatology 52:1934–1947

Rosendahl Huber S, van Beek J, de Jonge J, Luytjes W, van Baarle D (2014) T cell responses to viral infections–opportunities for peptide vaccination. Front Immunol 5:171

Seeger C, Mason WS (2000) Hepatitis B virus biology. Microbiol Mol Biol Rev 64:51e68

Seder RA, Darrah PA, Roederer M (2008) T-cell quality in memory and protection: implications for vaccine design. Nat Rev Immunol 8:247–258

Sheiko MA, Golden-Mason L, Giugliano S, Hurtado CW, Mack CL, Narkewicz MR, Rosen HR (2016) CD4+ and CD8+ T cell activation in children with hepatitis C. J Pediatr 170:142–148
Shrestha B, Diamond MS (2007) Fas ligand interactions contribute to CD8+ T-cell-mediated control of West Nile virus infection in the central nervous system. J Virol 81:11749–11757
Turtle L, Bali T, Buxton G, Chib S, Chan S, Soni M, Hussain M, Isenman H, Fadnis P, Venkataswamy MM, Satishkumar V (2016) Human T cell responses to Japanese encephalitis virus in health and disease. J Exp Med 213(7):1331–1352
Utzschneider DT, Charmoy M, Chennupati V, Pousse L, Ferreira DP, Calderon-Copete S, Danilo M, Alfei F, Hofmann M, Wieland D, Pradervand S (2016) T cell factor 1-expressing memorylike CD8+ T cells sustain the immune response to chronic viral infections. Immunity 45(2):415–427
Vande Pol SB, Klingelhutz AJ (2013) Papillomavirus E6 oncoproteins. Virology 445:115–137
Wehrens EJ, Wong K, Gupta A, Benedict C, Zuniga E (2016) IL-27 suppresses CD4 and CD8 T cell cytotoxicity and viral control during cytomegalovirus infection. J Immunol 196(1 Suppl):217.4
Weiskopf D, Angelo MA, de Azeredo EL, Sidney J, Greenbaum JA, Fernando AN et al (2013) Comprehensive analysis of dengue virus-specific responses supports an HLA-linked protective role for CD8+ T cells. Proc Natl Acad Sci U S A 110(22):E2046–E2053
Zhang Y, Zhang H, Ma W, Liu K, Zhao M, Zhao Y, Lu X, Zhang F, Li X, Gao GF, Liu WJ (2018) Evaluation of Zika virus-specific T-cell responses in immunoprivileged organs of infected Ifnar1−/−mice. J Vis Exp (140):e58110. https://doi.org/10.3791/58110
Zhao J, Zhao J, Perlman S (2010) T cell responses are required for protection from clinical disease and for virus clearance in severe acute respiratory syndrome coronavirus-infected mice. J Virol 84(18):9318–9325

Potentiality of Toll-Like Receptors (TLRS) in Viral Infections

10

A. M. V. N. Prathyusha, Prudhvi Lal Bhukya, and Pallaval Veera Bramhachari

Abstract

Living organisms exist in nature usually perceptible to infectious agents like viruses, bacteria, and many other pathogens. Immune system of host organism typically elicits both nonspecific innate and "specific" adaptive immune responses against foreign pathogens. Cells of innate immunity expresses a diversity of pattern recognition receptors (PRRs), which were developed during evolutionary process, recognizes conserved structures of distinct pathogens known as pathogen-associated molecular patterns (PAMPs). There are several classes of PRRs like Toll-like receptors (TLRS), RLRs, NLRs, and many DNA and RNA sensors. Among these PRRs, Toll-like receptors plays an important role in eliciting innate immunity, development of B and T cell responses and as well as pathogen-specific adaptive immune response. Immune responses in any viral infections are elicited by the recognition viral PAMPs like nucleic acids such as DNA and RNA, viral capsid proteins by the host PRR. TLRs sense these viral PAMPs and induce the antiviral response by inducing the immune active chemokines and cytokines. This chapter focuses on the responses of different TLRs with viral PAMPS, immune responses mediated by TLRS in viral infections, agonists of TLRs for treatment of viral infections.

Keywords

Viral Infections · PAMP · PRR · TLRs · Innate immune responses

A. M. V. N. Prathyusha · P. Veera Bramhachari (✉)
Department of Biotechnology, Krishna University, Machilipatnam, Andhra Pradesh, India

P. L. Bhukya
National Institute of Virology, Pune, Maharashtra, India

P. V. Bramhachari (ed.), *Dynamics of immune activation in viral diseases*,
https://doi.org/10.1007/978-981-15-1045-8_10

10.1 Introduction

Innate immune system expresses diverse pattern recognition receptors (PRRs) developed during the evolutionary process, which recognizes the distinct pathogens conserved region known as pathogen-associated molecular patterns (PAMPs). Eventually the cells of innate immune system detects the virus or viral components like viral nucleic acid (DNA or RNA), carbohydrate moiety and even viral proteins through germ line encoded receptors equipped on innate immune cells or in intracellular compartments of cells these receptors generally known as PRRs (Thompson et al. 2011). Several PRRs like Toll-like receptors (TLRs), RIG-I-like receptors (RIG-1 and MDA5), nucleotide-binding oligomerization domain (NLRs), and cyclic GMP-AMP synthase (cGAS) involve in regulation of viral infection. Among these, TLRs are most studied, well characterized, and also plays critical role in innate responses against complex viral pathogens, viz. herpes simplex virus (HSV), Herpes simplex encephalitis (HSE), influenza A virus, respiratory syncytial virus (RSV), coxsackie B virus, yellow fever virus, West Nile virus (WNV), Epstein Barr virus (EBV), human cytomegalovirus, Lymphocytic choriomeningitis virus (LCMV), Hepatitis B virus (HBV), Hepatitis C virus (HCV), enterovirus 71 (EV71), and coxsackievirus A16 (CA16).

Different TLRs recognize some viral components and modulates viral replication, cytokine production in vitro. For instance, TLR2, and TLR3 were reported to participate in cellular activation and cytokine production against Flavivirus. Furthermore, TLRs 2, 3, 4, 7, 8, and 9 are well characterized reported to be involved in activating inflammatory response against several viruses viz influenza A virus (Le Goffic et al. 2007), HSV (Sato et al. 2006), coxsackie B virus (Triantafilou et al. 2005), or RSV (Rudd et al. 2005). Multiple TLRs including TLR3, 7, 8, and 9 are significantly associated with yellow fever virus, moreover, TLR3 signaling was highlighted during WNV infection merely not for mounting cellular antiviral response nonetheless for shaping detrimental innate and adaptive immune response in vivo (Wang et al. 2004).

TLRs generate innate immune signals that orchestrate innate and also adaptive immune response. TLRs has the ability to employ diverse downstream signaling molecules furthermore cross talk with other immune pathways is an essential factor in determining type, magnitude, and duration of inflammatory response. Several single nucleotide polymorphisms (SNPs) in *TLR* genes are associated with altered susceptibility in viral infections and immune response which promotes viral replication (Medvedev 2017). This chapter emphasizes the role of multiple TLRs and their agonists in viral infections.

10.2 Toll-Like Receptors

TLRs are membrane sentinels belong to a class of type I transmembrane proteins helps in detection of microbial infection through conserved molecular pattern of pathogen proteins derived from DNA and RNA viruses, Gram-positive and -negative bacteria, fungi, and protozoan (Pasare and Medzhitov 2005; Dowling and

Dellacasagrande 2016). In human, there are 11 types of TLRs (TLR1–TLR11) whereas in mouse 12 TLRs (TLR1–TLR9 and TLR11–TLR13) have been identified (Gay and Gangloff 2007). Each TLR has specific location on innate immune cells either intra- or extracellular. Subgroup of TLRs like TLR1, TLR2, TLR4, and TLR5 are generally expressed on cells surface and recognizes PAMPS of viral proteins, extracellular fungal and bacterial cell wall portions. Another subgroup of TLRs like TLR3, TLR7, TLR8, and TLR9 localized in cellular compartments such as endosomes and senses nucleic acids, DNA and RNA (Gay and Gangloff 2007; Takeda and Akira 2004). Not all the TLRs function independently. For instance, TLR4 involves MD2, however, TLR2 requires TLR1 or TLR6, while ds RNA-mediated activation of TLR3 is enhanced by CD14 (Lee et al. 2006). Responses of TLR class proteins in combating viral PAMPs have observed both in vivo and in vitro studies.

10.3 TLR 1/TLR2 in Viral Infection

TLR1 and TLR2 regulate innate immune response by recognizing PAMPs-like glycoproteins and lipoproteins to activate immune cells (Ma et al. 2015). TLR2-dependent secretion of pro-inflammatory cytokines and cell surface molecules which activate promotion of T cells function in non-parenchymal liver cells like Kupffer cells (Lin et al. 2010). Preiss et al. (2008) reported that stimulation of TLR2 leads to activation of NF-κB and other cytokines like tumor necrosis factor alpha (TNF-α) and interleukin 8 (IL-8) in MyD88-dependent manner to suppress invading pathogens in both hepatoma cell lines and human primary hepatocytes. In an in vitro study carried out by Thompson et al. (2009) reported that when Hepatoma cell lines stimulated with IIL-1 receptor and TLR2, replication and capsids formation of HBV is inhibited. Secretions of pro-inflammatory cytokines viz. IL6, IL12p40, and TNF-α were observed in macrophages in a TLR2-dependent fashion when triggered with full-length HBV-HBc antigen (Cooper et al. 2005). Zhang et al. (2012) reported that TLR2 activates antiviral signalling pathways like NFKB, PI3/Akt, and MAPK producing various pro-inflammatory cytokines in hepatoma cell lines. In a similar study, it is observed that in HepG2.2.15 cells and primary woodchuck hepatocytes TLR2-mediated innate immune response that leads to decreased replication and gene expression of HBV and woodchuck hepatitis virus. Even though TLR2 inhibits replication and nucleocapsid formation of HBV while virus has evolved a clever strategy to counteract antiviral function of TLR2 by using its antigens like HBeAg, HBsAg, and by whole HBV virions to decrease expression level of TLR2, which ultimately affects amount of inflammatory cytokines. In many other DNA and RNA viruses, like EBV, HSV, human cytomegalovirus, HCV, RSV, and LCMV antiviral action mediated by TLR2 was observed. The antiviral response of TLR2 was observed in many cell types like glial cells of nervous system in LCMV virus infection, monocytes in EBV infection, microglial cells in HSV infection and leukocytes in HSV infection (Zhou et al. 2009; Gaudreault et al. 2007; Aravalli et al. 2008; Murawski et al. 2009). Transmembrane domain of HIV-1 envelope interacts with transmembrane domain of TLR2 and hinders secretion of TNF-α and MCP-1 in mouse macrophages.

10.4 TLR3 in Viral Infection

Toll-like receptor 3 (TLR3) is one of the important intracellular receptors present on intracellular membrane organelles like endosomes, lysosome, and endoplasmic reticulum (ER). Apart from these organelles other immune cells like natural killer cells (NK cells), mast cells, dendritic cells (DCs), monocytes, epithelial cells, and variety of antigen processing cells also express TLR3 on their cell surface (Latz et al. 2004; Nicodemus and Berek 2010). TLR3 is well known for detection of viral intracellular double-stranded RNA (ds RNA) and stimulates NF-κB activation through antiviral TRIF pathway (Karimi-Googheri and Arababadi 2014). When TLR3 interacts with its ligand ds RNA, the cytoplasmic domain of TLR3 known as Toll/interleukin 1 receptor (TIR) interacts with TRIF and further activates the downstream signalling molecules viz. TRAF6, RIP-1, and TBK1 which finally activates several antiviral transcription factors like NF-κB, AP-1, and IRF-3 which transcribes the genes of cytokines, chemokines, and several co stimulatory molecules (Karimi-Googheri and Arababadi 2014; Chang and Toledo-Pereyra 2012).

TLR3-dependent inflammatory responses permit WNV to affect CNS (central nervous system) prompting reversible breakdown of blood–brain barrier. TLR3 enhances WNV's ability to replicate in CNS triggering lethal encephalitis (Wang et al. 2004). TLR3 also initiates inflammatory response against HBV and HCV and affects chronicity of virus infection, and thus subsequent pathological changes. Several studies reported the correlation of genetic polymorphisms in *TLR3* gene with susceptibility or resistance to copious infectious and immune diseases. SNPs in *TLR3* gene may associate with changes in protein or gene expression, which influences function and efficacy of signal transduction. These changes in TLR3 cause an altered immune response. Previous reports, *TLR3* polymorphisms rs1879026, rs3775296, rs3775291, and rs5743305 have been associated with outcome of HCV and HBV infection, and development of consequential liver cirrhosis and HCC (hepatocellular carcinoma) primarily in Asian populations (Fischer et al. 2018). *TLR3* gene represents a good candidate approach for predicting progression of HCV and immune control over HBV infections (Al-Anazi et al. 2017).

10.5 TLR4 in Viral Infection

TLR4 is a receptor for sensing various microbial PAMPs such as lipopolysaccharides (LPS) and activates intracellular antimicrobial and antiviral signalling pathways. Apart from LPS, it also senses hyaluronan, heat shock protein (hsp60), free fatty acids, and adjuvant monophosphoyl lipid A (MPLA). Structure of TLR4 consists of three main domains such as extracellular domain that is rich in leucine repeats, hydrophobic transmembrane domain, and cytoplasmic TIR domains

(Yamamoto et al. 2004). In comparison with other TLRs, TLR4 has a unique property as it combines with two different signalling adaptors molecules like MyD88 and TRIF activates transcription of several pro-inflammatory cytokines like IL6, IL12, TNFα, and type 1 interferon (Yamamoto et al. 2004; Evans et al. 2003). In case of MyD88-dependent pathway, TLR4 which is present on cells' surface dimerizes with myeloid differentiation 2 proteins (MD-2), which further results in recruitment of two adaptor proteins known as TIRAP and MyD88. This series of events results in activation of pro-inflammatory transcription factors NF-κB, AP-1, IRF5, and production of pro-inflammatory cytokines (Rosadini et al. 2015). TLR4 is known to internalize into endosomes and recruits TRAM and TRIF adaptor proteins and activates transcription factor like IRF3 which induces production of type I interferon (Kawai and Akira 2011).

10.6 TLR5 in Viral Infection

TLR5 recognizes flagellin protein as a ligand from invading motile bacterial pathogens and mediates immune responses by the production of cytokines. When TLR5 receptor is activated by bacterial flagellin, TLR5 activates NF-κB and further stimulates TNFα production to elicit efficient immunity against invading pathogens (Srivastava et al. 2013). Increasing numbers of recent studies have demonstrated the efficacy of flagellin in adjuvant activity and also about bridging innate and adaptive immune responses via TLR5. Isogawa in 2005 reported that intravenous injection of TLR5-specific ligand flagellin from S*almonella muenchen* in HBV transgenic mice exhibited complete inhibition of HBV replication in liver non-cytopathically within 24 h in an enhanced α/β interferon dependent way (Isogawa et al. 2005). This study states that ligand specific for TLR5 can be used as therapeutic targets for inhibiting the replication of HBV. Zhao et al. developed a new vaccine with the adjuvant activity of fusion protein Flagellin (FliC) from *Salmonella abortus equi* and gD protein from EHV-1 virus. This novel vaccine significantly induced specific antibody responses and increased IFN-γ and IL-4 levels in host (Zhao et al. 2019).

10.7 TLR7 in Viral Infection

TLR7 plays a significant role in immunopathogenesis of HBV, influenza A virus, HCV, EV71, and CA16 viral infections. Single-stranded RNA (ssRNA) that is derived from invading pathogens viz. viruses, bacterium, fungi, and parasites act as a ligand for TLR7 receptors (Krieg 2007; Hayashi et al. 2012). TLR7 is expressed on intracellular membrane organelles like endosomes, endoplasmic reticulum (ER), and lysosome of different immune cells like NK cells (Hackstein et al. 2012; Bourquin et al. 2009), common dendritic cells (cDC), plasmacytoid dendritic cells (pDC) (Kader et al. 2013), and cytotoxic T lymphocytes and macrophages (Budimir et al. 2013; Gantier et al. 2008). Structurally N-terminal part of TLR7 contains leucine-rich repeats (LRRs), which is located on inner surface of endosomes.

Following this LRR domain, there are transmembrane region and cytoplasmic TIR receptor domain. Interaction of TLR7 with its ligand-like pathogens ssRNAs results in activation of MyD88-IRF7-dependent innate immune signalling pathways to elicit cytokine production (Sajadi et al. 2013), promote formation of autophagosomes which directly combat pathogen invasion (Song et al. 2018). Viruses evolved several strategies to combat TLR7 pathway-dependent autophagy and enhance their replication in host cells. EV71 and CA16 virus evade immune pathways by lowering production of INF-1, downstream signaling molecules in TLR7 pathway this results in reduced autophagosomes production and in turn autophagy (Song et al. 2018). This ultimately leads to successful replication of viruses in host cells.

10.8 TLR8 in Viral Infection

Despite other TLRs in viral infections, TLR8 recognizes the PAMPs expressed by infectious agents and stimulates innate immune responses for production of cytokines to combat invading pathogens (Sarvestani et al. 2012). TLR8 is present on endosomes which specifically recognizes GU and G-rich nucleotides in ssRNAs of pathogens. TLR8 recognizes many viral genomes like HCV and HIV. TLR8 mediates its antiviral action through MyD88-dependent manner (Heil et al. 2004; Zhang et al. 2016).

10.9 TLR9 in Viral Infection

TLR9 is a receptor that is generally present on extracellular surfaces and on membranes of intracellular organelles like endosomes and ER. TLR9 identifies unmethylated CpG DNA of invading pathogens like bacteria, viruses, and parasites. TLR9 belongs to type 1 transmembrane proteins and structurally TLR9 consists of LRRs repeats to its N terminal domain, followed by LRRs domain there are transmembrane regions and cytoplasmic TOLL/TIR domains are present. Expression of TLR9 receptor is constrained only to pDCs and B lymphocytes. Once TLR9 binds to its ligand, activates MyD88, forms complex with IRAK family protein employs signaling cascade that eventually triggers NF-κB and AP-1 transcription factors to elicit immune responses against invading pathogens (Hemmi et al. 2000; Latz et al. 2004).

Interestingly, few viruses viz. HIV, HCV, EBV, and HPV thought to inhibit TLR9 signaling to escape from immune response. HBV virus might escape TLR9 pathway either directly through activity of surface antigen (HBsAg), or indirectly through increased secretion of IL-10 by monocytes. HBV virus directly escapes immune system by binding HBsAg to C-type lectin receptor BDCA-2 upregulates SOCS-1, which inhibits TLR9–IRF7–IFN pathway thus reducing the secretion of IFN from pDC cells. HSV virus modulates NF-κB signaling via TLR pathway to directly benefit virus replication which simultaneously endows suppression of INF secretion. Nonetheless, omnipresent NF-κB signals trigger transcription of key modules of innate feedbacks to viral infection such as cytokines, chemokines, adhesion, as well as antiapoptotic proteins. Remarkably, HSV modulates the expression of

NF-κB through its numerous gene products. Therefore, HSV impairs both TLR-facilitated NF-κB pathway but also inhibits/activates NF-κB signaling through its own gene proteins in TLR independent manner to ensure viral replication and immune escape. Studies on HSV discovered that TLR2 and NF-κB-dependent pathways are harnessed for viral replication.

10.10 TLR Antagonists in Antiviral Therapy

Association of viral outbreaks with public health concern globally, there is a dire need for the development of safe and effective innate immune strategies to combat viral infections. Innate immunity delivers the first line of defense against viruses, especially TLRs, constitute critical components of innate immune pathways. Therefore, copious viruses evolved strategies to deploy TLR signaling to escape defensive responses of host. Agonist therapy might control viral infections during very early stage as they block virus attachment to cell surface via TLRs or by preventing TLR-mediated immune activation. Agonists are molecules that mimic the viral pattern recognition molecules and bind to TLRs to elicit both innate and adaptive immune response. Furthermore, agonist viral therapy may be useful for prophylactic or therapeutic treatment for viruses. Several agonists were successfully employed in preclinical and clinical settings against different viruses. Each TLR responds to diverse agonists, viz. bacterial and fungal components as well as viral RNA, except for TLR10, whose agonist is unknown (Lee et al. 2006). TLR3, 4, 7, 8, and 9 agonists were used in nonhuman primate models for DENV and hepatitis B virus (Chihab et al. 2019) and in murine models for HSV types-1 and -2. These agonists were discovered to reduce viral symptoms and replication. Conversely, TLR7 and TLR8 agonists were also reported to be used in treating viral infections along with inflammatory disorders. However, the first identified TLR guanosine analog Isatoribine has comparable chemical structural features in common with ribonucleoside ribavirin, component of current standard for HCV care therapy. TLR7 agonist Isatoribine may signify an important and unique approach for chronic HCV therapy (Xiang et al. 2007). Interestingly, both TLR7 (Imidazoquinoline) and TLR8 (R848 (resiquimod) agonists were used to treat HSV-2 induced psoriasis and genital lesions (Mark et al. 2007; Gotovtseva et al. 2008) and related imiquimod is used in treatment of genital warts caused by human papillomavirus.

Nonetheless, carboxy-terminal domain and amino-terminal amphipathic helix are indispensable for antiviral activity against HCV (Helbig et al. 2011) while SAM domain (Upadhyay et al. 2014) is obligatory for activity against TBEV and carboxy-terminal domain for DENV, respectively (Helbig et al. 2013). Notably, Viperin is another agonist that comprises carboxy-terminal domain, central domain with conserved CXXCXXC motif typical for radical SAM family of enzymes (SAM domain) and amino-terminal domain that mimics TLR ligands that acts as antiviral agents (Vanwalscappel et al. 2018). TLR7/8 agonist R848 is deemed as the most potent inhibitor and might be for prophylactic or therapeutic treatment of ZIKA. RNA-seq analysis recognized several genes that were strongly induced by R848 in monocytes. Viperin, an interferon-induced gene was identified to be active against several viruses, which also has the ability to inhibit ZIKV replication. Notably, the transduction of viperin lentiviral expression vector to microglial CHME3 cells rendered them resistant to ZIKA, preventing viral replication (synthesis of viral protein and RNA). CRISPR/Cas9 knockout of viperin in macrophages relieved block to infection, demonstrate the role of viperin. TLR agonists may be useful for prophylactic or therapeutic treatment for ZIKV (Vanwalscappel et al. 2018).

TLR9 agonists have been used for cancer and chronic infections for their ability to enhance both innate and adaptive antitumoral/antiviral responses. TLR9 Ligands agonists PF-3512676 and ISS-1018 have been tested in clinical trials as immune cancer therapeutics (Aillot et al. 2018). Low-molecular-weight mannogalactofucans (LMMGFs), a potent TLR2 agonist, block the interaction of viral components with TLR and suppress deleterious effects of TLR2 in response to HSV as seen in HSE cases (Jahanban-Esfahlan et al. 2019). GS-9688 is a potent and selective small molecule agonist of human TLR8, currently in Phase 2 trials for treatment of CHB (Chronic Hepatitis B infection) (Amin et al. 2019).

10.11 Conclusion and Future Perspectives

TLRs represent as primary sensors of innate immune response against viral pathogens. Priming of TLR results in the activation of cellular signaling pathways involved in both innate and adaptive immune responses. TLRs in mount protective immune system against infection and their cross talk among other PRRs in association with pathogen recognition. Triggering of TLRs 7, 8, and 9 prompts secretion of pro-inflammatory cytokines (IFN-α, IL-12, and TNF-α) that ensues in T_H1-like immune responses, important in antiviral and tumor immunity. TLR activation by topical administration of TLR7/8/9 agonists represents a powerful treatment modality for epithelial viral and neoplastic lesions. Currently, imiquimod represents the only TLR agonist licensed for the treatment of humans (e.g., CA, actinic keratosis, superficial basal cell carcinoma). Several other agonists of TLR7, 8, and 9 were shown to be effective therapeutic agents against infections, cancer, and allergic disorders, but are not yet approved for treatment in humans. Agonists of TLRs 4, 7, 8, and 9 were also used as adjuvants for prophylactic and therapeutic vaccination.

Acknowledgments The authors are grateful to NIV Pune and Krishna University, Machilipatnam for the support extended.

Conflict of Interest The authors declare that they have no competing interests.

References

Aillot L, Bonnin M, Ait-Goughoulte M, Bendriss-Vermare N, Maadadi S, Dimier L et al (2018) Interaction between toll-like receptor 9-CpG oligodeoxynucleotides and hepatitis B virus virions leads to entry inhibition in hepatocytes and reduction of alpha interferon production by plasmacytoid dendritic cells. Antimicrob Agents Chemother 62(4):e01741-17

Al-Anazi MR, Matou-Nasri S, Abdo AA, Sanai FM, Alkahtani S, Alarifi S et al (2017) Association of toll-like receptor 3 single-nucleotide polymorphisms and hepatitis C virus infection. J Immunol Res 2017:1590653

Amin OE, Colbeck E, Daffis S, Pattabiraman D, Sitali C, Rosenberg W et al (2019) In vitro modulation by TLR8 agonist GS-9688 of multiple regulatory cell types in patients with chronic hepatitis B. J Hepatol 70:E445

Aravalli RN, Hu S, Lokensgard JR (2008) Inhibition of toll-like receptor signaling in primary murine microglia. J Neuroimmune Pharmacol 3(1):5–11

Bourquin C, Schmidt L, Lanz AL, Storch B, Wurzenberger C, Anz D et al (2009) Immunostimulatory RNA oligonucleotides induce an effective antitumoral NK cell response through the TLR7. J Immunol 183(10):6078–6086

Budimir N, De Haan A, Meijerhof T, Waijer S, Boon L, Gostick E et al (2013) Critical role of TLR7 signaling in the priming of cross-protective cytotoxic T lymphocyte responses by a whole inactivated influenza virus vaccine. PLoS One 8(5):e63163

Chang WJ, Toledo-Pereyra LH (2012) Toll-like receptor signaling in liver ischemia and reperfusion. J Investig Surg 25(4):271–277

Chihab H, Zaidane I, Elhabazi A, Jadid FZ, El Fihri R, Elmessaoudi-Idrissi M, Chair M et al (2019) Toll-like receptor 9 polymorphisms and hepatitis B virus clearance in Moroccan chronic carriers. Gene 687:212–218

Cooper CL, Davis HL, Angel JB, Morris ML, Elfer SM, Seguin I et al (2005) CPG 7909 adjuvant improves hepatitis B virus vaccine seroprotection in antiretroviral-treated HIV-infected adults. AIDS 19(14):1473–1479

Dowling JK, Dellacasagrande J (2016) Toll-like receptors: ligands, cell-based models, and readouts for receptor action. Methods Mol Biol 1390:3–27

Evans JT, Cluff CW, Johnson DA, Lacy MJ, Persing DH, Baldridge JR (2003) Enhancement of antigen-specific immunity via the TLR4 ligands MPL™ adjuvant and Ribi. 529. Expert Rev Vaccines 2(2):219–229

Fischer J, Koukoulioti E, Schott E, Fülöp B, Heyne R, Berg T et al (2018) Polymorphisms in the toll-like receptor 3 (TLR3) gene are associated with the natural course of hepatitis B virus infection in Caucasian population. Sci Rep 8(1):12737

Gantier MP, Tong S, Behlke MA, Xu D, Phipps S, Foster PS et al (2008) TLR7 is involved in sequence-specific sensing of single-stranded RNAs in human macrophages. J Immunol 180(4):2117–2124

Gaudreault E, Fiola S, Olivier M, Gosselin J (2007) Epstein-Barr virus induces MCP-1 secretion by human monocytes via TLR2. J Virol 81(15):8016–8024

Gay NJ, Gangloff M (2007) Structure and function of toll receptors and their ligands. Annu Rev Biochem 76:141–165

Gotovtseva EP, Kapadia AS, Smolensky MH, Lairson DR (2008) Optimal frequency of imiquimod (aldara) 5% cream for the treatment of external genital warts in immunocompetent adults: a meta-analysis. Sex Transm Dis 35(4):346–351

Hackstein H, Hagel N, Knoche A, Kranz S, Lohmeyer J, Von Wulffen W et al (2012) Skin TLR7 triggering promotes accumulation of respiratory dendritic cells and natural killer cells. PLoS One 7(8):e43320

Hayashi T, Yao S, Crain B, Chan M, Tawatao RI, Gray C et al (2012) Treatment of autoimmune inflammation by a TLR7 ligand regulating the innate immune system. PLoS One 7(9):e45860

Heil F, Hemmi H, Hochrein H, Ampenberger F, Kirschning C, Akira S et al (2004) Species-specific recognition of single-stranded RNA via toll-like receptor 7 and 8. Science 303(5663):1526–1529

Helbig KJ, Eyre NS, Yip E, Narayana S, Li K, Fiches G, McCartney EM, Jangra RK, Lemon SM, Beard MR (2011) The antiviral protein viperin inhibits hepatitis C virus replication via interaction with nonstructural protein 5A. Hepatology 54(5):1506–1517

Helbig KJ, Carr JM, Calvert JK, Wati S, Clarke JN, Eyre NS, Narayana SK, Fiches GN, McCartney EM, Beard MR (2013) Viperin is induced following dengue virus type-2 (DENV-2) infection and has anti-viral actions requiring the C-terminal end of viperin. PLoS Negl Trop Dis 7(4):e2178

Hemmi H, Takeuchi O, Kawai T, Kaisho T, Sato S, Sanjo H et al (2000) A toll-like receptor recognizes bacterial DNA. Nature 408(6813):740

Isogawa M, Robek MD, Furuichi Y, Chisari FV (2005) Toll-like receptor signaling inhibits hepatitis B virus replication in vivo. J Virol 79(11):7269–7272

Jahanban-Esfahlan R, Seidi K, Majidinia M, Karimian A, Yousefi B, Nabavi SM et al (2019) Toll-like receptors as novel therapeutic targets for herpes simplex virus infection. Rev Med Virol 29(4):e2048

Kader M, Smith AP, Guiducci C, Wonderlich ER, Normolle D, Watkins SC et al (2013) Blocking TLR7-and TLR9-mediated IFN-α production by plasmacytoid dendritic cells does not diminish immune activation in early SIV infection. PLoS Pathog 9(7):e1003530

Karimi-Googheri M, Arababadi MK (2014) TLR3 plays significant roles against hepatitis B virus. Mol Biol Rep 41(5):3279–3286

Kawai T, Akira S (2011) Toll-like receptors and their crosstalk with other innate receptors in infection and immunity. Immunity 34(5):637–650

Krieg AM (2007) The toll of too much TLR7. Immunity 27(5):695–697

Latz E, Schoenemeyer A, Visintin A, Fitzgerald KA, Monks BG, Knetter CF et al (2004) TLR9 signals after translocating from the ER to CpG DNA in the lysosome. Nat Immunol 5(2):190

Le Goffic R, Pothlichet J, Vitour D, Fujita T, Meurs E, Chignard M et al (2007) Cutting edge: influenza a virus activates TLR3-dependent inflammatory and RIG-I-dependent antiviral responses in human lung epithelial cells. J Immunol 178(6):3368–3372

Lee HK, Dunzendorfer S, Soldau K, Tobias PS (2006) Double-stranded RNA-mediated TLR3 activation is enhanced by CD14. Immunity 24(2):153–163

Lin SC, Lo YC, Wu H (2010) Helical assembly in the MyD88–IRAK4–IRAK2 complex in TLR/IL-1R signalling. Nature 465(7300):885

Ma Z, Zhang E, Yang D, Lu M (2015) Contribution of toll-like receptors to the control of hepatitis B virus infection by initiating antiviral innate responses and promoting specific adaptive immune responses. Cell Mol Immunol 12(3):273

Mark KE, Corey L, Meng TC, Magaret AS, Huang ML, Selke S, Slade HB, Tyring SK, Warren T, Sacks SL, Leone P, Bergland VA, Wald A (2007) Topical resiquimod 0.01% gel decreases herpes simplex virus type 2 genital shedding: a randomized, controlled trial. J Infect Dis 195(9):1324–1331

Medvedev R (2017) Impact of oxidative stress by hepatitis C virus-mediated Nrf2 inhibition on autophagy and viral particle release (Doctoral dissertation, Johann Wolfgang Goethe-Universität Frankfurt am Main)

Murawski MR, Bowen GN, Cerny AM, Anderson LJ, Haynes LM, Tripp RA et al (2009) Respiratory syncytial virus activates innate immunity through toll-like receptor 2. J Virol 83(3):1492–1500

Nicodemus CF, Berek JS (2010) TLR3 agonists as immunotherapeutic agents. Immunotherapy 2(2):137–140

Pasare C, Medzhitov R (2005) Control of B-cell responses by toll-like receptors. Nature 438(7066):364

Preiss S, Thompson A, Chen X, Rodgers S, Markovska V, Desmond P, Visvanathan K, Li K, Locarnini S, Revill P (2008) Characterization of the innate immune signalling pathways in hepatocyte cell lines. J Viral Hepat 15(12):888–900

Rosadini CV, Zanoni I, Odendall C, Green ER, Paczosa MK, Philip NH et al (2015) A single bacterial immune evasion strategy dismantles both MyD88 and TRIF signaling pathways downstream of TLR4. Cell Host Microbe 18(6):682–693

Rudd BD, Burstein E, Duckett CS, Li X, Lukacs NW (2005) Differential role for TLR3 in respiratory syncytial virus-induced chemokine expression. J Virol 79(6):3350–3357

Sajadi SMA, Mirzaei V, Hassanshahi G, Khorramdelazad H, Daredor HY, Hosseini SMH et al (2013) Decreased expressions of toll-like receptor 9 and its Signaling molecules in chronic hepatitis B virus–infected patients. Arch Pathol Lab Med 137(11):1674–1679

Sarvestani ST, Williams BR, Gantier MP (2012) Human toll-like receptor 8 can be cool too: implications for foreign RNA sensing. J Interf Cytokine Res 32(8):350–361

Sato A, Linehan MM, Iwasaki A (2006) Dual recognition of herpes simplex viruses by TLR2 and TLR9 in dendritic cells. Proc Natl Acad Sci 103(46):17343–17348

Song J, Hu Y, Li J, Zheng H, Wang J, Guo L et al (2018) Suppression of the toll-like receptor 7-dependent type I interferon production pathway by autophagy resulting from enterovirus 71 and coxsackievirus A16 infections facilitates their replication. Arch Virol 163(1):135–144

Srivastava T, Sharma M, Yew KH, Sharma R, Duncan RS, Saleem MA et al (2013) LPS and PAN-induced podocyte injury in an in vitro model of minimal change disease: changes in TLR profile. J Cell Commun Signal 7(1):49–60

Takeda K, Akira S (2004) TLR signaling pathways. Semin Immunol 16(1):3–9

Thompson AJ, Colledge D, Rodgers S, Wilson R, Revill P, Desmond P et al (2009) Stimulation of the interleukin-1 receptor and toll-like receptor 2 inhibits hepatitis B virus replication in hepatoma cell lines in vitro. Antivir Ther 14(6):797–808

Thompson MR, Kaminski JJ, Kurt-Jones EA, Fitzgerald KA (2011) Pattern recognition receptors and the innate immune response to viral infection. Viruses 3(6):920–940

Triantafilou K, Orthopoulos G, Vakakis E, Ahmed MAE, Golenbock DT, Lepper PM et al (2005) Human cardiac inflammatory responses triggered by Coxsackie B viruses are mainly toll-like receptor (TLR) 8-dependent. Cell Microbiol 7(8):1117–1126

Upadhyay AS, Vonderstein K, Pichlmair A, Stehling O, Bennett KL, Dobler G, Guo JT, Superti-Furga G, Lill R, Overby AK, Weber F (2014) Viperin is an iron-sulfur protein that inhibits genome synthesis of tick-borne encephalitis virus via radical SAM domain activity. Cell Microbiol 16(6):834–848

Vanwalscappel B, Tada T, Landau NR (2018) Toll-like receptor agonist R848 blocks Zika virus replication by inducing the antiviral protein viperin. Virology 522:199–208

Wang T, Town T, Alexopoulou L, Anderson JF, Fikrig E, Flavell RA (2004) Toll-like receptor 3 mediates West Nile virus entry into the brain causing lethal encephalitis. Nat Med 10(12):1366

Xiang AX, Webber SE, Kerr BM, Rueden EJ, Lennox JR, Haley GJ et al (2007) Discovery of ANA975: an oral prodrug of the TLR-7 agonist isatoribine. Nucleosides Nucleotides Nucleic Acids 26(6–7):635–640

Yamamoto M, Takeda K, Akira S (2004) TIR domain-containing adaptors define the specificity of TLR signaling. Mol Immunol 40(12):861–868

Zhang X, Kraft A, Broering R, Schlaak JF, Dittmer U, Lu M (2012) Preclinical development of TLR ligands as drugs for the treatment of chronic viral infections. Expert Opin Drug Discovery 7(7):597–611

Zhang Y, El-Far M, Dupuy FP, Abdel-Hakeem MS, He Z, Procopio FA et al (2016) HCV RNA activates APCs via TLR7/TLR8 while virus selectively stimulates macrophages without inducing antiviral responses. Sci Rep 6:29447

Zhao Y, Chang J, Zhang B, Tong P, Wang C, Ran D et al (2019) TLR-5 agonist Salmonella abortus equi flagellin FliC enhances FliC-gD-based DNA vaccination against equine herpesvirus 1 infection. Arch Virol 164(5):1371–1382

Zhou Y, Ye L, Wan Q, Zhou L, Wang X, Li J et al (2009) Activation of toll-like receptors inhibits herpes simplex virus-1 infection of human neuronal cells. J Neurosci Res 87(13):2916–2925

Activation of Complement System During Viral Infections: Prospects and Future Challenges

11

Prudhvi Lal Bhukya and Pallaval Veera Bramhachari

Abstract

The complement system is homeostatic system evolved to remain constant check on pathogen but as per the recent knowledge it is involved in many biological processes including complementing adaptive immunity. It gets activated in most of the viral infections and leads to neutralization of virus via opsonization of C3b, aggregation, phagocytosis, membrane attack complex (MAC) mediated lysis of virus or virus-infected cells. Primary work of complement in viral diseases clears the virus by MAC or by opsonization nonetheless it offers favorable milieu locally during localized infection. It increases vascular permeability, generates edema, recruits phagocytes by chemotaxis, mediates release of cytokines depending on the cell type. This acute inflammation generated due to local activation of complement in initial stage of infection is crucial. C3a and C5a anaphylatoxins modulate adaptive immunity generation via modulating priming of T cells and enhancing Th1 immunity. In the absence of certain complement components and its receptors, some viruses become more pathogenic than in general, denote complement functions at more than one step in different viruses of their pathogenesis.

Keywords

Complement components · Complement activation · Viral infection · Pathogenesis

P. L. Bhukya (✉)
Hepatitis Division, National Institute of Virology, Pune, Maharashtra, India

P. V. Bramhachari
Department of Biotechnology, Krishna University,
Machilipatnam, Andhra Pradesh, India

P. V. Bramhachari (ed.), *Dynamics of immune activation in viral diseases*,
https://doi.org/10.1007/978-981-15-1045-8_11

11.1 Introduction

Complement system is the first line of defense system presumed to be evolved in invertebrates by gene duplication and diversions (Smith et al. 1999). Conventionally it is considered as an innate immune system as it keeps constant vigilance on pathogens and killing of pathogens in body fluids. However, recent research studies highlighted its function as a link between innate and adaptive immune systems (Dunkelberger and Song 2010). It also acts as critical mediator for clearance of immune complexes and injured cells. Complement system is versatile and works apart from immune functions like maturation of neuronal synapse, angiogenesis, tissue repair and regeneration, immune complex clearance, and in lipid metabolism. This versatile function might be because it coevolved with the immunity and other cellular pathways (Ricklin et al. 2010).

Indeed complement contributes to pathogen clearance, moreover aiding adaptive immunity against viruses, especially C3d fragment linked with antigen increases antibody response basically by mounting enhanced germinal center assembly. Complement also augments a good $CD4^+$ and $CD8^+$ T cell response against viruses through signaling of complement receptors on these cells and enhanced antigen presentation function by Antigen Presenting cells (APCs) (Kolev et al. 2014). However, on the other side viruses either cleared by complement or viruses use complement for invading inside cells through complement receptors moreover some viruses suppress synthesis of complement or code/acquire host complement regulatory proteins (Agrawal et al. 2017). This chapter emphasizes on the protective role of complement during viral infections.

11.2 Complement System

Complement system is a group of soluble and cell surface proteins that starts the series of activation cascade of proteins against the inflammatory signal. These proteins are present in all vertebrates in inactive form zymogen while activated form is serine proteases. However, a host cell does not activate complement due to presence of complement regulatory proteins present on their surface. There are mainly three major pathways by which complement system gets activated. These are classical pathway, alternate pathway, lectin pathway; converge at single step of C3 convertase formation, and cleavage of C3 component.

Classical pathway (CP) is antibody dependent since it needs antibody for its activation, upon binding of IgG or IgM to the antigen C1 complex can bind to antibody. C1q multimeric protein possesses globular heads that bind to Fc portion of antibody while C1r and C1s subsequently get activated and further cleaves C4 and C2. The activation product C4bC2a forms classical pathway C3 convertase leaving away C4a and C2b. The alternate pathway (AP) activated by tick over mechanism of C3, that is, spontaneous hydrolysis of C3 into $C3H_2O$ further binds to foreign surface. Factor B binds to bound $C3H_2O$ followed by factor D that cleaves factor B into Ba and Bb, this generates alternate pathway C3 convertase C3bBb. Properdin

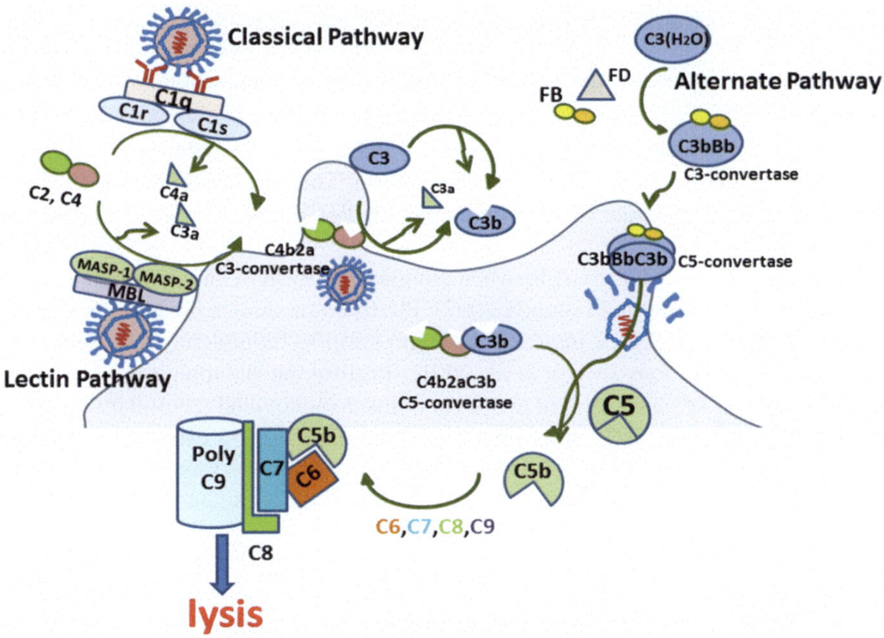

Fig. 11.1 The complement system activation during viral infection, classical alternate, and lectin pathway

binding stabilizes alternate pathway convertase. Alternate pathway loop continues C3b formation for coating pathogen leading either neutralization or phagocytosis by phagocytic cells (Fig. 11.1).

The lectin pathway (LP) activated in the presence of carbohydrate moieties such as mannose, complement proteins mannose-binding lectin and ficolin. Although there are four types of MASPs (MBL-associated proteins) MASP1, 2, 3, and MAP 19.

Only MASP2 is capable to cleave C4 and C2when it is associated with MBL. C3 convertase is the same for classical and lectin pathway C4bC2a (Fig. 11.1). Research reports on the role of MASP1, MASP3, and MAP 19 are not clearly documented. However only MASP1 can cleave C2 for the matter of fact that, this step increases activation of lectin pathway in presence of pathogen carbohydrate moieties.

The C3 convertase formed by CP, LP, and AP cleaves C3 into C3a and C3b. C3b generated can bind on pathogen surface for clearance while few C3b molecules bind to C3 convertase to form C5 convertase (C3bBbC3b) AP or (C4bC2aC3b) CP and LP. C5 convertase cleaves C5 into C5a and C5b, which binds away from C5 convertase. C3a and C5a generated during complement activation act as anaphylatoxins. C6, C7, and C8 bind to C5b to form C6C7C8 complex also referred as a nascent MAC (Membrane Attack Complex), binding of poly C9 completes the MAC. Around 22 C9 molecules bound to C5bC6C7C8 forms pore of approx. 120-A^0 size on the cell membrane of pathogen leads to lysis of pathogen (Dudkina et al. 2016).

11.3 Downstream Effects of Complement Activation

Activation of complement increases vascular permeability helps the recruitment of immune cells, as well as chemotactic cells by secretion of anaphylatoxins during virus infection in a local milieu. Infected cells and immune cells generate a number of different cytokines also called a cytokine storm. This acute inflammation generated due to local activation of complement in the initial stage of infection is crucial.

Reciprocal interactions exist between complement system and proinflammatory cytokines (Markiewski and Lambris 2007). Plerthora of studies have reported that, proinflammatory cytokines increase the expression of complement receptors in inflammatory conditions (Mäck et al. 2001). In different pathophysiological scenarios, the effect of generation of anaphylatoxins does not lead to similar outcome, for example, in hepatitis B virus infection generation of C5a eventually leads to fulminant hepatitis condition while in the case of influenza virus infection, C3 is necessary for recovery from infection, whereas c5a does not seem to have a protective role in this case (Xu et al. 2014).

11.4 Cross Talk Between Complement and Adaptive Immunity

C3b when cleaved into iC3b and C3dg by factor I (Atkinson et al. 2018), C3dg binding to antigen interacts with cr2 receptor on B cells which is B cell co-receptor and it has very strong costimulatory effect on B cells (Roozendaal and Carroll 2007) CR2-mediated signaling helps B cells to survive in germinal center (81) it helps DC for long retention of antigen thereby it contributes to enhanced B cell memory (Fischer et al. 1998) C3a and C5a has a regulatory effects on B cells such as C3a inhibits polyclonal response and C5a promotes migration of B cells to the site of complement activation (Burg et al. 1995; Fischer and Hugli 1997; Ottonello et al. 1999).

In the absence of C3a-C3aR signaling, DCs fail to induce potent CD4 T cell response against antigen (Sacks 2010) anaphylatoxins provide a co-stimulatory and survival signal to T cells (Strainic et al. 2008). Anaphylatoxins' interaction with their respective receptors has a very important role which decides the consequence of APC and T cell interaction, induction of Th1, Th2, Th17, and T reg cells (Strainic et al. 2013). Especially it has been shown that complement enhances CD4 and CD8 T cell response against virus infection through anaphylatoxin receptors (Kolev et al. 2014). In addition, complement also enhances CD8 T cell immunity in lymphocytic choriomeningitis virus infection. (Fang et al. 2007). These mechanisms composed of direct engagement between complement proteins with CRs on T cells, while indirect regulation via APC engagement, and alteration of cytokine profiles through CR–TLR cross talk.

11.5 Conclusions and Future Perspectives

Complement pathway is reported and accepted to a mediator/connecting link among adaptive and innate immunity. However, C3d–CR2 role in monitoring Innate immunity is long known, while recent studies have established complex approach that complement drives T cell responses. To date, it was expected that complement pathway occurs only extracellular milieu, therefore, it is able for opsonization of viruses only outside cell. However, it is now clear that major complement activation proteins like C3 and C5 can be cleaved inside cell in a noncanonical way, which suggests viruses opsonization with C3b may occur even inside the cell. Recent studies elegantly documented that cleavage of C3 and C5 proteins inside T cell, subsequently engages respective receptors for their cleaved products in an autocrine fashion control T_H1 induction. Therefore, intracellularly generated C3a and C5a fragments are anticipated to play predominant role in producing effective defensive response against viruses. However, numerous subversion mechanisms of virus evade complement-mediated adaptive immunity have been depicted. Conversely, many evasion strategies of viruses remain undiscovered. This raises the question—how do T cell tropic viruses manipulate the noncanonical complement activation inside the cell for their survival? And more importantly, in general, how viruses regulate intracellular complement activation? Better knowledge on unknown evasion mechanisms devised by viruses would not only add to our knowledge yet prerequisite for capitalizing rational vaccine design and therapeutic potential.

Acknowledgments The authors are grateful to NIV, Pune and Krishna University, Machilipatnam for the support extended.

Conflict of Interest The authors declare that they have no competing interests.

References

Agrawal P, Nawadkar R, Ojha H, Kumar J, Sahu A (2017) Complement evasion strategies of viruses: an overview. Front Microbiol 8:1117. https://doi.org/10.3389/fmicb.2017.01117

Atkinson JP, Du Clos TW, Mold C, Kulkarni H, Hourcade D, Wu X (2018) 21 - the human complement system: basic concepts and clinical relevance. In: Clinical Immunology, 5th edn, pp 299–317.e1. https://doi.org/10.1016/B978-0-7020-6896-6.00021-1

Burg M, Martin U, Rheinheimer C, Köhl J, Bautsch W, Böttger EC, Klos A (1995) IFN-gamma up-regulates the human C5a receptor (CD88) in myeloblastic U937 cells and related cell lines. J Immunol 155(9):4419–4426

Dudkina NV, Spicer BA, Reboul CF, Conroy PJ, Lukoyanova N, Elmlund H, Law RH, Ekkel SM, Kondos SC, Goode RJ, Ramm G, Whisstock JC, Saibil HR, Dunstone MA (2016) Structure of the poly-C9 component of the complement membrane attack complex. Nat Commun 7:10588. https://doi.org/10.1038/ncomms10588

Dunkelberger JR, Song WC (2010) Complement and its role in innate and adaptive immune responses. Cell Res 20(1):34–50. https://doi.org/10.1038/cr.2009.139

Fang C, Miwa T, Shen H, Song WC (2007) Complement-dependent enhancement of CD8+ T cell immunity to lymphocytic choriomeningitis virus infection in decay-accelerating factor-deficient mice. J Immunol 179(5):3178–3186. Erratum in: J Immunol. 2007 Nov 15;179(10):7185
Fischer WH, Hugli TE (1997) Regulation of B cell functions by C3a and C3a(desArg): suppression of TNF-alpha, IL-6, and the polyclonal immune response. J Immunol 159(9):4279–4286
Fischer MB, Goerg S, Shen L, Prodeus AP, Goodnow CC, Kelsoe G, Carroll MC (1998) Dependence of germinal center B cells on expression of CD21/CD35 for survival. Science 280(5363):582–585
Kolev M, Le Friec G, Kemper C (2014) Complement--tapping into new sites and effector systems. Nat Rev Immunol 14(12):811–820. https://doi.org/10.1038/nri3761
Mäck C, Jungermann K, Götze O, Schieferdecker HL (2001) Anaphylatoxin C5a actions in rat liver: synergistic enhancement by C5a of lipopolysaccharide-dependent alpha(2)-macroglobulin gene expression in hepatocytes via IL-6 release from Kupffer cells. J Immunol 167(7):3972–3979
Markiewski MM, Lambris JD (2007) The role of complement in inflammatory diseases from behind the scenes into the spotlight. Am J Pathol 171(3):715–727
Ottonello L, Corcione A, Tortolina G, Airoldi I, Albesiano E, Favre A, D'Agostino R, Malavasi F, Pistoia V, Dallegri F (1999) rC5a directs the in vitro migration of human memory and naive tonsillar B lymphocytes: implications for B cell trafficking in secondary lymphoid tissues. J Immunol 162(11):6510–6517
Ricklin D, Hajishengallis G, Yang K, Lambris JD (2010) Complement: a key system for immune surveillance and homeostasis. Nat Immunol 11(9):785–797. https://doi.org/10.1038/ni.1923
Roozendaal R, Carroll MC (2007) Complement receptors CD21 and CD35 in humoral immunity. Immunol Rev 219:157–166
Sacks SH (2010) Complement fragments C3a and C5a: the salt and pepper of the immune response. Eur J Immunol 40(3):668–670. https://doi.org/10.1002/eji.201040355
Smith LC, Azumi K, Nonaka M (1999) Complement systems in invertebrates. The ancient alternative and lectin pathways. Immunopharmacology 42(1–3):107–120
Strainic MG, Liu J, Huang D, An F, Lalli PN, Muqim N, Shapiro VS, Dubyak GR, Heeger PS, Medof ME (2008) Locally produced complement fragments C5a and C3a provide both costimulatory and survival signals to naive CD4+ T cells. Immunity 28(3):425–435. https://doi.org/10.1016/j.immuni.2008.02.001
Strainic MG, Shevach EM, An F, Lin F, Medof ME (2013) Absence of signaling into $CD4^+$ cells via C3aR and C5aR enables autoinductive TGF-β1 signaling and induction of $Foxp3^+$ regulatory T cells. Nat Immunol 14(2):162–171. https://doi.org/10.1038/ni.2499
Xu GL, Chen J, Yang F, Li GQ, Zheng LX, Wu YZ (2014) C5a/C5aR pathway is essential for the pathogenesis of murine viral fulminant hepatitis by way of potentiating Fgl2/fibroleukin expression. Hepatology 60(1):114–124. https://doi.org/10.1002/hep.27114

MicroRNAs and Their Role in Viral Infection

12

Divya Tiraki

Abstract

MicroRNAs (miRNAs) are small noncoding RNA molecules that act as central regulators of gene expression. miRNAs have diverse functions in physiology. As regulatory molecules, miRNAs are involved in the manifestation of various viral diseases in humans. Viruses can take advantage of host miRNAs to facilitate their survival and replication while, for host cells, miRNAs can serve as direct antiviral entities. miRNAs modulate viral pathogenesis either by directly binding to viral genomes or by modulating cellular antiviral responses. Circulating miRNAs are increasingly being recognized as an emerging class of disease biomarkers. These miRNAs exhibit a consistent expression profile among healthy individuals. Another level of complexity is added by the miRNAs encoded by the viral genomes using the host RNAi machinery. These vir-miRNAs help in the viral infection establishment. In this chapter, we discuss in brief about what are miRNAs, their functional role, different viral miRNAs, and cellular miRNAs involved in the pathogenesis of classical viral infections.

Keywords

MicroRNAs · Virus · Replication · Cellular miRNA · Viral miRNA

12.1 What Are miRNAs?

MicroRNAs (miRNAs) are 21–23 nucleotide small noncoding RNA molecules that post-transcriptionally regulate gene expression by binding to complementary sequences on target mRNA transcripts (mRNAs) resulting in translational repression

D. Tiraki (✉)
Department of Communicable Diseases, Interactive Research School for Health Affairs (IRSHA), Bharati Vidyapeeth Deemed University, Pune, Maharashtra, India

P. V. Bramhachari (ed.), *Dynamics of immune activation in viral diseases*,
https://doi.org/10.1007/978-981-15-1045-8_12

or target degradation (Girard et al. 2008; Bartel 2009). miRNAs have attracted immense attention during the last decade as tiny regulators with profound impact on eukaryotic gene expression (Scaria et al. 2007). During the last decade, miRNA research has gone from discovering the existence of miRNAs in mammals to exploring their therapeutic applications in numerous diseases.

miRNAs are well conserved in eukaryotic organisms and are vital and evolutionarily ancient component of genetic regulation (Tanzer and Stadler 2004). miRNAs can specifically bind to 3′UTRs (Grimson et al. 2007), 5′UTR (Lytle et al. 2007), or the coding region of mRNA transcripts (Friedman et al. 2008). miRNAs have diverse functions in physiology; ranging from cell differentiation, proliferation, apoptosis to the endocrine system, hematopoiesis, limb morphogenesis, fat metabolism, etc. (Girard et al. 2008). They display different expression profiles from tissue to tissue, reflecting diversity in cellular phenotypes. A direct role of miRNAs in immune response also has been demonstrated (Gantier et al. 2007). miRNAs are generally tissue specific and cell associated, however, they have also been found in the extracellular milieu (circulating miRNAs). Circulating miRNAs in serum and plasma are increasingly being recognized as an emerging class of potentially useful disease biomarkers readily detectable in the blood (Wang et al. 2009; Zhang et al. 2010a, b).

12.2 Biogenesis of miRNAs in Animals

The biogenesis of miRNA is a complex multistep process, which begins in the nucleus, goes through several post-transcriptional modifications, and terminates in the cytoplasm. A majority of miRNAs are derived from long double-stranded RNAs (dsRNAs), which are sequentially cleaved into shorter intermediates by specialized ribonuclease III (RNase III) enzymes that partner with dsRNA-binding proteins (Axtell et al. 2011; Kim et al. 2009). Mature miRNAs are found as one arm of a hairpin structure within RNA polymerase II (Pol II) transcripts called pri-miRNAs, which are characteristically longer, capped, and polyadenylated (Cai et al. 2004; Lee et al. 2004).

RNA Pol II generates long pri-miRNAs either from introns of protein-coding genes or independent genomic transcription units (Cai et al. 2004; Corcoran et al. 2009; Lee et al. 2004). Drosha, a nuclear RNase III enzyme, together with its dsRNA-binding domain (dsRBD) partner DGCR8 further processes the pri-miRNAs into pre-miRNAs (~70 nt stem–loop structures) (Denli et al. 2004; Han et al. 2004; Lee et al. 2002, 2003). A RanGTP-dependent Exportin5 transporter then exports this pre-miRNA hairpin to the cytoplasm (Bohnsack et al. 2004; Yi et al. 2003; Zeng and Cullen 2004). In the cytoplasm, second enzymatic processing is brought about by Dicer (RNase III enzyme) in association with its dsRBD partner TRBP, liberating a ~22 nt dsRNA duplex (miRNA–miRNA*) from the pre-miRNA (Bernstein et al. 2001; Hutvágner et al. 2001; Ketting et al. 2001).

These duplexes contain the mature miRNA bound to its complement, miRNA*. The dsRNA duplex is subsequently loaded onto an Argonaute (Ago 1–4) protein, forming an effector complex known as the RNA-induced silencing complex (RISC) (Hammond et al. 2000; Mourelatos et al. 2002). The orientation of loading onto the RISC is determined by the thermodynamic stability present at the 5′ end of the mature miRNA (Khvorova et al. 2003). The strand which is relatively more unstable base pairs at the 5′end. This helps in the preferential usage of the mature miRNA strand whereas the passenger strand miRNA*, which remains unused is degraded. In the effector complex, miRNAs bind the host mRNAs through the sequence complementarity at the 3′ UTR.

12.3 Interplay of miRNAs with Their Targets

A single miRNA may control the expression of more than one target mRNAs. The specificity required for the binding interactions is attributed to the 5′ region of miRNA. The target interactions are restricted to a region confined to ~6–8 nt sequence at the 5′ end of the miRNA, known as the "seed sequence" (Wee et al. 2012). This seed sequence is highly conserved among different species and change in the sequence alters the target binding (Ameres et al. 2007). The loops formed in the miRNA:mRNA duplexes may also have a profound effect on the efficiency of miRNA-mediated gene regulation (Ye et al. 2008). The properties associated with the target:miRNA interactions such as sequence, structure-associated free energy, and evolutionary conservation are used in the current target prediction algorithms (Doench and Sharp 2004; Martin et al. 2007).

12.4 Effector Mechanisms of miRNAs

miRNAs direct the RISC to downregulate gene expression by one of the two post-transcriptional mechanisms; mRNA cleavage or translational repression. The degree of miRNA: mRNA complementarity determines which mechanism is followed. Higher degree of complementarity directs toward the Ago-catalyzed degradation of target miRNA sequences through the mRNA cleavage mechanism. Conversely, a central mismatch bypasses degradation and facilitates the translational repression process.

12.5 Types of miRNAs

miRNAs are often tissue specific and cell-associated, however, they are also found in the extracellular milieu. Hence, based on their location in the body, they can be classified as (a) cell-associated miRNAs and (b) circulating miRNAs (cell-free miRNAs). Based on their origin, miRNAs are also classified as (a) host-encoded miRNAs and (b) virus-encoded miRNAs.

12.5.1 Cell-Associated miRNAs

All miRNAs originate in particular cells/tissues and play a vital role in the regulation of cellular processes there. For example, miR-122, a liver-specific miRNA is abundantly present in the liver and constitutes ~70% of total miRNA pool. Cell-associated miRNAs play a role in the pathogenesis of diseases, which is attributed to their dysregulation in the tissue of their origin. Thus, these miRNAs can act as biomarkers or indicators of diseases. For example, miR-122 has been reported as an indicator of liver disease progression (Girard et al. 2008), miR-208 acts as a predictive marker for myocardial injury (Ji et al. 2009).

12.5.2 Circulating miRNAs

Cell-free miRNAs that are found circulating in the extracellular milieu are called as circulating miRNAs. In 2008, Lawrie et al. first reported the presence of miRNAs in serum (Lawrie et al. 2008). Cell-free miRNAs are not only detected in plasma and serum, but also in body fluids such as urine, tears, breast milk, amniotic fluid, pleural fluid, cerebrospinal fluid, and saliva (Chen et al. 2008; Weber et al. 2010). Due to their remarkable stability and ease of access, circulating miRNAs serve as an important tool as a biomarker of various diseases (Mitchell et al. 2008). These cell-free miRNAs act as mediators for cell–cell communication and immune regulation (Valadi et al. 2007). Altered levels of circulating miRNAs have been reported in different pathological conditions (Laterza et al. 2009; Mitchell et al. 2008; Mo et al. 2012; Nogales-Gadea et al. 2014; Zhang et al. 2010a, b). The origin of miRNAs is diverse and microRNAs are released in circulation by blood cells (Hunter et al. 2008); organs of the body (Laterza et al. 2009) and tumors (Mo et al. 2012). Circulating miRNAs are widely used as biomarkers in a variety of pathological conditions and few of them have been summarized in Table 12.1. As biomarkers, these miRNAs help in various ways like tracking the progression of the disease, as a prognostic tool, and as a diagnostic tool.

Table 12.1 Circulating miRNAs as biomarkers

miRNA	Pathological condition	References
miR-122 and miR-155	Inflammatory biomarkers for different liver injuries	(Wang et al. 2009; Bala et al. 2012)
miR-25 and miR-223	Lung cancer	(Chen et al. 2008)
miR-184	Squamous cell carcinoma	(Wong et al. 2008)
miR-92a	Leukemia	(Tanaka et al. 2009)
miR-20 and miR-92a	Predictive biomarkers in HCV liver disease	(Shrivastava et al. 2013)
miR-21	B lymphoma	(Lawrie et al. 2008)
miR-208	Myocardial injury	(Ji et al. 2009)

12.5.3 Host-Encoded miRNAs

This class of miRNAs includes all the miRNAs that originate in the mammalian system through the canonical miRNA biogenesis pathways. These include both the cell-associated and circulating miRNAs discussed above.

12.5.4 Virus-Encoded miRNAs

Several studies have reported the existence of virus-encoded miRNAs (vmiRNAs) in DNA and RNA viruses. The biogenesis of these miRNAs is mainly dependent on host proteins and enzymes. Most of the viruses generate viral pri-miRNAs using host RNA polII, which are further processed to pre-miRNAs by host microprocessor complex. Viral miRNAs sharing homology with host miRNAs are capable of cheating host cells and directly takeover the regulatory pathways of host miRNAs. The host–virus interactions directed by viral miRNAs have been studied extensively in DNA viruses, like Human cytomegalovirus (HCMV) (Grey et al. 2007), Herpes simplex virus-1 (HSV-1) (Flores et al. 2013; Umbach et al. 2008). HCMV-encoded miR-UL112-1 targets the 3′UTR of MICB, vital for NK cell-mediated destruction of virus infected cells. HSV-1-encoded miR-H2-3p and miR-H3 target ICP0 and ICP34.5 genes, thereby maintaining the latency. The examples of RNA viruses that encode miRNAs are West Nile virus (WNV): KUN-miR-1 (Hussain et al. 2012), Dengue virus (DENV): DENV-vsRNA-5 (Hussain and Asgari 2014), enterovirus 71: vsRNA1 (Weng et al. 2014) and Hepatitis A virus (HAV): hav-miR-1-5p, hav-miR-2-5p, and hav-miR-N1-3p (Shi et al. 2014a, b).

12.6 miRNAs in Viral Infections

A growing body of evidence suggests a significant role of cellular miRNA-mediated RNAi in intricate networks of host–virus interactions. Lecellier et al. reported the first antiviral miRNA in human cells capable of restricting accumulation of reovirus primate foamy virus type 1 (PFV-1) (Lecellier et al. 2005). Cellular as well as viral miRNAs have been reported to be involved in the pathogenesis of viruses such as hepatitis viruses (Hepatitis A virus (HAV), Hepatitis B virus (HBV), Hepatitis C virus (HCV), Hepatitis E virus (HEV)), West Nile Virus (WNV), Measles virus (MeV), Influenza A virus (IAV), Human Immunodeficiency Virus-1 (HIV-1), Dengue (DENV), and Herpes Simples virus-1 (HSV-1) (Bogerd et al. 2014).

Multiple lines of evidence have confirmed the ability of miRNAs to modulate antiviral activity through two mechanisms:

1. *miRNAs directly exert an antiviral function by interacting with the genome.* Computational and experimental approaches predict UTRs, particularly 3′UTR to be a more common miRNA target. However, few reports have shown that miRNAs can bind to 5′UTR as well. Few examples for virus–miRNA interactions

are 5 IFN-inducible miRNAs, miR-196,-296, -351, -431, and -448 target HCV genome and inhibit replication (Pedersen et al. 2007), miR-23 inhibits PRRSV replication by targeting the PRRSV RNAs (Zhang et al. 2014a, b), let 7c inhibits H1N1 by targeting the 3′UTR of viral M1 gene (Ma et al. 2012).
2. *miRNAs modulate host factors to provide a less conducive environment for virus replication.* Examples are miR-146 that is involved in the pathogenesis of Dengue virus (Wu et al. 2013), Chikungunya (Selvamani et al. 2014), and VSV (Hou et al. 2009), miR-23 that inhibits PRRSV replication by upregulating type 1 IFNs (Zhang et al. 2014a, b), miR-30e*, which suppresses DENV replication through NF-κB-dependent IFN production (Zhu et al. 2014a), miR-21 attenuates human cytomegalovirus (HCMV) replication by targeting Cdc25a, a cell cycle regulator (Fu et al. 2014), miR-23a facilitates HSV-1 replication by suppressing IFN regulatory factor 1 (IRF1) (Ru et al. 2014), miR-15b modulates HBV replication by targeting HNF1α (Dai et al. 2014). Qian et al. have reported the role of miR-122 in Borna virus-induced disease. miR-122 exerted a direct antiviral function by translation and replication inhibition and also by indirectly acting through IFN to increase the host immunity (Qian et al. 2010).

12.6.1 Regulation of Viral Infection by Viral miRNAs

Viral miRNAs have been reported in Human Immunodeficiency Syndrome (HIV), Polyomaviruses, Herpesviruses, and Adenoviruses. Over 90% of these miRNAs are found to be conserved between related viral species. Some miRNAs make use of the cellular miRNA pathway to regulate the viral gene expression and few others also have cellular targets.

12.6.1.1 HIV miRNAs

Owing to the presence of secondary structures in the viral genome, there is a high chance that HIV-1 may encode miRNAs. Omoto et al. showed that HIV-1 encodes miR-N367 in the Nef/LTR overlapping region, which targets the negative response element of the LTR U3 region and inhibits the LTR's promoter activity (Omoto et al. 2004). Another miRNA, miR-H1 downregulates the expression of apoptosis antagonizing transcription factor (AATF) (Kaul et al. 2009). Next, Ouellet et al. reported miR-TAR-5p/3p in HIV-1 infected cell lines to be involved in the asymmetrical processing of the HIV-1 TAR element by Dicer (Ouellet et al. 2008). They also prevented apoptosis of infected cells thereby promoting survival and replication by targeting host genes nucleophosmin 1 (NPM), caspase 8 (*CASP8*), Ikaros family Zinc Finger 1 (*IKZF1*), and Ikaros family Zinc Finger 3 (*Aiolos*). Further, miR-H3 targeting TATA box in the viral 5′ LTR promoter was reported (Zhang et al. 2014a, b).

12.6.1.2 Polyomavirus miRNAs

These viruses are capable of establishing persistent infections in humans and transforming the infected cells. SV40, a monkey polyomavirus encodes miRNAs that regulate viral pathogenesis. SV40 encodes a single functional pre-miRNA which is antisense to the SV40 early mRNA. It is involved in evasion of immune responses by cleaving the large T-antigen mRNA (Sullivan et al. 2009). Several similar miRNAs have been reported in other polyomaviruses like BK polyomavirus (BKPyV), JC polyomavirus (JCPyV), simian agent 12 (SA12), Murine polyomavirus (MPyV), and Markel cell polyomavirus (MCPyV) (Seo et al. 2008, 2009; Sullivan et al. 2009).

12.6.1.3 Herpesvirus miRNAs

This is a DNA virus causing persistent infection. Few of these herpesviruses and their encoded miRNAs are discussed in brief in below sections.

12.6.1.3.1 Epstein-Barr Virus (EBV)

It encodes five nonhomologous miRNAs as identified in the B cell line latently infected with EBV (Pfeffer et al. 2004). These are located in two major regions: BHFR1 (developmental stage specific) and BART (tissue specific) genes. These may target chemokines, cytokines, apoptotic, and cell growth control genes.

12.6.1.3.2 Kaposi's Sarcoma-Associated Herpesvirus (KSHV)

A cluster of 11 Kaposi's Sarcoma-Associated Herpesvirus (KSHV) miRNAs have been identified using RACE-like cloning from a single 4.5-kb region of the genome (Cai et al. 2006; Pfeffer et al. 2005; Samols et al. 2007). Out of these 11, miR-K12-10 codes for Kaposi's family of proteins, which affect cell growth, enhance cytokine production in infected cells. Computational predictions suggest that these miRNA might play a pivotal role in regulating viral pathogenesis and replication.

12.6.1.3.3 Human Cytomegalovirus

Human cytomegalovirus (HCMV) encodes for nine miRNAs, which are dispersed throughout the viral genome. Of these, three are from complementary strands of known ORFs, five in intergenic regions and one within intron (Pfeffer et al. 2005). These miRNAs are associated with the autoregulation of virus and are involved in the latent stage of the life cycle.

12.6.1.4 Adenovirus miRNAs

Adenovirus (ADV), a DNA virus encodes a 160-nt noncoding RNA, VA1pre-miRNA during the late lytic replication. This miRNA could inhibit the RNAi expression in vivo by inhibiting the nuclear transport of shRNA through its competition with exportin 5 (Cullen 2006).

12.6.1.5 Hepatitis A

Using computational analysis and experimental validation, Shi et al. reported the production of HAV-encoded miRNA on the antigenome, hav-miR-N1-3p, in KMB17 and HEK 293 T cells (Shi et al. 2014a). Parallelly in the same year, the same group reported identification of two more HAV-encoded miRNAs, hav-miR-1-5p and hav-miR-2-5p using deep sequencing and in silico approaches in KMB17 cells (Shi et al. 2014b). Further in 2016, once again Shi et al. reported the mechanism of these HAV vmiRNAs wherein they proposed that these vmiRNAs serve as a self-mediated feedback regulator during HAV infection by attenuating the accumulation of viral genomic RNAs (Shi et al. 2016).

12.6.2 Cellular miRNAs in Viral Infection

Several cellular miRNAs are known to regulate viral pathogenesis by altering the host mRNA/gene expression. Several other cellular miRNAs involved in the host immune system are modulated by the viruses thereby affecting the course of infection. In the section below, we shed light upon the effect of viruses on cellular miRNAs and cellular miRNAs that regulate the viral life cycle in several human viral infections.

12.6.2.1 Hepatitis B

HBV infection can cause acute or chronic hepatitis B, liver cirrhosis, and hepatocellular carcinoma (Kiyosawa et al. 1990). In an attempt to identify host miRNAs involved in HBV replication, both miRNAs inhibiting and promoting HBV replication have been identified.

12.6.2.1.1 miRNAs Inhibiting HBV Replication

Wang et al. addressed the effect of miR-122 on HBV and found that miR-122 mimics inhibit HBV replication whereas anti-miR-122 increased HBV replication in hepatoma cells (Wang et al. 2012). Several other studies showed that HBV infection reduced miR-122 levels and viral load inversely correlated with the miR-122 levels in HBV-infected patients (Chen et al. 2011; Fan et al. 2011; Qiu et al. 2010; Wang et al. 2012). Furthermore, it was also shown that HBx could bind PPARγ and inhibit miR-122 transcription (Song et al. 2013b). miR-122 was involved in regulating HBx protein. miR-122 induced downregulation of SOCS3 thereby enhancing IFN-mediated inhibition of HBV (Gao et al. 2015). Pederson et al. showed that upregulated IFNs result in lowering of miR-122 both in vitro and in vivo (Pedersen et al. 2007). Wang et al. have reported a correlation of circulating miR-122 levels with the prognosis of CHB related liver failure (Wang et al. 2016).

Zhang et al. observed inhibition of HBV replication by miR-199a-3p and miR-210 by binding to S protein and preS1 coding regions respectively (Zhang et al. 2010a, b). Similarly, miR-125a-5p inhibited translation of S transcript (Potenza et al. 2011). Few additional miRNAs, let-7, miR-196b, -205, -345, -433, and -511 target HBV genome (Wu et al. 2011). miR-99a targets IGF-1R and induces cell cycle arrest (Li et al. 2011). miR-141 inhibits HBV replication by downregulating PPARα required for transcription of HBV genome (Hu et al. 2012). Downregulation of miR-22, a tumor suppressor, was seen in HBV-related HCC (Shi and Xu 2013). Kohno et al. observed suppression of HBV replication by overexpression of mi-1231 without affecting the interferon-stimulated genes (ISGs) (Kohno et al. 2014). Recently, Wang et al. showed that HBV expression dramatically downregulates miR-101 whereas overexpression of miR-101 suppresses HBV replication by targeting FOXO1 (Wang and Tian 2017).

12.6.2.1.2 miRNAs Promoting HBV Replication

miR-1 upregulates farnesoid X receptor α (FXRα), a nuclear receptor required in HBV transcription (Zhang et al. 2011). Serum miR-122 levels correlated well with ALT levels as well as HBV and HBsAg (Bala et al. 2012). Jin et al. have shown downregulation of HBX1P, an HBV inhibitor, by miR-501 (Jin et al. 2013). Few other studies have shown upregulation of miR-99a and -125 in HBV patients (Akamatsu et al. 2015; Giray et al. 2014; Hayes et al. 2012). In 2016, one more study reported enhanced HBV replication by miR-449a, an epigenetically regulated miRNA by targeting CREBP5 (Zhang et al. 2016). Lin et al. showed that miR-99 family promoted autophagy through mTOR/ULK1 signaling which in turn enhanced HBV replication (Lin et al. 2017). Another study by Qiu et al. reported that miRNA let-7a promotes HBV replication in HCC tissues (Qiu et al. 2017). And previously, Deng et al. demonstrated sequestration of let-7a by HBV (Deng et al. 2016). Recently, Deng et al. have reported that miR-125b-5p could target the LIN28B/let-7 axis and stimulate HBV replication (Deng et al. 2017b). This is the first evidence for posttranscriptional miRNA regulation.

12.6.2.2 Hepatitis C

HCV infection disrupts multiple pathways regulated by miRNAs, like antigen processing, cell cycle, immune response, proteasome, and lipid metabolism (Ura et al. 2009). miRNAs act either by controlling target gene expression or by directly targeting the HCV genome (Kerr et al. 2011; Pedersen et al. 2007; Peng et al. 2009).

12.6.2.2.1 miRNAs Promoting HCV Replication

Jopling and colleagues, for first the time, reported a positive role of miR-122 in HCV infection (Jopling et al. 2005). This study demonstrated a reduction of HCV RNA levels by sequestration of miR-122 in subgenomic replicon (SGR) cells. It has been observed that miR-122 expression facilitates an efficient replication in HepG2 (Narbus et al. 2011), as well as many non-hepatic cells upon infection with HCV (Chang et al. 2008; Fukuhara et al. 2012). miR-122 enhances HCV replication by binding directly to two closely spaced target sites in the 5′UTR of the viral genome, resulting in increased accumulation and expression of the viral RNA (Jangra et al. 2010; Jopling et al. 2008; Machlin et al. 2011; Shimakami et al. 2012; Pang et al. 2012). These target sites are conserved across HCV genotypes and hence are a potential therapeutic candidate.

The beneficial role of miR-122 seen in vitro does not translate into a correlation with HCV viral load in patients. This comment is supported by the finding that nonresponders to standard IFN-cased therapy have low miR-122 pre-treatment levels (Kamo et al. 2014; Sarasin-Filipowicz et al. 2009; Su et al. 2013). In fact pretreatment miR-122 levels might serve to be biomarker for predicting the therapeutic outcome. Circulating miR-122 serves as a biomarker to predict therapeutic efficacy in HCV-infected patients (Jiao et al. 2017). Koberle et al. showed that IFN therapy did not lower miR-122 in HCV-infected patients (Köberle et al. 2013). Hepatic miR-122 expression levels were significantly increased in HCV-3-infected patients (Oliveira et al. 2016). Recently, van der Ree et al. demonstrated miravirsen dosing decreased miR-122 levels in CHC patients (van der Ree et al. 2016). Insufficient clinical data is available and hence more studies are required to understand the regulatory role of miR-122 in various states of HCV infection. Also, the results for the clinical trials for miravirsen are awaited.

miRNAs like miR-141, -192, -491, and -215 are reported to facilitate HCV replication (Banaudha et al. 2011; Ishida et al. 2011). miR-199a-5p facilitated HCV replication by regulating pro-survival pathways through PI3K/Akt, Ras/ERK, and Wnt/β-catenin (Wang et al. 2015). Recently, Bandiera et al. reported facilitation of HCV infection by upregulating miR-146a-5p in hepatocytes (Bandiera et al. 2016).

12.6.2.2.2 miRNAs Inhibiting HCV Replication

Several studies have identified miRNAs with anti-HCV activity, miR-199, -196, -29, -130, -27, and let-7b (Bandyopadhyay et al. 2011; Bhanja Chowdhury et al. 2012; Cheng et al. 2012; Hou et al. 2010; Murakami et al. 2009). Pederson et al. have reported induction of antiviral miRNAs like miR-196, -296, -431, -351 and -441 by IFNβ (Pedersen et al. 2007). Mukherjee et al. have recently demonstrated that miR-181c binds to E1 and NS5A regions of HCV genome, enhanced HOXA1 expression (required for normal development and differentiation of mammalian tissue) (Mukherjee et al. 2014). Further, they showed that overexpression of miR-181c inhibited viral replication indicating it to be a therapeutic target. Recently, Chou et al. demonstrated that let-7g inhibits HCV replication by cooperating with IFN/RBV through p38/AP-1 signaling (Chou et al. 2016).

12.6.2.3 Hepatitis A

A few studies have reported the involvement of cellular miRNAs in HAV infection. Profiling and characterization of cellular miRNAs were carried out in HAV infected human fibroblasts (Shi et al. 2016). The alteration of miRNA profile was suggested to be related to the establishment of persistent HAV infection (Shi et al. 2016). In another study, Weseslindtner et al. have shown upregulation of miR-106a and -197 in patients with acute viral hepatitis (caused by HAV, HBV, HCV, and HEV). This report, however, does not differentiate between these etiologies (Weseslindtner et al. 2016).

12.6.2.4 Hepatitis E

Cheng et al. showed downregulation of miR-221 and -222 in swine HEV-4 ORF3 expressing HEK 293 cells by regulating the $p27^{kip1}$ expression (Cheng et al. 2013). Another study by Trehanpati et al. reported for the first time, the role of miRNA signatures in predicting ALF in HE during pregnancy. They demonstrated that the interaction of distinct miRNAs in particular immune pathways was responsible for the diverse outcomes of HE infection including inflammatory responses, liver failure, or death (Trehanpati et al. 2017). As mentioned in the earlier section,

Weseslindtner et al. have reported the upregulation of miR-106a, -122, and -197 in patients with acute viral hepatitis (caused by HAV, HBV, HCV, and HEV) with no differentiation in the different etiologies (Weseslindtner et al. 2016). Recently, positive regulation of Hepatitis E virus replication by interacting with the RdRp region of the viral genome has been reported by Haldipur et al. (2018).

12.6.2.5 Influenza A Virus

Several cellular miRNAs are involved in regulating the Influenza A virus (IAV) life cycle. These miRNAs act either as diagnosis markers or by targeting IAV viral RNA or by controlling the IAV inflammatory responses.

12.6.2.5.1 microRNAs as Diagnostic Markers

PBMCs from critically ill patients with swine-origin pandemic H1N1 had increased levels of miR-148 and decreased levels of miR-29a and -31 (Song et al. 2013a). Higher levels of miR-34c-3p and lower levels of miR-29a-3p, -181a-5p, and -30c-5p were reported in throat swabs of H1N1 patients (Peng et al. 2016). Sera showed significantly high levels of miR-150 in critically ill patients as compared to healthy controls and those with mild disease (Moran et al. 2015). miRNA microarray profiling of PBMCs from H1N1/H3N2-infected patients revealed dysregulation of 14 miRNAs (increased miR-229-5p, -335, -1260, and -664 levels and decreased miR-26a, -18a, 185, -30a, -34b, -628-3p, -576-3p, -665, -765, and -1285 levels). H&N9-infected sera showed upregulation of miR-17, -20a, -106a, and -376c (Zhu et al. 2014b).

12.6.2.5.2 microRNAs Targeting IAV RNA

PB1 mRNA of (1) H1N1 is destabilized by miR-491, -323, and -654 (Song et al. 2010); (2) H5N1 is targeted by miR-485 (Ingle et al. 2015); and (3) H3N2 is targeted by miR-3145 (Khongnomnan et al. 2015). Next, Ma et al. showed that let 7c precursor binds to 3′UTR of M1 mRNA and inhibits viral replication (Ma et al. 2012). miR-33a indirectly diminishes replication by binding to 3′UTR of Archain (ARCN1) thereby suppressing NP and M1 expression (Hu et al. 2016). Another miRNA, miR-21 inhibits the virus by targeting NP, PB1, PB2, NA, and HA of H1N1 (Waring et al. 2018).

12.6.2.5.3 microRNAs Controlling IAV Inflammatory Response

Chen et al. have reported that increased expression of IRF5 gene correlated with reduced levels of miR-302a in PBMCs and throat swabs of IAV patients as compared to healthy individuals (Chen et al. 2017). miR-146a directly targets TRAF6 in H3N2-infected human nasal epithelial cells (Deng et al. 2017a). miR-4776 binds to 3′UTR of NFkB and leads to its reduction thereby inhibiting viral replication (Othumpangat et al. 2017). H3N2 infection decreases miR-302c expression which targets NFkB inducing kinase (NIK) (Gui et al. 2015). miR-132, -46a, and -1275 reduced IRAK and MAPK3 expression in H1N1/H3N2-infected cells (Buggele et al. 2012). Yet another study by Dong et al. reported that miR-9 promoted viral replication by suppressing MCP1P1 (Dong et al. 2017).

Several other miRNAs like miR-302c, -449b, -125a/b, -136, -194, -483-3p, -132, and -26a have been shown to target cytokine production (Chen et al. 2017; Gui et al. 2015; Buggele et al. 2013; Hsu et al. 2017; Zhao et al. 2015; Wang et al. 2017; Maemura et al. 2018; Lagos et al. 2010; Gao et al. 2017).

12.6.2.6 Dengue (DENV)

Several studies have reported increased levels of let-7c and miR-30e* in DENV-infected HeLa; Huh7, and U937-DC-SIGN cells (Escalera-Cueto et al. 2015; Zhu et al. 2014a). Let-7c inhibits viral replication by activating TLR2 via NS1 and leads to IL6 and TNFα production (Chen et al. 2015). miR-30e* inhibits viral replication by targeting NFkB thereby increasing NFkB-dependent IFNβ expression. miR-126-5p is lowered during DENV infection (Kakumani et al. 2013) whereas miR-126-5p mimics decreased both viral RNA levels and viremia in supernatants (Kakumani et al. 2016). DENV induces miR-146a in THP-1 cells and primary monocytes thereby facilitating replication through impairment of IFNβ production (Wu et al. 2013). miR-147a (a negative regulator of pro-inflammation) was significantly increased in PBMCs infected with DENV (Tolfvenstam et al. 2011).

12.6.2.7 Human Immunodeficiency Virus-1

Involvement of RNAi machinery in Human Immunodeficiency Virus-1 (HIV-1) came into limelight when Dicer and Drosha were knockeddown in HIV-1-infected PBMCs. Since then several miRNAs have been studied. miR-29a and -29b (expressed in PBMCs) targeted sites in the HIV *nef* region and affected the HIV-1 life cycle (Ahluwalia et al. 2008). An experiment by Triboulet et al. using HIV-1-infected Jurkat cells showed downregulation of miR-17/92 cluster, which targeted P300/CBP-associated factor (PCAF), required for HIV-1 Tat (Triboulet et al. 2007). miR-28, -125b, -15o, -223, and -383 (anti-HIV) were under expressed activated CD4+ cells. These miRNAs targeted 3′end of HIV-1 mRNA (Huang et al. 2007). miR-29a regulates the viral replication by targeting both the 3′end of HIV-1 transcripts and HIV-1 *nef* (Nathans et al. 2009; Ahluwalia et al. 2008; Hariharan et al. 2005). miRNAs also play a role in the HIV-1-induced encephalitis (HIVE). miR-129* and -130a (targeting caspase-6 mRNA) and miR-146a (targeting MCP-2) are under expressed in the brains of HIVE patients (Noorbakhsh et al. 2010; Rom et al. 2010). On the other hand, miR-21 which targets MEF2C is overexpressed in the brains of HIVE patients (Yelamanchili et al. 2010).

12.6.2.8 Herpes Simplex Virus-1 (HSV-1)

Unlike several other viral infections discussed above, cellular miRNAs are also involved in Herpes Simplex Virus (HSV-1) life cycle. miR-101 targets 3′UTR of mitochondrial ATP synthase subunit β (ATP5B) which is required for HSV-1 replication (Zheng et al. 2011), thereby contributing to viral latency. On the other hand, miR-23a targets IRF-1 and inhibits interferon pathway, thus, facilitating HSV-1 replication (Ru et al. 2014). Further, HSV-1 induces miR-146a (a pro-inflammatory miRNA) that targets complement factor H and induces arachidonic cascade, contributing to Alzheimer-like neuropathological change (Hill et al. 2009).

12.7 Conclusion

As discussed in the sections above, it is increasingly becoming apparent that viruses can make use of the host RNAi machinery in several ways so as to regulate the viral life cycle and host response. Viral regulation of cellular miRNA profile seems to be a complex yet widely used mechanism affecting host cell environment and viral life cycle by variations in host miRNA targets. Another level of complexity is added by the viral-encoded miRNAs. This is another strategy used by the viruses to establishing themselves in the host by counterattacking the host response. Here we have touched upon a few examples of viral infections as discussing all the viral infections is beyond the scope of this chapter. Most of the miRNAs serve as potential diagnostic and prognostic biomarkers for several viral infections. Others which act on either the host or the viral components required for the replication may act as potential therapeutic targets. A classical example is the anti-miR-122 (Miravirsen) against HCV replication, which is being in clinical trial and results are promising. It would be very interesting to utilize the potential therapeutic miRNAs reported against viral infections and develop strategies for management of such viral infections.

References

Ahluwalia JK et al (2008) Human cellular microRNA hsa-miR-29a interferes with viral nef protein expression and HIV-1 replication. Retrovirology 5:117

Akamatsu S et al (2015) Differences in serum microRNA profiles in hepatitis B and C virus infection. J Infect 70(3):273–287

Ameres SL et al (2007) Molecular basis for target RNA recognition and cleavage by human RISC. Cell 130(1):101–112

Axtell MJ et al (2011) Vive la différence: biogenesis and evolution of microRNAs in plants and animals. Genome Biol 12(4):221

Bala S et al (2012) Circulating microRNAs in exosomes indicate hepatocyte injury and inflammation in alcoholic, drug-induced, and inflammatory liver diseases. Hepatology 56(5):1946–1957

Banaudha K et al (2011) MicroRNA silencing of tumor suppressor DLC-1 promotes efficient hepatitis C virus replication in primary human hepatocytes. Hepatology 53(1):53–61

Bandiera S et al (2016) HCV-induced up-regulation of miR-146a-5p in hepatocytes promotes viral infection and deregulates metabolic pathways associated with liver disease pathogenesis. J Virol 90(14):6387–6400

Bandyopadhyay S et al (2011) Hepatitis C virus infection and hepatic stellate cell activation downregulate miR-29: miR-29 overexpression reduces hepatitis C viral abundance in culture. J Infect Dis 203(12):1753–1762

Bartel DP (2009) MicroRNAs: target recognition and regulatory functions. Cell 136(2):215–233

Bernstein E et al (2001) Role for a bidentate ribonuclease in initiation step of RNA interference. Nature 409(6818):363–366

Bhanja Chowdhury J et al (2012) Hepatitis C virus infection modulates expression of interferon stimulatory gene IFITM1 by upregulating miR-130a. J Virol 86(18):10221–10225

Bogerd H et al (2014) Replication of many human viruses is refractory to inhibition by endogenous cellular microRNAs. J Virol 88(14):8065–8076

Bohnsack MT et al (2004) Exportin 5 is a RanGTP-dependent dsRNA-binding protein that mediates nuclear export of pre-miRNAs. RNA 10(2):185–191

Buggele WA et al (2012) Influenza A virus infection of human respiratory cells induces primary microRNA expression. J Biol Chem 287:31027–31040
Buggele WA et al (2013) Small RNA profiling of influenza A virus-infected cells identifies miR-449b as a regulator of histone deacetylase 1 and interferon beta. PLoS One 8:e76560
Cai X et al (2004) Human microRNAs are processed from capped, polyadenylated transcripts that can also function as mRNAs. RNA 10(12):1957–1966
Cai X et al (2006) Human papillomavirus genotype 31 does not express detectable microRNA levels during latent or productive virus replication. J Virol 80:10890–10893
Chang J et al (2008) Liver-specific microRNA miR-122 enhances replication of hepatitis C virus in nonhepatic cells. J Virol 82(16):8215–8223
Chen X et al (2008) Characterization of microRNAs in serum: a novel class of biomarkers for diagnosis of cancer and or diseases. Cell Res 18(10):997–1006
Chen Y et al (2011) A liver-specific microRNA binds to a highly conserved RNA sequence of hepatitis B virus and negatively regulates viral gene expression and replication. FASEB J 25(12):4511–4521
Chen J et al (2015) Activation of TLR2 and TLR6 by dengue NS1 protein and its implications in the immunopathogenesis of dengue virus infection. PLoS Pathog 11:e1005053
Chen X et al (2017) MicroRNA-302a suppresses influenza A virus-stimulated interferon regulatory factor-5 expression and cytokine storm induction. J Biol Chem 292:21291–21303
Cheng J-C et al (2012) Let-7b is a novel regulator of hepatitis C virusreplication. Cell Mol Life Sci 69(15):2621–2633
Cheng Y et al (2013) Identification of miR-221 and -222 as important regulators in genotype IV swine hepatitis E virus ORF3-expressing HEK. Virus Genes 47:49–55
Chou W et al (2016) MicroRNA let-7g cooperates with interferon/ribavirin to repress hepatitis C virus replication. J Mol Med 94(3):311–320
Corcoran DL et al (2009) Features of mammalian microRNA promoters emerge from polymerase II chromatin immunoprecipitation data. PLoS One 4(4):e5279
Cullen BR (2006) Viruses and microRNAs. Nat Genet 38:S25–S30
Dai X et al (2014) Modulation of HBV replication by microRNA-15b through targeting hepatocyte nuclear factor 1α. Nucleic Acids Res 42(10):6578–6590
Deng M et al (2016) Hepatitis B virus mRNAs functionally sequester let-7a and enhance hepatocellular carcinoma. Cancer Lett 383(1):62–72
Deng Y et al (2017a) MicroRNA-146a induction during influenza H3N2 virus infection targets and regulates TRAF6 levels in human nasal epithelial cells (hNECs). Exp Cell Res 352:184–192
Deng W et al (2017b) MicroRNA-125b-5p mediates post-transcriptional regulation of hepatitis B virus replication via LIN28B/let-7 axis. RNA Biol 14(10):1389–1398
Denli AM et al (2004) Processing of primary microRNAs by microprocessor complex. Nature 432(7014):231–235
Doench JG, Sharp PA (2004) Specificity of microRNA target selection in translational repression. Genes Dev 18(5):504–511
Dong C et al (2017) Modulation of influenza A virus replication by microRNA-9 through targeting MCPIP1. J Med Virol 89:41–48
Escalera-Cueto M et al (2015) Let-7c overexpression inhibits dengue virus replication in human hepatoma Huh-7 cells. Virus Res 196:105–112
Fan CG et al (2011) miR-122 inhibits viral replication and cell proliferation in hepatitis B virus-related hepatocellular carcinoma and targets NDRG3. Oncol Rep 26(5):1281–1286
Flores O et al (2013) Mutational inactivation of herpes simplex virus 1 microRNAs identifies viral mRNA targets and reveals phenotypic effects in culture. J Virol 87(12):6589–6603
Friedman RC et al (2008) Most mammalian mRNAs are conserved targets of microRNAs. Genome Res 19(1):92–105
Fu Y-R et al (2014) miR-21 attenuates human cytomegalovirus replication in neural cells by targeting Cdc25a. J Virol 89:1070–1082
Fukuhara T et al (2012) Expression of microRNA miR-122 facilitates an efficient replication in nonhepatic cells upon infection with hepatitis C virus. J Virol 86(15):7918–7933

Gantier MP et al (2007) Fine-tuning of innate immune response by microRNAs. Immunol Cell Biol 85:458–462
Gao D et al (2015) Down-regulation of suppressor of cytokine signaling 3 by miR-122 enhances interferon-mediated suppression of hepatitis B virus. Antivir Res 118:20–28
Gao S et al (2017) Influenza A virus-induced downregulation of miR-26a contributes to reduced IFNalpha/beta production. Virol Sin 32:261–270
Girard M et al (2008) miR-122, a paradigm for role of microRNAs in liver. J Hepatol 48(4):648–656
Giray BG et al (2014) Profiles of serum microRNAs; miR-125b-5p and miR223-3p serve as novel biomarkers for HBV-positive hepatocellular carcinoma. Mol Biol Rep 41(7):4513–4519
Grey F et al (2007) A human cytomegalovirus-encoded microRNA regulates expression of multiple viral genes involved in replication. PLoS Pathog 3(11):e163
Grimson A et al (2007) MicroRNA targeting specificity in mammals: determinants beyond seed pairing. Mol Cell 27(1):91–105
Gui S et al (2015) Mir-302c mediates influenza A virus-induced IFNbeta expression by targeting NF-kappaB inducing kinase. FEBS Lett 589:4112–4118
Haldipur B et al (2018) Positive regulation of hepatitis E virus replication by MicroRNA-122. J Virol 92(11):e01999-17
Hammond SM et al (2000) An RNA-directed nuclease mediates post-transcriptional gene silencing in Drosophila cells. Nature 404(6775):293–296
Han J et al (2004) Drosha-DGCR8 complex in primary microRNA processing. Genes Dev 18(24):3016–3027
Hariharan M et al (2005) Targets for human encoded microRNAs in HIV genes. Biochem Biophys Res Commun 337:1214–1218
Hayes CN et al (2012) Hepatitis B virus-specific miRNAs and Argonaute2 play a role in viral life cycle. PLoS One 7(10):e47490
Hill JM et al (2009) HSV-1 infection of human brain cells induces miRNA-146a and Alzheimer-type inflammatory signaling. Neuroreport 20:1500–1505
Hou J et al (2009) MicroRNA-146a feedback inhibits RIG-I-dependent type I IFN production in macrophages by targeting TRAF6, IRAK1, and IRAK2. J Immunol 183(3):2150–2158
Hou W et al (2010) MicroRNA-196 represses Bach1 protein and hepatitis C virus gene expression in human hepatoma cells expressing hepatitis C viral proteins. Hepatology 51(5):1494–1504
Hsu AC et al (2017) MicroRNA- 125a and -b inhibit A20 and MAVS to promote inflammation and impair antiviral response in COPD. JCI Insight 2:e90443
Hu W et al (2012) MicroRNA-141 represses HBV replication by targeting PPARA. PLoS One 7(3):e34165
Hu Y et al (2016) MicroRNA-33a disturbs influenza A virus replication by targeting ARCN1 and inhibiting viral ribonucleoprotein activity. J Gen Virol 97:27–38
Huang J et al (2007) Cellular microRNAs contribute to HIV-1 latency in resting primary CD4R T lymphocytes. Nat Med 13:1241–1247
Hunter MP et al (2008) Detection of microRNA expression in human peripheral blood microvesicles. PLoS One 3(11):e3694
Hussain M, Asgari S (2014) MicroRNA-like viral small RNA from dengue virus 2 autoregulates its replication in mosquito cells. PNAS 111(7):2746–2751
Hussain M et al (2012) West Nile virus encodes a microRNA-like small RNA in 3' untranslated region which up-regulates GATA4 mRNA and facilitates virus replication in mosquito cells. Nucleic Acids Res 40(5):2210–2223
Hutvágner G et al (2001) A cellular function for RNA-interference enzyme dicer in maturation of let-7 small temporal RNA. Science 293(5531):834–838
Ingle H et al (2015) The microRNA miR-485 targets host and influenza virus transcripts to regulate antiviral immunity and restrict viral replication. Sci Signal 8:126
Ishida H et al (2011) Alterations in microRNA expression profile in HCV-infected hepatoma cells: involvement of miR-491 in regulation of HCV replication via PI3 kinase/Akt pathway. Biochem Biophys Res Commun 412(1):92–97

Jangra RK et al (2010) Regulation of hepatitis C virus translation and infectious virus production by microRNA miR-122. J Virol 84(13):6615–6625
Ji X et al (2009) Plasma miR-208 as a biomarker of myocardial injury. Clin Chem 55(11):1944–1949
Jiao X et al (2017) Serum and exosomal miR-122 and miR-199a as a biomarker to predict therapeutic efficacy of hepatitis C patients. J Med Virol 89(9):1597–1605
Jin J et al (2013) MicroRNA-501 promotes HBV replication by targeting HBXIP. Biochem Biophys Res Commun 430(4):1228–1233
Jopling CL et al (2005) Modulation of hepatitis C virus RNA abundance by a liver-specific microRNA. Science 1:1577–1581
Jopling CL et al (2008) Position-dependent function for a tandem MicroRNA miR-122-binding site located in hepatitis C virus RNA genome. Cell Host Microbe 4(1):77–85
Kakumani PK et al (2013) Role of RNA interference (RNAi) in dengue virus replication and identification of NS4B as an RNAi suppressor. J Virol 87:8870–8883
Kakumani PK et al (2016) Role of human GRP75 in miRNA mediated regulation of dengue virus replication. Gene 586:7–11
Kamo Y et al (2014) Significance of micro RNA-122 in chronic hepatitis C patients with serotype 1 on interferon therapy. Hepatol Res 45(1):88–96
Kaul D et al (2009) HIV-1 genome-encoded hiv1-mir-H1 impairs cellular responses to infection. Mol Cell Biochem 323(1–2):143–148
Kerr TA et al (2011) MicroRNAs and liver disease. Transl Res 157(4):241–252
Ketting RF et al (2001) Dicer functions in RNA interference and in synsis of small RNA involved in developmental timing in C. elegans. Genes Dev 15(20):2654–2659
Khongnomnan K et al (2015) Human miR-3145 inhibits influenza A viruses replication by targeting and silencing viral PB1 gene. Exp Biol Med 240:1630–1639
Khvorova A et al (2003) Functional siRNAs and miRNAs exhibit strand bias. Cell 115(2):209–216
Kim VN et al (2009) Biogenesis of small RNAs in animals. Nat Rev Mol Cell Biol 10(2):126–139
Kiyosawa K et al (1990) Interrelationship of blood transfusion, non-a, non-B hepatitis and hepatocellular carcinoma: analysis by detection of antibody to hepatitis C virus. Hepatology 12(4 Pt 1):671–675
Köberle V et al (2013) Serum microRNA-122 kinetics in patients with chronic hepatitis C virus infection during antiviral therapy. J Viral Hepat 20(8):530–535
Kohno T et al (2014) Human microRNA hsa-miR-1231 suppresses hepatitis B virus replication by targeting core mRNA. J Viral Hepat 21(9):e89–e97
Lagos D et al (2010) miR-132 regulates antiviral innate immunity through suppression of the p300 transcriptional co-activator. Nat Cell Biol 12:513–U232
Laterza OF et al (2009) Plasma MicroRNAs as sensitive and specific biomarkers of tissue injury. Clin Chem 55(11):1977–1983
Lawrie CH et al (2008) Detection of elevated levels of tumour-associated microRNAs in serum of patients with diffuse large B-cell lymphoma. Br J Haematol 141(5):672–675
Lecellier C-H et al (2005) A cellular microRNA mediates antiviral defense in human cells. Science 308(5721):557–560
Lee Y et al (2002) MicroRNA maturation: stepwise processing and subcellular localization. EMBO J 21(17):4663–4670
Lee Y et al (2003) nuclear RNase III Drosha initiates microRNA processing. Nature 425(6956):415–419
Lee Y et al (2004) MicroRNA genes are transcribed by RNA polymerase II. EMBO J 23(20):4051–4060
Li D et al (2011) MicroRNA-99a inhibits hepatocellular carcinoma growth and correlates with prognosis of patients with hepatocellular carcinoma. J Biol Chem 286(42):36677–36685
Lin Y et al (2017) microRNA-99 family modulates hepatitis B virus replication by promoting IGF-1R/PI3K/Akt/mTOR/ULK1 signaling-induced autophagy. Cell Microbiol 19(5):e12709
Lytle JR et al (2007) Target mRNAs are repressed as efficiently by microRNA-binding sites in 5' UTR as in 3' UTR. PNAS 104(23):9667–9672

Ma YJ et al (2012) Cellular microRNA let-7c inhibits M1 protein expression of the H1N1 influenza A virus in infected human lung epithelial cells. J Cell Mol Med 16:2539–2546
Machlin ES et al (2011) Masking 5′ terminal nucleotides of hepatitis C virus genome by an unconventional microRNA-target RNA complex. PNAS 108(8):3193–3198
Maemura T et al (2018) Lung-derived exosomal miR-483-3p regulates the innate immune response to influenza virus infection. J Infect Dis 217(9):1372–1382
Martin G et al (2007) Prediction and validation of microRNA targets in animal genomes. J Biosci 32(6):1049–1052
Mitchell PS et al (2008) Circulating microRNAs as stable blood-based markers for cancer detection. PNAS 105(30):10513–10518
Mo M-H et al (2012) Cell-free circulating miRNA biomarkers in cancer. J Cancer 3:432–448
Moran J et al (2015) Circulating levels of miR-150 are associated with poorer outcomes of A/H1N1 infection. Exp Mol Pathol 99:253–261
Mourelatos Z et al (2002) miRNPs: a novel class of ribonucleoproteins containing numerous microRNAs. Genes Dev 16(6):720–728
Mukherjee A et al (2014) Transcriptional suppression of miR-181c by hepatitis C virus enhances homeobox A1 expression. J Virol 88(14):7929–7940
Murakami Y et al (2009) Regulation of hepatitis C virus genome replication by miR-199a*. J Hepatol 50(3):453–460
Narbus CM et al (2011) HepG2 cells expressing miR-122 support entire hepatitis C virus life cycle. J Virol 85(22):12087–12092
Nathans R et al (2009) Cellular microRNA and P bodies modulate host-HIV-1 interactions. Mol Cell 34:696–709
Nogales-Gadea G et al (2014) Analysis of serum miRNA profiles of myasnia gravis patients. PLoS One 9(3):1–11
Noorbakhsh F et al (2010) MicroRNA profiling reveals new aspects of HIV neurodegeneration: caspase-6 regulates astrocyte survival. FASEB J 24:1799–1812
Oliveira KG et al (2016) Increased hepatic expression of miRNA-122 in patients infected with HCV genotype 3. Med Microbiol Immunol 205(2):111–117
Omoto S et al (2004) HIV-1 nef suppression by virally encoded microRNA. Retrovirology 1:44
Othumpangat S et al (2017) Upregulation of miRNA- 4776 in influenza virus infected bronchial epithelial cells is associated with downregulation of NFKBIB and increased viral survival. Viruses 9(5):E94
Ouellet DL et al (2008) Identification of functional microRNAs released through asymmetrical processing of HIV-1 TAR element. Nucleic Acids Res 36:2353–2365
Pang PS et al (2012) Structural map of a MicroRNA-122:hepatitis C virus complex. J Virol 86(2):1250–1254
Pedersen IM et al (2007) Interferon modulation of cellular microRNAs as an antiviral mechanism. Nature 449(7164):919–922
Peng X et al (2009) Computational identi025ication of hepatitis C virus associated microRNA-mRNA regulatory modules in human livers. BMC Genomics 10(1):373
Peng F et al (2016) Identification of microRNAs in throat swab as the biomarkers for diagnosis of influenza. Int J Med Sci 13:77–84
Pfeffer S et al (2004) Identification of virusencoded microRNAs. Science 304:734–736
Pfeffer S et al (2005) Identification of microRNAs of the herpesvirus family. Nat Methods 2:269–276
Potenza N et al (2011) Human microRNA hsa-miR-125a-5p interferes with expression of hepatitis B virus surface antigen. Nucleic Acids Res 39(12):5157–5163
Qian J et al (2010) Modulation of miR-122 on persistently Borna disease virus infected human oligodendroglial cells. Antivir Res 87(2):249–256
Qiu L et al (2010) MiR-122-induced down-regulation of HO-1 negatively affects miR-122-mediated suppression of HBV. Biochem Biophys Res Commun 398(4):771–777
Qiu D et al (2017) Expression of microRNA let-7a positively correlates with hepatitis B virus replication in hepatocellular carcinoma tissues. Exp Biol Med 242(9):939–944

Rom S et al (2010) CCL8/MCP-2 is a target for mir-146a in HIV-1-infected human microglial cells. FASEB J 24:2292–2300
Ru J et al (2014) MiR-23a facilitates replication of HSV-1 through suppression of interferon regulatory factor 1. PLoS One 9(12):e114021
Samols MA et al (2007) Identification of cellular genes targeted by KSHV-encoded microRNAs. PLoS Pathog 3:611–618
Sarasin-Filipowicz M et al (2009) Decreased levels of microRNA miR-122 in individuals with hepatitis C responding poorly to interferon therapy. Nat Med 15(1):31–33
Scaria V et al (2007) Host-virus genome interactions: macro roles for microRNAs. Cell Microbiol 9(12):2784–2794
Selvamani SP et al (2014) Chikungunya virus exploits miR-146a to regulate NF-kB pathway in human synovial fibroblasts. PLoS One 9(8):19–22
Seo G et al (2008) Evolutionarily conserved function of a viral microRNA. J Virol 82:9823–9828
Seo G et al (2009) Merkel cell polyomavirus encodes a microRNA with the ability to autoregulate viral gene expression. Virology 383:183–187
Shi C, Xu X (2013) MicroRNA-22 is down-regulated in hepatitis B virus-related hepatocellular carcinoma. Biomed Pharmacother 67(5):375–380
Shi J et al (2014a) Identification and validation of a novel microRNA-like molecule derived from a cytoplasmic RNA virus antigenome by bioinformatics and experimental approaches. Virol J 11(1):1–14
Shi J et al (2014b) Novel microRNA-like viral small regulatory RNAs arising during human hepatitis A virus infection. FASEB J 28(10):4381–4393
Shi J et al (2016) Hepatitis A virus-encoded miRNAs attenuate accumulation of viral genomic RNAs in infected cells. Virus Genes 52(3):317–324
Shimakami T et al (2012) Base pairing between hepatitis C virus RNA and microRNA 122: 3' of its seed sequence is essential for genome stabilization and production of infectious virus. J Virol 86(13):7372–7383
Shrivastava S et al (2013) Up-regulation of circulating miR-20a is correlated with hepatitis C virus-mediated liver disease progression. Hepatology 58(3):863–871
Song L et al (2010) Cellular microRNAs inhibit replication of the H1N1 influenza A virus in infected cells. J Virol 84:8849–8860
Song H et al (2013a) Microarray analysis of microRNA expression in peripheral blood mononuclear cells of critically ill patients with influenza A (H1N1). BMC Infect Dis 13:257
Song K et al (2013b) Epigenetic regulation of MicroRNA-122 by peroxisome proliferator activated receptor-gamma and hepatitis b virus X protein in hepatocellular carcinoma cells. Hepatology 58(5):1681–1692
Su T-H et al (2013) Serum microRNA-122 level correlates with virologic responses to pegylated interferon therapy in chronic hepatitis C. PNAS 110(19):7844–7849
Sullivan CS et al (2009) Murine polyomavirus encodes a microRNA that cleaves early RNA transcripts but is not essential for experimental infection. Virology 387:157–167
Tanaka M et al (2009) Down-regulation of miR-92 in human plasma is a novel marker for acute leukemia patients. PLoS One 4(5):e5532
Tanzer A, Stadler PF (2004) Molecular evolution of a MicroRNA cluster. J Mol Biol 339(2):327–335
Tolfvenstam T et al (2011) Characterization of early host responses in adults with dengue disease. BMC Infect Dis 11:209
Trehanpati N et al (2017) miRNA signatures can predict acute liver failure in hepatitis E infected pregnant females. Heliyon 3(4):e00287
Triboulet R et al (2007) Suppression of microRNA-silencing pathway by HIV-1 during virus replication. Science 315:1579–1582
Umbach JL et al (2008) MicroRNAs expressed by herpes simplex virus 1 during latent infection regulate viral mRNAs. Nature 454(7205):780–783
Ura S et al (2009) Differential microRNA expression between hepatitis B and hepatitis C leading disease progression to hepatocellular carcinoma. Hepatology 49(4):1098–1112

Valadi H et al (2007) Exosome-mediated transfer of mRNAs and microRNAs is a novel mechanism of genetic exchange between cells. Nat Cell Biol 9(6):654–659
van Der Ree MH et al (2016) Miravirsen dosing in chronic hepatitis C patients results in decreased microRNA-122 levels without affecting or microRNAs in plasma. Aliment Pharmacol Ther 43(1):102–113
Wang Y, Tian H (2017) miR-101 suppresses HBV replication and expression by targeting FOXO1 in hepatoma carcinoma cell lines. Biochem Biophys Res Commun 487(1):167–172
Wang K et al (2009) Circulating microRNAs, potential biomarkers for drug-induced liver injury. PNAS 106(11):4402–4407
Wang S et al (2012) Loss of microRNA 122 expression in patients with hepatitis B enhances hepatitis B virus replication through cyclin G 1-modulated P53 activity. Hepatology 55(3):730–741
Wang H et al (2015) Inhibition of microRNA-199a-5p reduces replication of HCV via regulating pro-survival pathway. Virus Res 208:7–12
Wang WJ et al (2016) Correlation between circulating miR-122 and prognosis of chronic HBV-related liver failure. J Dig Dis 17(5):334–339
Wang K et al (2017) miR-194 inhibits innate antiviral immunity by targeting FGF2 in influenza H1N1 virus infection. Front Microbiol 8:2187
Waring BM et al (2018) MicroRNA-based attenuation of influenza virus across susceptible hosts. J Virol 92:e01741–e01717
Weber JA et al (2010) microRNA spectrum in 12 body fluids. Clin Chem 56(11):1733–1741
Wee LM et al (2012) Argonaute divides its RNA guide into domains with distinct functions and RNA-binding properties. Cell 151(5):1055–1067
Weng K-F et al (2014) A cytoplasmic RNA virus generates functional viral small RNAs and regulates viral IRES activity in mammalian cells. Nucleic Acids Res 42(20):12789–12805
Weseslindtner L et al (2016) Micro RNAs mir-106a, mir-122 and mir-197 are increased in severe acute viral hepatitis with coagulopathy. Liver Int 36(3):353–360
Wong T-SS et al (2008) Mature miR-184 as potential oncogenic microRNA of squamous cell carcinoma of tongue. Clin Cancer Res 14(9):2588–2592
Wu F et al (2011) Targets for human encoded microRNAs in HBV genes. Virus Genes 42(2):157–161
Wu S et al (2013) miR-146a facilitates replication of dengue virus by dampening interferon induction by targeting TRAF6. J Infect 67(4):329–341
Ye W et al (2008) Effect of central loops in miRNA:MRE duplexes on efficiency of miRNA-mediated gene regulation. PLoS One 3(3):e1719
Yelamanchili SV et al (2010) MicroRNA-21 dysregulates the expression of MEF2C in neurons in monkey and human SIV/HIV neurological disease. Cell Death Dis 1:e77
Yi R et al (2003) Exportin-5 mediates nuclear export of pre-microRNAs and short hairpin RNAs. Genes Dev 17(24):3011–3016
Zeng Y, Cullen BR (2004) Structural requirements for pre-microRNA binding and nuclear export by Exportin 5. Nucleic Acids Res 32(16):4776–4785
Zhang G et al (2010a) Suppression of hepatitis B virus replication by microRNA-199a-3p and microRNA-210. Antivir Res 88(2):169–175
Zhang Y et al (2010b) Plasma microRNA-122 as a biomarker for viral-, alcohol-, and chemical-related hepatic diseases. Clin Chem 56(12):1830–1838
Zhang X et al (2011) Modulation of hepatitis B virus replication and hepatocyte differentiation by MicroRNA-1. Hepatology 53(5):1476–1485
Zhang Q et al (2014a) MicroRNA-23 inhibits PRRSV replication by directly targeting PRRSV RNA and possibly by upregulating type I interferons. Virology 450–451:182–195
Zhang Y et al (2014b) A novel HIV-1-encoded microRNA enhances its viral replication by targeting the TATA box region. Retrovirology 11:23
Zhang X et al (2016) Epigenetically regulated miR-449a enhances hepatitis B virus replication by targeting cAMP-responsive element binding protein 5 and modulating hepatocytes phenotype. Sci Rep 6:25389

Zhao L et al (2015) Identifcation of cellular microRNA-136 as a dual regulator of RIG-I-mediated innate immunity that antagonizes H5N1 IAV replication in A549 cells. Sci Rep 5:14991

Zheng S et al (2011) MiR-101 regulates HSV-1 replication by targeting ATP5B. Antivir Res 89:219–226

Zhu X et al (2014a) MicroRNA-30e∗ suppresses dengue virus replication by promoting NF-κB-dependent IFN production. PLoS Negl Trop Dis 8(8):e3088

Zhu Z et al (2014b) Comprehensive characterization of serum microRNA profile in response to the emerging avian influenza A (H7N9) virus infection in humans. Viruses 6:1525–1539

Evolution, Distribution, and Diversity of Immunodeficiency Viruses

13

Harika Sai Vemuri, Surekha Challa, and Nageswara Rao Reddy Neelapu

Abstract

Immunodeficiency viruses infect the host and primarily affect the immune system of an organism, i.e., host. Till date Human Immunodeficiency Virus (HIV), Simian Immunodeficiency Virus (SIV), Feline Immunodeficiency Virus (FIV), Bovine Immunodeficiency Virus (BIV), and Dog Immunodeficiency Virus (DIV) were reported in the literature. These viruses belong to phylum—Incertae sedis, family—Retroviridae, genus—*Lentivirus*, and order—Ortervirales. The review discusses about evolution, distribution, and diversity of immunodeficiency viruses, which helps in understanding the biology of HIV and how to develop a vaccine to the most harmful and dreadful diseases.

Keywords

Bovine immunodeficiency virus (BIV) · Dog immunodeficiency virus (DIV) · Feline immunodeficiency virus (FIV) · Human immunodeficiency virus (HIV) · Immunodeficiency virus · Simian immunodeficiency virus (SIV)

13.1 Immunodeficiency Viruses

Viruses that infect the host and affect the immune system of the host upon infection are generally known as immunodeficiency viruses (IV). The host then acquires a disease known as Acquired Immunodeficiency Disease Syndrome (AIDS). These immunodeficiency viruses belong to phylum—Incertae sedis, family—Retroviridae, genus—*Lentivirus*, and order—Ortervirales. The different types of

H. S. Vemuri · S. Challa · N. R. R. Neelapu (✉)
Department of Biochemistry and Bioinformatics, Institute of Science, Gandhi Institute of Technology and Management (GITAM) (Deemed to be University), Rushikonda, Visakhapatnam, Andhra Pradesh, India

P. V. Bramhachari (ed.), *Dynamics of immune activation in viral diseases*,
https://doi.org/10.1007/978-981-15-1045-8_13

immunodeficiency viruses are Human Immunodeficiency Virus (HIV) (Barré-Sinoussi et al. 1983; Clavel et al. 1986), Simian Immunodeficiency Virus (SIV) (Daniel et al. 1985), Feline Immunodeficiency Virus (FIV) (Pedersen et al. 1987), Bovine Immunodeficiency Virus (BIV) (Van Der Maaten et al. 1972a), and Dog Immunodeficiency Virus (DIV) (Safran et al. 1992). Among these viruses, HIV and SIV are widely studied and there are also reports on other immunodeficiency viruses like FIV, BIV, and DIV.

13.2 Human Immunodeficiency Virus

HIV is a retrovirus that infects human and affects the immune system of the human upon infection (Fig. 13.1). The host then acquires a disease known as AIDS. Worldwide statistics as of 2017 for HIV and AIDS are mentioned in Table 13.1. In 1983, HIV called HIV1 was isolated by "... researchers at the Pasteur Institute in France ..." which was known to cause AIDS (Barré-Sinoussi et al. 1983). In 1986 "... HIV-2, was isolated from AIDS patients in West Africa ..." (Clavel et al. 1986).

13.2.1 Evolution, Distribution, and Diversity of HIV

Two types of HIVs exist, HIV1 and HIV2; among them, HIV1 was discovered first and HIV2 was identified later (Fig. 13.2). HIV1 is distributed worldwide (Fig. 13.3), whereas HIV2 is observed mainly in western Africa (Vidal et al. 2000). The reason for widespread HIV1 can due to the ancestor of HIV1, which might have mutated at a much faster rate and traveled along with *Homo sapiens* population generating diversity in HIV1 (Vidal et al. 2000; Sharp and Hahn 2008). HIV1 group might have stemmed from strain SIVcpz of SIV (Keele et al. 2006), whereas HIV2 group might have stemmed from strain SIVsmm of SIV (Gao et al. 1992, 1994). HIV1 is further classified into group M (main), group O (outlier), group N (non-M/non-O), and group P. Group M is further subdivided into ten subtypes A–K, CRF's (Vidal et al. 2000; Sharp and Hahn 2008). Lemey et al. (2004) classified HIV2 into groups A and B. Further, HIV2 was classified

Table 13.1 Worldwide statistics as of 2017 for HIV and AIDS (Global Statistics 2017)

S. no.	Parameter	Statistics
1	People living with HIV	36.9 million (31.1–43.9 million)
2	People on antiretroviral therapy	21.7 million (19.1–22.6 million)
3	People newly infected with HIV	1.8 million (1.4–2.4 million)
4	People died with HIV infection	940,000 (670,000–1.3 million)
5	People living with HIV since epidemic	77.3 million (59.9–100 million)
6	People died with HIV infection since epidemic	35.4 million (25.0–49.9 million)

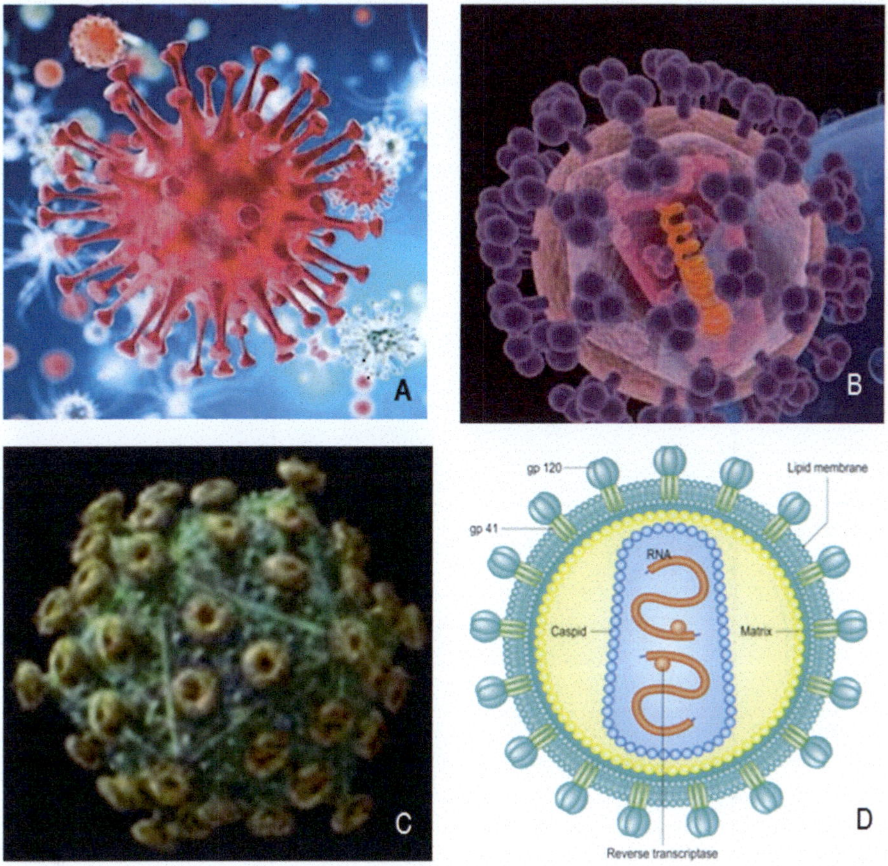

Fig. 13.1 (**a**) HIV infects humans and causes disease known as AIDS, (**b**) SIV infects nonhuman primates and causes disease known as SAIDS, (**c**) FIV infects cats and causes disease known as FAIDS, and (**d**) BIV infects cattle and causes disease known as BAIDS

phylogenetically into hypothetical groups such as group C–G, and 96FR12034 (Damond et al. 2004). Epidemiological, physiological, clinical, and phylogenetic evidence favored that HIV1 and HIV2 are due to "… several cross-species transmission of HIV from chimpanzee to humans …" (Castro-Nallar et al. 2012; Huet et al. 1990; Hahn et al. 2000; Plantier et al. 2009; Van Heuverswyn and Peeters 2007). Intensive studies were carried out on the evolution and divergence of HIV1 and HIV2 using phylogeny. The divergence time of HIV1, subtype A of HIV2 and subtype B of HIV2 dated to 1920s (Worobey et al. 2008), 1940 ± 16 (Lemey et al. 2003) and 1945 ± 14 (Gilbert et al. 2007), respectively. HIV1 was introduced into North America in 1968 (1966–1970) (Pérez-Losada et al. 2010; Surekha and Neelapu 2019).

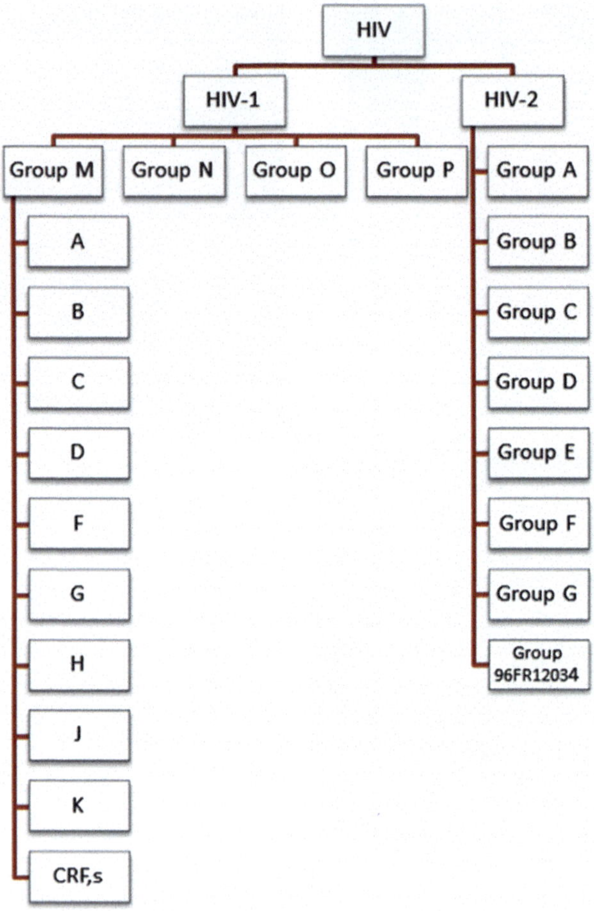

Fig. 13.2 Classification of HIV into two major types HIV1 and HIV2

13.2.2 Pathogenesis of HIV

HIV1 advances into the host cells using two molecules of HIV envelope (glycoprotein—gp120 and gp41). This ingress of glycoproteins stimulates intracellular signal cascades and facilitates replication of virus. The glycoproteins gp120 and gp41 form spikes on the surface of virion's. The protein gp120 binds to the $CD4^+$ receptor and attaches to the host cell membrane. A marked decrease in $CD4^+$ T cells (both activated and memory) is the characteristic feature of infection and disease. The virus then interacts with receptors (CCR5, CXCR4) leading to structural changes of the protein. The viral and the cellular membranes of the host are fused and form a viral pore. This allows the transfer of the viral core into the cytoplasm of the host. After the core disassembles, the RNA is reverse transcribed into DNA by the virus' with the help of enzyme reverse transcriptase. The DNA is then integrated into the

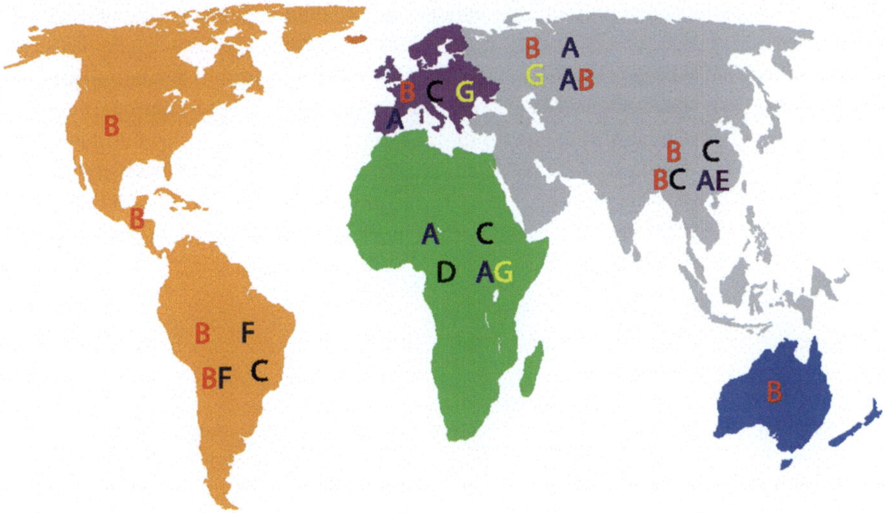

Fig. 13.3 Worldwide representation of HIV1 and its subtypes across the continents (Source: Castro-Nallar et al. (2012))

gene rich and transcriptionally active domains of the host's genome with the help of the viral protein integrase, DNA repair enzymes of the host and LEDGF/p75 (lens epithelium-derived growth factor—integrase binding host factor). Once the host cell is transformed into a potential virus producer, transcription of viral genetic material occurs. Later viral proteins are transported and assembled at the cell membrane with the help of vesicular sorting pathway (ESCRT-I, -II, and -III). TSG101 domain, a short sequence motif in p6 of Gag I is used for protein sorting. Cleavage of the Gag-Pol polyprotein by the viral protease produces mature infectious virions …" (Naif 2013).

13.3 Simian Immunodeficiency Virus

Simian Immunodeficiency Virus (SIV) infects "non-human primates" and affects the immune system of the "non-human primates" upon infection (Fig. 13.1). The host then acquires a disease known as Simian Acquired Immunodeficiency Disease Syndrome (SAIDS). SIV was isolated in 1985 from captive rhesus macaques suffering from SAIDS (Daniel et al. 1985).

13.3.1 Evolution, Distribution, and Diversity of SIV

SIVs infect nearly 45 different species of "non-human primates." Different types of SIV infecting "non-human primates" are mentioned in Table 13.2 (Gordon et al. 2005). SIV may be present at least 32,000 years ago in monkeys and apes (Worobey

Table 13.2 Major types of SIVs and their corresponding nonhuman primate hosts infected with SIV

S. no.	SIV type	Nonhuman primate infected with SIV	S. no.	SIV type	Nonhuman primate infected with SIV
1	SIVcpz	Chimpanzee	19	SIVsun	Sun-tailed monkey
2	SIVgor	Gorillas	20	SIVprg	Preuss's guenon
3	SIVagm	African green monkeys	21	SIVwrc	Western red colobus
4	SIVsmm	Sooty mangabeys	22	SIVolc	Olive colobus
5	SIVrcm	Red-capped mangabeys	23	SIVkrc	Kibale red colobus
6	SIVgsn	Greater spot-nosed monkeys	24	SIVtrc	Tshuapa red colobus
7	SIVmus	Mustached guenons	25	SIVsyk	Sykes' monkey
8	SIVmon	Mona monkeys	26	SIVdeb	De Brazza's monkey
9	SIVagi	Agile mangabey	27	SIVden	Dent's mona monkey
10	SIVmnd 2	Mandrill	28	SIVwol	Wolf's mona monkey
11	SIVdrl	Drill monkey	29	SIVgsn/SIVmon/ SIVmus 1/SIVmus 2 clade, SIVtal	Northern talapoin
12	SIVmac	Rhesus macaque	30	SIVasc	Red-tailed guenon
13	SIVmne	Pig-tailed macaque	31	SIVbkm	Black mangabey
14	SIVstm	Stump-tailed macaque	32	SIVreg [formerly SIVery]	Red-eared guenon
15	SIVsab, SIVver, SIVgri	Grivet monkey	33	SIVblu	Blue monkey
16	SIVtan	Tantalus monkey	34	SIVcol	Colobus guereza
17	SIVmnd 1	Mandrill	35	SIVkcol 1	Black-and-white colobus
18	SIVlho	L'hoest's monkey	36	SIVkcol 2, SIVblc (formerly SIVbcm)	Black colobus

et al. 2010). Primate's infection with SIV dated back to 14 Ma, if the virus and host were coevolved then it dates to 85 Ma (Compton et al. 2013). The "non-human primates" include species of gorilla, chimpanzees, and monkeys. The species of gorilla are *Gorilla gorilla* (western lowland gorillas), and *Gorilla beringei* (eastern Grauer's gorillas). The species of western gorilla are subdivided into *Gorilla gorilla diehli* (Cross River gorilla) and *Gorilla gorilla gorilla* (western lowland gorilla). The eastern species are classified into *Gorilla beringei graueri* (Grauer's gorilla), *Gorilla beringei beringei* (mountain gorilla), and *Gorilla beringei* subspecies (Bwindi gorilla). The chimpanzee species include *Pan troglodytes verus* (western Africa), *Pan troglodytes* ellioti (Nigerian), *Pan troglodytes troglodytes* (central), and *Pan troglodytes schweinfurthii* (eastern). The monkey species include African green monkeys, sooty mangabeys, *Cercocebus torquatus* (red-capped mangabeys),

Cercopithecus nictitans (greater spot-nosed monkeys), *Cercopithecus cephus* (mustached guenons), and *Cercopithecus mona* (mona monkeys). The association of SIV with apes, gorillas, and monkeys can be approximately 32,000 years or even much longer (Worobey et al. 2010). It is hypothesized that "... several cross-species transmission events ..." of SIVcpz and SIVsmm might have resulted in evolution of HIV1 and HIV2 (Huet et al. 1990; Hahn et al. 2000; Plantier et al. 2009; Van Heuverswyn and Peeters 2007). SIVcpz from chimpanzees "... crossed the species barrier ..." and migrated to humans giving rise to HIV1. Similarity, SIVsmm from sooty mangabeys "... crossed the species barrier ..." and migrated to humans giving rise to HIV-2.

13.3.2 Pathogenesis of SIV

SIV infects $CD4^+$ T cells and the SIV-infected cells undergo apoptosis within 1 day of infection. The immune system of simians is not able to replace the cells at the same pace leading to the deterioration of immune function. Subsequently, the host acquires immunodeficiency syndrome leaving the animal susceptible to fatigue and life-threatening coinfections (Fig. 13.4a). "... The interaction of SIV with the host's immune system triggers innate immune responses followed by virus-specific adaptive cellular and humoral immune responses. Rapidly occurring mutations lead to immune evasion. Chronic immune activation contributes to the functional impairment of the immune system. As the disease progresses, adaptive immune responses are unsuccessful in containing the virus replication, and overt signs of a chronic immune suppression become evident ..." (Schmitz and Korioth-Schmitz 2013). SIV infection leads to changes in mucosal tissues (cervicovaginal, gastrointestinal, and penile tissues) and all organ systems (brain, lung, and heart) (Haase 2011). The severity of the infection has an impact and functionally impairs the organs (Alammar et al. 2011; Kelly et al. 2012). Successful infection of SIV also results in loss of $CD4^+$ T cells. SIV uses glycoproteins to bind $CD4^+$ receptors of T cells and interacts with co-receptor CXCR4 leading to conformational changes of the protein. "... The viral and the cellular membranes of the host are fused allowing the transfer of the SIV genome into the cytoplasm of the host. The RNA is reverse transcribed into DNA by the virus with the help of enzyme reverse transcriptase. Then this DNA is integrated into the host cell's genome, transforming the host cell into a potential virus producer. The virus then uses host cell's replication machinery to transcribe its DNA back into RNA. The copies of RNA are packed into virus particles of about 80–100 nm in diameter and bud off or free from the host cell infecting more cells. SIV infections are non-pathogenic in their natural African simian primates. However, if the virus infects an Asian or Indian rhesus macaque, these non-African simian primates will develop simian AIDS (SAIDS). SIVs nef gene down-regulates expression of CD3, CD4, and MHC class I and therefore do not induce immunodeficiency. Whereas, nef gene in HIV-1 lost its ability to down-regulate CD3, which results in the immune activation and apoptosis ..." (Swigut et al. 2004).

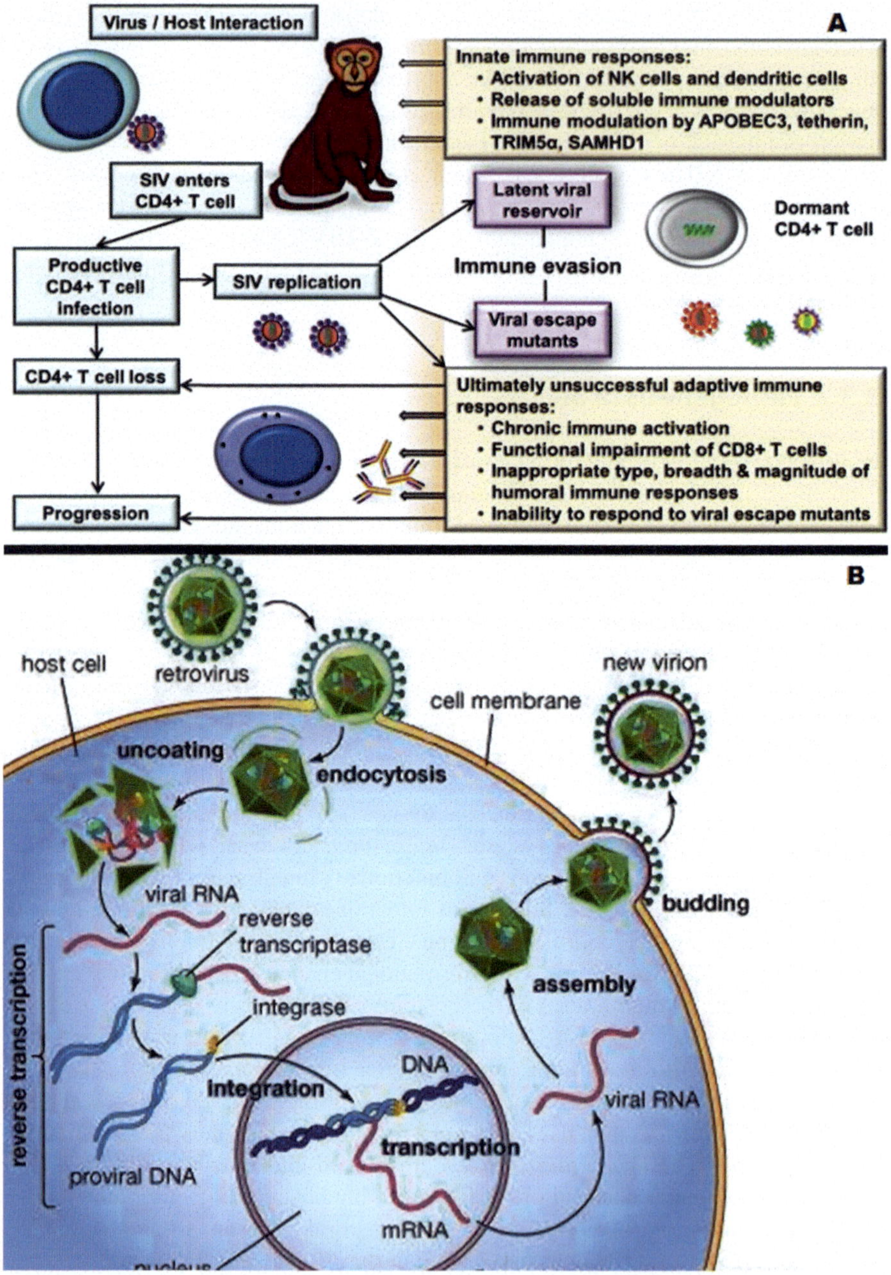

Fig. 13.4 (**a**) Consequences and immunopathogenesis in simian when infected with a simian immunodeficiency virus (SIV) (Source: Schmitz and Korioth-Schmitz (2013)). (**b**) Replication of SIV in the host simian (Source: Encyclopaedia Britannica Inc. https://www.britannica.com/science/SIV)

13.4 Feline Immunodeficiency Virus

Feline Immunodeficiency Virus (FIV) infects cats and affects the immune system of the cats upon infection (Fig. 13.1). The cat then acquires a disease known as Feline Acquired Immunodeficiency Disease Syndrome (FAIDS). FIV was first isolated from a cat exhibiting an immunodeficiency-like syndrome (Pedersen et al. 1987).

13.4.1 Evolution, Distribution, and Diversity of FIV

FIV is endemic in Felidae species (Biek et al. 2003, 2006; Brown et al. 1994; Carpenter et al. 1996, 1998; Carpenter and O'Brien 1995; Driciru et al. 2006; Hofmann-Lehmann et al. 1996; Munson et al. 2004; Olmsted et al. 1992; Troyer et al. 2004; Troyer et al., 2005), from free-ranging lions (Pecon-Slattery et al. 2004) to domestic cats (Olmsted et al. 1989). FIV occurs in *Felis catus* (domestic cats), *Puma concolor* (pumas) (Langley et al. 1994), *Panthera leo* (lions), *Otocolobus manul* (pallas cat) (Barr et al. 1997), *Panthera pardus* (puma leopard), *Acinonyx jubatus* (cheetah), *Leopardus pardalis* (ocelot), *Panthera onca* (jaguar), *Leopardus weidii* (margay), *Leopardus tigrina* (tigrina), large African lions, and *Crocuta crocuta* (spotted hyaena) (Troyer et al. 2005). The different types of FIVs like FIV_{Ple}, FIV_{Fca}, FIV_{Pco}, FIV_{Oma}, FIV_{Lpa}, FIV_{Aju}, FIV_{Ccr}, FIV_{Pon}, FIV_{Lwe}, and FIV_{Lti} were reported in literature (Fig. 13.5). The major types of FIVs and their corresponding felines infected with FIV are listed in Table 13.3.

Currently, isolates of FIV are classified into six subtypes (A, B, C, D, E, and F) (Weaver 2010). Weaver (2010) performed a detailed phylogenetic analysis of FIV and classified FIV into six subtypes (A, B, C, D, E, and F) (Fig. 13.4). Sodora et al. (1994) initially classified FIV into three subtypes (A, B, and C). Sodora et al. (1994) used phylogenetic analysis to classify FIV isolates based on FIV *env* (envelope sequence). The geographical location of the subtype A is California and Europe; subtype B is Japan and the central and eastern USA; and Subtype C is southwestern Canada. Nishimura et al. (1998), later classified FIV into five subtypes (A, B, C, D,

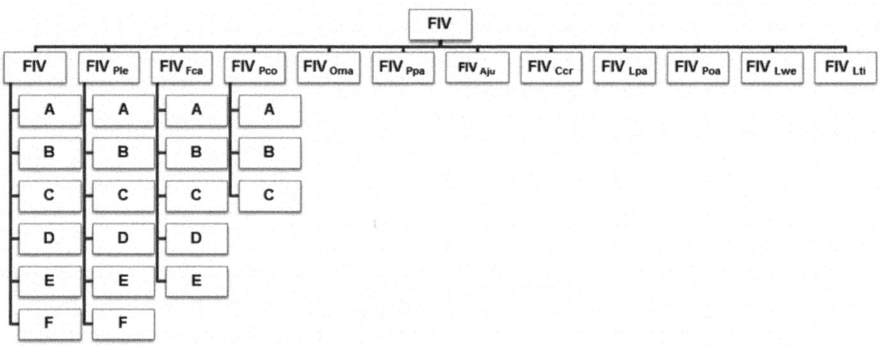

Fig. 13.5 Classification of FIV and their subtypes

Table 13.3 Major types of FIVs and their corresponding felines infected with FIV

S. no.	FIV type	Felines infected with FIV
1	FIV_{Ple}	*Panthera leo* (lions)
2	FIV_{Fca}	*Felis catus* (domestic cats)
3	FIV_{Pco}	*Puma concolor* (pumas)
4	FIV_{Oma}	*Otocolobus manul* (pallas cat)
5	FIV_{Ppa}	*Panthera pardus* (puma leopard)
6	FIV_{Aju}	*Acinonyx jubatus* (cheetah)
7	FIV_{Ccr}	*Crocuta crocuta* (spotted hyaena)
8	FIV_{Lpa}	*Leopardus pardalis* (ocelot)
9	FIV_{Pon}	*Panthera onca* (jaguar)
10	FIV_{Lwe}	*Leopardus weidii* (margay)
11	FIV_{Lti}	*Leopardus tigrina* (tigrina)

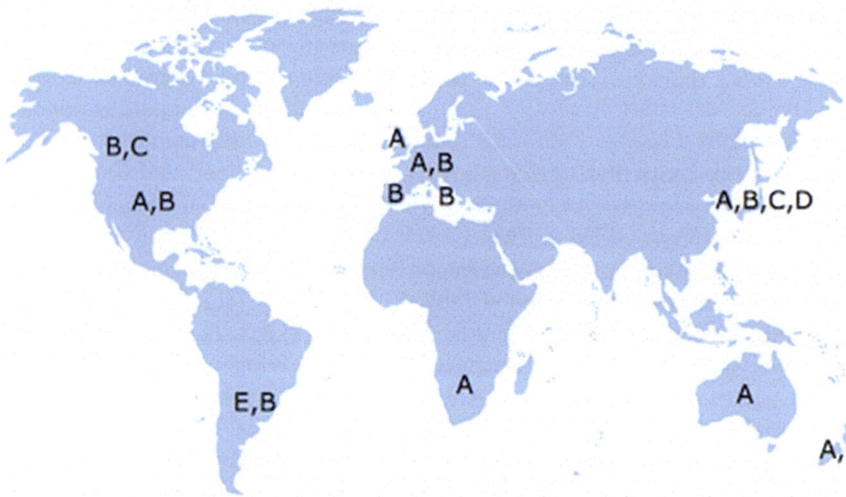

Fig. 13.6 Worldwide distribution of FIV and its subtypes across the continents (Source: Teixeira et al. (2012))

and E) based on FIV *env* (envelope sequence). The geographical location of the subtype A is California and Northern Europe; subtype B is central and eastern USA and southern European countries; subtype C is California and British Columbia; subtype D is Japan; and subtype E is Argentina. FIV_{Ple} has diverged into six subtypes A–F, each with distinct geographic areas of endemicity (Brown et al. 1994; Troyer et al. 2004; O'Brien et al. 2006). FIV_{Fca} has diverged into five subtypes, A–E (Sodora et al. 1994; Kakinuma et al. 1995; Pecoraro et al. 1996), where subtypes A, B, and C are widespread worldwide. Subtype D is found in Japan and Vietnam (Kakinuma et al. 1995; Nakamura et al. 2003), whereas subtype E is only found in Argentina (Pecon-Slattery et al. 2008). The four subtypes A–D are found in cat populations from Japan (Nishimura et al. 1998; Kakinuma et al. 1995). FIV_{Pco} has diverged into three subtypes A, B, and C (Pecon-Slattery et al. 2008) (Fig. 13.6).

13.4.2 Pathogenesis of FIV

FIV uses receptor CD9 (Poeschla and Looney 1998) for entry and then infects $CD4^+$ T (Brown et al. 1991), FIV can also infect astroglial cells, $CD8^+$ T lymphocytes, macrophages, and kidney cells (Pedersen et al. 1989; Brunner and Pedersen 1989; Phillips et al. 1990; Zenger et al. 1995). "… This leads to a significant drop in cells which have critical roles in the immune system. Low levels of $CD4^+$ and other affected immune system cells cause the cat to be susceptible to opportunistic diseases once the disease progresses to feline acquired immune deficiency syndrome (FAIDS) …" (Bendinelli et al. 1995). Symptoms of immunodeficiency associated with FIV are chronic lesions, opportunistic infections, neurological abnormalities, and weight loss (Yamamoto et al. 2010).

CD134 is mostly present on T cells which are activated and binds to OX40 ligand, causing T cell activation, stimulation, proliferation, and apoptosis. The virus enters the host's cells using the glycoprotein env and interacts with surface receptors. The glycoprotein SU binds to receptor CD134 of the host cell leading to conformational changes of protein SU. These changes facilitate interaction between SU and the chemokine receptor CXCR4; and fuses viral membrane and cellular membrane of the host (Hu 2012). "… This allows the transfer of the viral RNA into the cytoplasm of the host. Then RNA is reverse transcribed and integrated into the genome of the host through non-homologous recombination. Once viral RNA is integrated into the host genome, the virus can be dormant in the asymptomatic stage without being detected by the immune system of the host or can cause lysis of the host cell …" (Lecollinet and Richardson 2008; Hartmann 2011).

13.5 Bovine Immunodeficiency Virus

Bovine Immunodeficiency Virus (BIV) infects cattle and affects the immune system of the cattle upon infection (Fig. 13.1). The cattle then acquires a disease known as Bovine Acquired Immunodeficiency Disease Syndrome (BAIDS). BIV strain "… R-29 was originally isolated from an 8-year-old dairy cow suspected of having bovine lymphosarcoma …" (Van Der Maaten et al. 1972a).

13.5.1 Evolution, Distribution, and Diversity of BIV

BIV is widespread in dairy and beef cattle in Australia (Forman et al. 1992), Canada (McNab et al. 1994), France (Polack et al. 1996), Germany (Muluneh 1994), Italy (Cavirani et al. 1998), Japan (Hirai et al. 1996; Meas et al. 1998), Korea (Cho et al. 1999), Netherlands (Horzinek et al. 1991a), New Zealand (Horner 1991), the UK (Clayton 1994), the USA (Amborski et al. 1989; Cockerell et al. 1992; St. Cyr Coats et al. 1994), and buffaloes in Pakistan (Meas et al. 2000). The prevalence of BIV in dairy cattle is higher than beef cattle (Amborski et al. 1989; Cho et al. 1999). The different BIV strains reported in literature are BIV R-29 (isolated from cow) (Van

Der Maaten et al. 1972b), BIV-106, and BIV-127 (variants of BIV R-29) (Braun et al. 1988; Garvey et al. 1990), BIV CR1 (BIV strain from Costa Rica (Hidalgo et al. 1995), BIV FL112 (Hirari et al. 1996), and FL491 (Suarez et al. 1993) (strain from Florida, USA).

13.5.2 Pathogenesis of BIV

BIV infects nondividing host cells and uses two phases—entry phase (first phase) and replication phase (second phase) to replicate itself (Berkowitz et al. 2001). In the first phase, glycoprotein of the virus envelope interacts and binds with the specific receptor of the cell. "… Then virus uses either of the ways, receptor mediated endocytosis or viral envelope-cell membrane fusion to enter into the host cell. The virus then dissembles in the cell and reverse-transcribes the RNA genome into DNA with the help of enzyme reverse transcriptase (Gonda, 1992). The activity of reverse transcriptase is higher at lower concentrations of Mn^{2+} ions when compared to Mg^{2+} ions (Horzinek et al. 1991b). The DNA is then transported to nucleus and is integrated into the genome of the host cell. In the second phase, the integrated viral DNA is transcribed and the transcript (viral mRNA) is transported and translated in the cytoplasm …." The translated viral structural proteins are then assembled into virus particles and form a viral complex (virus buds) along with the viral RNA at the plasma membrane. Then viral proteases process the viral buds and are released from the cell in the mature stages, which are ready to infect another cell (Gonda 1992).

13.6 Dog Immunodeficiency Viruses

Dog Immunodeficiency Virus (DIV) is a retrovirus which infects dog and affects the immune system of the dogs upon infection. The dog then acquires a disease known as DAIDS. Safran et al. (1992) isolated canine immunodeficiency virus, (CIV)—(canine lentivirus) from a leukemic dog. The ultrastructure and morphogenesis of CIV is strikingly similar to that displayed by other lentiviruses, while the immunological relatedness close to any other lentivirus or oncovirus was not established. These findings suggest that this canine retrovirus, fits in the subfamily lentivirus and not related to other known members. Additional, investigations on DIV (CIV) are essential to establish the biology, immunopathogenesis of virus; and how immunodeficiency is acquired by dogs.

13.7 Conclusions and Future Perspectives

Immunodeficiency viruses infect the host and primarily affect the immune system of an organism i.e. host. HIV, SIV, FIV, BIV, and DIV were known to infect humans, simians, cats, cattle, and dogs respectively. The ultrastructure and morphogenesis of IVs like HIV, SIV, FIV, BIV, and DIV are strikingly similar. At the same time immunological relatedness of the lentiviruses like HIV, SIV, FIV and BIV are close with exception in DIV. Therefore, there is a need to establish the immunological relatedness of the DIV. The similarities in the ultrastructure, morphogenesis and immunological relatedness of lentivirus provide us with an opportunity for better understanding of the immunodeficiency in different hosts. This information further can be used to develop a vaccine to the most harmful and dreadful disease.

Acknowledgements HSV, SC, and NRRN are grateful to GITAM (Deemed to be University) for providing necessary facilities to carry out the research work and for extending constant support.

Authors Contribution HSV, CS and NNRR initiated the review, participated in writing and revised the manuscript.

Conflict of Interest Statement The authors declare that there is no potential conflict of interest.

References

Alammar L, Gama L, Clements JE (2011) Simian immunodeficiency virus infection in the brain and lung leads to differential type I IFN signaling during acute infection. J Immunol 186:4008–4018

Amborski GF, Lo J, Seger CL (1989) Serological detection of multiple retroviral infections in cattle: bovine leukemia virus, bovine syncytial virus and bovine visna virus. Vet Microbiol 20:247–253

Barr MC, Zou L, Long F, Hoose WA, Avery RJ (1997) Proviral organization and sequence analysis of feline immunodeficiency virus isolated from a Pallas' cat. Virology 228:84–91

Barré-Sinoussi F, Chermann JC, Rey F, Nugeyre MT, Chamaret S, Gruest J, Dauguet C, Axler-Blin C, Vézinet-Brun F, Rouzioux C, Rozenbaum W, Montagnier L (1983) Isolation of a T-lymphotropic retrovirus from a patient at risk for acquired immune deficiency syndrome (AIDS). Science 220:868–871

Bendinelli M, Pistello M, Lombardi S, Poli A, Garzelli C, Matteucci D, Ceccherini-Nelli L, Malvaldi G, Tozzini F (1995) Feline immunodeficiency virus: an interesting model for AIDS studies and an important cat pathogen. Clin Microbiol Rev 8:87–112

Berkowitz R, Ilves H, Lin WY, Eckert K, Coward A, Tamaki S, Veres G, Plavec I (2001) Construction and molecular analysis of gene transfer systems derived from bovine immunodeficiency virus. J Virol 75(7):3371–3382

Biek R, Rodrigo AG, Holley D, Drummond A, Anderson CR Jr, Ross HA, Poss M (2003) Epidemiology, genetic diversity, and evolution of endemic feline immunodeficiency virus in a population of wild cougars. J Virol 77:9578–9589

Biek R, Ruth TK, Murphy KM, Anderson CR Jr, Johnson M, DeSimone R, Gray R, Hornocker MG, Gillin CM, Poss M (2006) Factors associated with pathogen seroprevalence and infection in Rocky Mountain cougars. J Wildl Dis 42:606–615

Braun MJ, Lahn S, Boyd AL, Kost TA, Nagashima K, Gonda MA (1988) Molecular cloning of biologically active proviruses of bovine immunodeficiency-like virus. Virology 167:515–523

Brown EW, Yuhki N, Packer C, O'Brien SJ (1994) A lion lentivirus related to feline immunodeficiency virus: epidemiologic and phylogenetic aspects. J Virol 68:5953–5968

Brown WC, Bissey L, Logan KS, Pedersen NC, Elder JH, Collisson EW (1991) Feline immunodeficiency virus infects both CD4+ and CD8+ T lymphocytes. J Virol 65:3359–3364

Brunner D, Pedersen NC (1989) Infection of peritoneal macrophages in vitro and in vivo with feline immunodeficiency virus. J Virol 63:5483–5488

Carpenter MA, Brown EW, Culver M, Johnson WE, Pecon-Slattery J, Brousset D, O'Brien SJ (1996) Genetic and phylogenetic divergence of feline immunodeficiency virus in the puma (*Puma concolor*). J Virol 70:6682–6693

Carpenter MA, Brown EW, MacDonald DW, O'Brien SJ (1998) Phylogeographic patterns of feline immunodeficiency virus genetic diversity in the domestic cat. Virology 251:234–243

Carpenter MA, O'Brien SJ (1995) Coadaptation and immunodeficiency virus: lessons from the Felidae. Curr Opin Genet Dev 5:739–745

Castro-Nallar E, Perez-Losada M, Burton GF, Crandall KA (2012) The evolution of HIV: inferences using phylogenetics. Mol Phylogenet Evol 62:777–792

Cavirani S, Donofrio G, Chiocco D, Foni E, Martelli P, Allegri G, Cabassi CS, De Iaco B, Flammini CF (1998) Seroprevalence to bovine immunodeficiency virus and lack of association with leukocyte counts in Italian dairy cattle. Prev Vet Med 37:147–157

Cho KO, Meas S, Park NY, Kim YK, Endoh D, Lee SI, Ohashi K, Sugimoto C, Onuma M (1999) Seroprevalence of bovine immunodeficiency virus in dairy and beef cattle herds in Korea. J Vet Med Sci 61:549–551

Clavel F, Guetard D, Brun-Vezinet F, Chamaret S, Rey MA, Santos-Ferreira MO, Laurent AG, Dauguet C, Katlama C, Rouzioux C, Klatzmann D, Champalimaud JL, Montagnier L (1986) Isolation of a new human retrovirus from West African patients with AIDS. Science 233:343–346

Clayton J (1994) Spectre of AIDS haunts reports of sick cows. Nat (Lond) 367:585

Cockerell GL, Jensen WA, Rovnak J, Ennis WH, Gonda MA (1992) Seroprevalence of bovine immunodeficiency-like virus and bovine leukemia virus in a dairy cattle herd. Vet Microbiol 31:109–116

Compton AA, Malik HS, Emerman M (2013) Host gene evolution traces the evolutionary history of ancient primate lentiviruses. Philos Trans R Soc Lond Ser B Biol Sci 68(1626):20120496. https://doi.org/10.1098/rstb.2012.0496

Damond F, Worobey M, Campa P, Farfara I, Colin G, Matheron S, Brun-Vézinet F, Robertson DL, Simon F (2004) Identification of a highly divergent HIV type 2 and proposal for a change in HIV type 2 classification. AIDS Res Hum Retrovir 20(6):666–672

Daniel MD, Letvin NL, King NW, Kannagi M, Sehgal PK, Hunt RD, Kanki PJ, Essex M, Desrosiers RC (1985) Isolation of T-cell tropic HTLV-III-like retrovirus from macaques. Science 228(4704):1201–1204

Driciru M, Siefert L, Prager KC, Dubovi E, Sande R, Princee F, Friday T, Munson L (2006) A serosurvey of viral infections in lions (*Panthera leo*), from Queen Elizabeth National Park, Uganda. J Wildl Dis 42:667–671

Forman AJ, Gibson CA, Rodwell BJ (1992) Serological evidence for the presence of bovine lentivirus infection in Australia. Aust Vet J 69:337

Gao F, Yue L, Robertson DL, Hill SC, Hui H, Biggar RJ, Neequaye AE, Whelan TM, Ho DD, Shaw GM (1994) Genetic diversity of human immunodeficiency virus type 2: evidence for distinct sequence subtypes with differences in virus biology. J Virol 68(11):7433–7447

Gao F, Yue L, White AT, Pappas PG, Barchue J, Hanson AP, Greene BM, Sharp PM, Shaw GM, Hahn BH (1992) Human infection by genetically diverse SIVSM-related HIV-2 in west Africa. Nature 358(6386):495–499

Garvey KJ, Obsrste MS, Esler JE, Braun MJ, Gonda MA (1990) Nucleotide sequence and genome organization of biologically active proviruses of the bovine immunodeficiency-like virus. Virology 175:391–409

Gilbert MTP, Rambaut A, Wlasiuk G, Spira TJ, Pitchenik AE, Worobey M (2007) The emergence of HIV/ AIDS in the Americas and beyond. Proc Natl Acad Sci U S A 104:18566–18570

Global Statistics (2017). https://www.hiv.gov/hiv-basics/overview/data-and-trends/global-statistics. Accessed 09 July 2017

Gonda MA (1992) Bovine immunodeficiency virus. AIDS 6(8):759–760

Gordon S, Pandrea I, Dunham R, Apetrei C, Silvestri G (2005) The call of the wild: what can be learned from studies of SIV infection of natural hosts? HIV Seq Compend 2004:2–29

Haase AT (2011) Early events in sexual transmission of HIV and SIV and opportunities for interventions. Annu Rev Med 62:127–139

Hahn BH, Shaw GM, De Cock KM, Sharp PM (2000) AIDS as a zoonosis: scientific and public health implications. Science 287:607–614

Hartmann K (2011) Clinical aspects of feline immunodeficiency and feline leukemia virus infection. Vet Immunol Immunopathol 143(3–4):190–201

Hidalgo G, Flores M, Bonilla JA (1995) Detection and isolation of bovine immunodeficiency-like virus (BIV) in dairy herds of Costa Rica. Zentralbl Veterinarmed B 42(3):155–161

Hirai N, Kabeya H, Ohashi K, Sugimoto C, Onuma M (1996) Immunomodulative effects of bovine immunodeficiency-like virus (BIV) infection and mixed infection of BIV and bovine leukemia virus on sheep. J Vet Med Sci 58:455–457

Hirari N, Kabeya Ohashi K, Sugimoto C, Onuma M (1996) Detection of antibodies against bovine immunodeficiency like virus in dairy cattle in Hokkaido. J Vet Med Sci 58(5):455–457

Hofmann-Lehmann R, Fehr D, Grob M, Elgizoli M, Packer C, Martenson JS, O'Brien SJ, Lutz H (1996) Prevalence of antibodies to feline parvovirus, calicivirus, herpesvirus, coronavirus, and immunodeficiency virus and of feline leukemia virus antigen and the interrelationship of these viral infections in free-ranging lions in east Africa. Clin Diagn Lab Immunol 3:554–562

Horner GW (1991) Serologic evidence of bovine immunodeficiency-like virus and bovine syncytial virus in New Zealand. Surveillance 18:9

Horzinek M, Keldermans L, Stuuman T, Black J, Her-Rewegh A, Sillekens P, Koolen M (1991a) Bovine immunodeficiency virus: immunochemical characterization and serological survey. J Gen Virol 72:2923–2928

Horzinek M, Keldermans L, Stuurman T, Black J, Herrewegh A, Sillekens P, Koolen M (1991b) Bovine immunodeficiency virus: immunochemical characterization and serological survey. J Gen Virol 72(12):2923–2928

Hu Q-Y (2012) Mapping of receptor binding interactions with the FIV surface glycoprotein (SU); implications regarding immune survelliance and cellular targets of infection. Retrovirology 1(11):1–11

Huet T, Cheynier R, Meyerhans A, Roelants G, Wain-Hobson S (1990) Genetic organization of a chimpanzee lentivirus related to HIV-1. Nature 345:356–359

Kakinuma S, Motokawa K, Hohdatsu T, Yamamoto JK, Koyama H, Hashimoto H (1995) Nucleotide sequence of feline immunodeficiency virus: classification of Japanese isolates into two subtypes which are distinct from non-Japanese subtypes. J Virol 69:3639–3646

Keele BF, Van Heuverswyn F, Li Y, Bailes E, Takehisa J, Santiago ML, Bibollet-Ruche F, Chen Y, Wain LV, Liegeois F, Loul S, Ngole EM, Bienvenue Y, Delaporte E, Brookfield JF, Sharp PM, Shaw GM, Peeters M, Hahn BH (2006) Chimpanzee reservoirs of pandemic and nonpandemic HIV-1. Science 313(5786):523–526

Kelly KM, Tarwater PM, Karper JM, Kelly KM, Tarwater PM, Karper JM, Bedja D, Queen SE, Tunin RS, Adams RJ, Kass DA, Mankowski JL (2012) Diastolic dysfunction is associated with myocardial viral load in simian immunodeficiency virus-infected macaques. AIDS 26:815–823

Langley RJ, Hirsch VM, O'Brien SJ, Adger-Johnson D, Goeken RM, Olmsted RA (1994) Nucleotide sequence analysis of a puma lentivirus (PLV-14): genomic organization and relationship to other lentiviruses. Virology 202:853–864

Lecollinet S, Richardson J (2008) Vaccination against the feline immunodeficiency virus: the road not taken. Comp Immunol Microbiol Infect Dis 31(2–3):167–190

Lemey P, Pybus OG, Rambaut A, Drummond AJ, Robertson DL, Roques P, Worobey M, Vandamme AM (2004) The molecular population genetics of HIV-1 group O. Genetics 167(3):1059–1068

Lemey P, Pybus OG, Wang B, Saksena NK, Salemi M, Vandamme A-M (2003) Tracing the origin and history of the HIV-2 epidemic. Proc Natl Acad Sci U S A 100:6588–6592

McNab WB, Jacobs RM, Smith HE (1994) A serological survey for bovine immunodeficiency-like virus in Ontario dairy cattle and associations between test results, production records and management practices. Can J Vet Res 58:36–41

Meas S, Kabeya H, Ohashi K, Matsuki S, Mikami Y, Sugimoto C, Onuma M (1998) Seroprevalence and field isolation of bovine immunodeficiency virus. J Vet Med Sci 60:1195–1202

Meas S, Seto J, Sugimoto C, Bakhsh M, Riaz M, Sato T, Neem K, Ohashi K, Onuma M (2000) Infection of bovine immunodeficiency virus and bovine leukemia virus in water buffalo and cattle populations in Pakistan. J Vet Med Sci 62:329–331

Muluneh A (1994) Seroprevalence of bovine immunodeficiency-virus (BIV) antibodies in the cattle population in Germany. J Veterinary Med Ser B 41:679–684

Munson L, Marker L, Dubovi E, Spencer JA, Evermann JF, O'Brien SJ (2004) Serosurvey of viral infections in free-ranging Namibian cheetahs (*Acinonyx jubatus*). J Wildl Dis 40:23–31

Naif HM (2013) Pathogenesis of HIV infection. Infect Dis Rep 5(Suppl 1):e6. https://doi.org/10.4081/idr.2013.s1.e6

Nakamura K, Suzuki Y, Ikeo K, Ikeda Y, Sato E, Nguyen NTP, Gojobori T, Mikami T, Miyazawa T (2003) Phylogenetic analysis of Vietnamese isolates of feline immunodeficiency virus: genetic diversity of subtype C. Arch Virol 148:783–791

Nishimura Y, Goto Y, Pang H, Endo Y, Mizuno T, Momoi Y, Watari T, Tsujimoto H, Hasegawa A (1998) Genetic heterogeneity of env gene of feline immunodeficiency virus obtained from multiple districts in Japan. Virus Res 57:101–112

O'Brien SJ, Troyer JL, Roelke M, Marker L, Pecon-Slattery J (2006) Plagues and adaptation: lessons from the Felidae models for SARS and AIDS. Biol Conserv 131:255–267

Olmsted RA, Hirsch VM, Purcell RH, Johnson PR (1989) Nucleotide sequence analysis of feline immunodeficiency virus: genome organization and relationship to other lentiviruses. Proc Natl Acad Sci U S A 86:8088–8092

Olmsted RA, Langley R, Roelke ME, Goeken RM, Adger-Johnson D, Goff JP, Albert JP, Packer C, Laurenson MK, Caro TM, Scheepers L, Wildt DE, Bush M, Martenson J, O'Brien SJ (1992) Worldwide prevalence of lentivirus infection in wild feline species: epidemiologic and phylogenetic aspects. J Virol 66:6008–6018

Pecon-Slattery J, Pearks Wilkerson AJ, Murphy WJ, O'Brien SJ (2004) Phylogenetic assessment of introns and SINEs within the Y chromosome using the cat family felidae as a species tree. Mol Biol Evol 21(12):2299–2309

Pecon-Slattery J, Troyer JL, Johnson WE, O'Brien SJ (2008) Evolution of feline immunodeficiency virus in Felidae: implications for human health and wildlife ecology. Vet Immunol Immunopathol 123(1–2):32–44

Pecoraro MR, Tomonaga K, Miyazawa T, Kawaguchi Y, Sugita S, Tohya Y, Kai C, Etcheverrigaray ME, Mikami T (1996) Genetic diversity of argentine isolates of feline immunodeficiency virus. J Gen Virol 77:2031–2035

Pedersen NC, Ho EW, Brown ML, Yamamoto JK (1987) Isolation of a T-lymphotropic virus from domestic cats with an immunodeficiency-like syndrome. Science 235:790–793

Pedersen NC, Yamamoto JK, Ishida T, Hansen H (1989) Feline immunodeficiency virus infection. Vet Immunol Immunopathol 21:111–129

Pérez-Losada M, Jobes DV, Sinangil F, Crandall KA, Posada D, Berman PW (2010) Phylodynamics of HIV-1 from a phase-III AIDS vaccine trial in North America. Mol Biol Evol 27:417–425

Phillips TR, Talbott RL, Lamont C, Muir S, Lovelace K, Elder JH (1990) Comparison of two host cell range variants of feline immunodeficiency virus. J Virol 64:4605–4613

Plantier J-C, Leoz M, Dickerson JE, De Oliveira F, Cordonnier F, Lemee V, Damond F, Robertson DL, Simon F (2009) A new human immunodeficiency virus derived from gorillas. Nat Med 15:871–872

Poeschla EM, Looney DJ (1998) CXCR4 is required by a nonprimate lentivirus: heterologous expression of feline immunodeficiency virus in human, rodent, and feline cells. J Virol 72:6858–6866

Polack B, Schwartz I, Berthelemy M, Belloc C, Manet G, Vuillaume A, Baron T, Gonda MA, Levy D (1996) Serologic evidence for bovine immunodeficiency virus infection in France. Vet Microbiol 48:165–173
Safran N, Perk K, Eyal O, Dahlberg JE (1992) Isolation and preliminary characterisation of a novel retrovirus isolated from a leukaemic dog. Res Vet Sci 52(2):250–255
Schmitz JE, Korioth-Schmitz B (2013) Immunopathogenesis of simian immunodeficiency virus infection in nonhuman primates. Curr Opin HIV AIDS 8(4):273–279
Sharp PM, Hahn BH (2008) The evolution of HIV-1 and the origin of AIDS. Philos Trans R Soc Lond Ser B Biol Sci 365(1552):2487–2494
Sodora DL, Shpaer EG, Kitchell BE, Dow SW, Hoover EA, Mullins JI (1994) Identification of three feline immunodeficiency virus (FIV) env gene subtypes and comparison of the FIV and human immunodeficiency virus type 1 evolutionary patterns. J Virol 68(4):2230–2238
St. Cyr Coats K, Pruett SB, Nash JW, Cooper CR (1994) Bovine immunodeficiency virus: incidence of infection in Mississippi dairy cattle. Vet Microbiol 42:181–189
Suarez DL, Van Der Maaten MJ, Wood C (1993) Isolation and characterization of new wild type isolates of bovine lentivirus. J Virol 67:5051–5055
Surekha C, Neelapu NRR (2019) Phylogenetic trees: applications, construction, and assessment. In: Shaik NA, Hakeem KR, Banaganapalli B, Elango R (eds) Essentials of bioinformatics, understanding bioinformatics: genes to proteins, vol III. Springer, Cham, pp 167-192
Swigut T, Alexander L, Morgan J, Lifson J, Mansfield KG, Lang S, Johnson RP, Skowronski J, Desrosiers R (2004) Impact of Nef-mediated downregulation of major histocompatibility complex class I on immune response to simian immunodeficiency virus. J Virol 78(23):13335–13344
Teixeira BM, Hagiwara MK, Cruz JC, Hosie MJ (2012) Feline immunodeficiency virus in South America. Viruses 4(3):383–396
Troyer JL, Pecon-Slattery J, Roelke ME, Black L, Packer C, O'Brien SJ (2004) Patterns of feline immunodeficiency virus multiple infection and genome divergence in a free-ranging population of African lions. J Virol 78:3777–3791
Troyer JL, Pecon-Slattery J, Roelke ME, Johnson W, VandeWoude S, Vazquez-Salat N, Brown M, Frank L, Woodroffe R, Winterbach C, Winterbach H, Hemson G, Bush M, Alexander KA, Revilla E, O'Brien SJ (2005) Seroprevalence and genomic divergence of circulating strains of feline immunodeficiency virus among Felidae and Hyaenidae species. J Virol 79:8282–8294
Van Der Maaten MJ, Booth AD, Segar CL (1972a) Isolation of a virus from cattle with persistent lymphocytosis. J Natl Cancer Inst 49:1649–1657
Van Der Maaten MJ, Boothe AD, Seger CL (1972b) Isolation of a virus from cattle with persistent lymphocytosis. J Natl Cancer Inst 49:1649–1657
Van Heuverswyn F, Peeters M (2007) The origins of HIV and implications for the global epidemic. Curr Infect Dis Rep 9:338–346
Vidal N, Peeters M, Mulanga-Kabeya C, Nzilambi N, Robertson D, Ilunga W, Sema H, Tshimanga K, Bongo B, Delaporte E (2000) Unprecedented degree of human immunodeficiency virus type 1 (HIV-1) group M genetic diversity in the Democratic Republic of Congo suggests that the HIV-1 pandemic originated in Central Africa. J Virol 74(22):10498–10507
Weaver EA (2010) A detailed phylogenetic analysis of FIV in the United States. PLoS One 5(8):e12004. https://doi.org/10.1371/journal.pone.0012004
Worobey M, Gemmel M, Teuwen DE, Haselkorn T, Kunstman K, Bunce M, Muyembe J-J, Kabongo J-MM, Kalengayi RM, Van Marck E, Gilbert MTP, Wolinsky SM (2008) Direct evidence of extensive diversity of HIV-1 in Kinshasa by 1960. Nature 455:661–664
Worobey M, Telfer P, Souquière S, Hunter M, Coleman CA, Metzger MJ, Reed P, Makuwa M, Hearn G, Honarvar S, Roques P, Apetrei C, Kazanji M, Marx PA (2010) Island biogeography reveals the deep history of SIV. Science 329(5998):1487. https://doi.org/10.1126/science.1193550
Yamamoto JK, Sanou MP, Abbott JR, Coleman JK (2010) Feline immunodeficiency virus model for designing HIV/AIDS vaccines. Curr HIV Res 8(1):14–25
Zenger E, Collisson EW, Barhoumi R, Burghardt RC, Danave IR, Tiffany-Castiglioni E (1995) Laser cytometric analysis of FIV-induced injury in astroglia. Glia 13:92–100

Role of Immune Cells in Hepatitis B Infection

14

Prakriti Sinha and Parul Sahu

Abstract

Hepatitis B virus infection is a common cause of liver cirrhosis and hepatocellular carcinoma leading to about one million deaths annually. The disease pathogenesis is complex and incompletely understood, especially when it turns chronic. Among various factors known to be responsible for the disease progression, the immune system has the most important role to play. Viral clearance is largely T cells dependent, and progression to a chronic state is due to insufficient T cells response, partly due to T-regulatory cells. In chronic infections, functions of important immune cells like natural killer cells and dendritic cells are impaired, which enables the viral persistence and promotes the pathogenesis in the host body. Similarly, the virus is reported to modulate functions of monocytes, macrophages, and Kupffer cells to promote an immune-tolerant local environment in the liver. Though B cells are expected to be central to the humoral immune response that clears the virus, its role in chronic HBV is still obscure. This chapter would elaborate on the significant roles of immune cells in pathogenesis and in clearance of the hepatitis B virus.

Keywords

Hepatitis B virus · Innate immunity · Adaptive immunity · T cells · Immunopathogenesis

P. Sinha (✉) · P. Sahu
National Institute of Immunology, New Delhi, India

P. V. Bramhachari (ed.), *Dynamics of immune activation in viral diseases*,
https://doi.org/10.1007/978-981-15-1045-8_14

14.1 Hepatitis B: The Disease

Hepatitis B virus infection is the most common cause of liver pathologies like cirrhosis and hepatocellular carcinoma leading to about one million deaths annually and around 257 million people are chronically infected globally (WHO report 2017). Most adults clear an acute viral infection without any symptoms, but 5–10% of them become chronic patients due to persisting presence of virus, while most neonates born to mothers with an active infection develop chronic hepatitis B infection. The infection could lead to a range of pathologies depending on the virus and host immune system interplay, from acute self-contained infection to chronic hepatitis, leading to cirrhosis, hepatocellular carcinoma (HCC), and death of the patient. Most acute infections are mild and asymptomatic with only a small proportion developing fatigue, nausea, jaundice, and rarely, acute liver failure. In chronic infections, the disease manifests itself to variable degree in each patient depending on various factors. Some patients remain asymptomatic or have nonspecific symptoms; some show symptoms similar to that in acute infection, while more severe cases have a more morbid prognosis with a 50% 5-year survival rate (Liang 2009).

Current treatment includes pegylated-interferon (IFN)-α and five nucleos(t)ide analogs (NAs). Though each of these drugs show some benefit in controlling the virus and symptoms, they do not completely clear the virus. Antiviral resistance and side effects limit usage of these drugs. Hence, therapy is advised for only those who are likely to develop serious complications (Koumbi 2015).

14.2 The Virus

The hepatitis B virus belongs to the *Hepadnaviridae* family of hepatotropic DNA viruses. Similar viruses are found to infect other apes, woodchucks, squirrels, ducks, and herons (Liang 2009). These viruses majorly infect the liver, but trace amounts of their DNA have been reported to be found in pancreas, kidney, and peripheral blood mononuclear cells (Murakami et al. 2004).

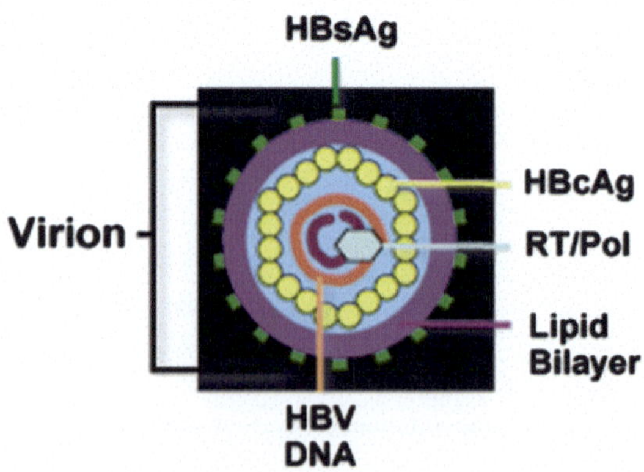

Fig. 14.1 Schematic depiction of a hepatitis B virion (Liang 2009)

The hepatitis B virus is 42 nm in diameter and consists of an outer host-derived lipid membrane, in which the hepatitis B surface antigen (HBsAg) is embedded. The inner nucleocapsid is composed of hepatitis B core antigen (HBcAg) which encapsulates the virally encoded polymerase and a partially double-stranded viral DNA genome (Fig. 14.1) (Liang 2009). An important feature of this virus is its replication strategy employing an RNA intermediate and the error-prone reverse transcriptase enzyme resulting in high mutation rate (Trepo et al. 2014).

The virus binds to host hepatocytes through its surface antigen (HBsAg) and Sodium taurocholate cotransporting polypeptide (NTCP) receptor, found on the surface of hepatocytes (Yan et al. 2012). After fusion with the host membrane, the core enters the cytosol and is transported to the host nucleus. With the help from host molecular machineries, the viral DNA is replicated, via an RNA intermediate, the viral proteins are expressed and new virions are assembled and released (Fig. 14.2) (Mohd-Ismail et al. 2019). Some of the viral proteins like HbsAg and HBeAg are synthesized and released in excess and are known to have immunomodulatory roles (Liang 2009).

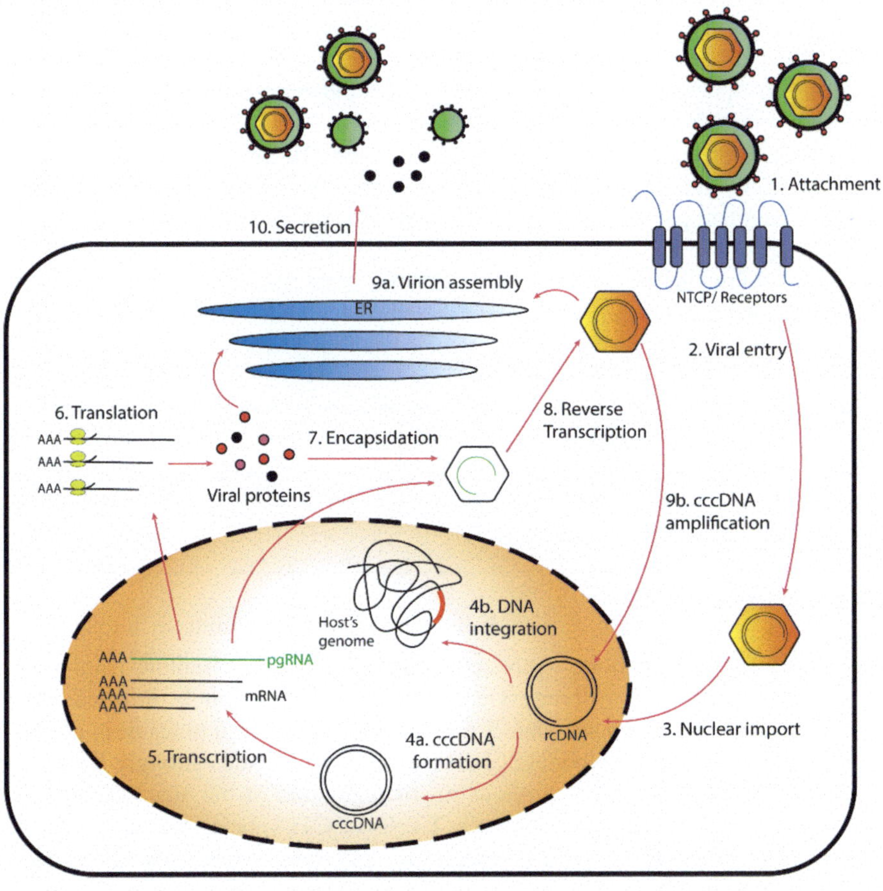

Fig. 14.2 Schematic representation of the Hepatitis B virus life cyce (Mohd-Ismail et al. 2019)

14.3 Natural History and Pathogenesis

HBV reaches high titers in the blood and this is the major route of transmission. On infection there is minimal immune response and hence the virus spreads rapidly in the hepatocytes. Immune activation within the first few weeks leads to inflammation in the liver and antiviral response. Cytotoxic T cells are the major cells responsible for killing infected hepatocytes as well as arresting viral replication by secretion of antiviral cytokines like INF-γ and TNF-α. Extensive cell death in the liver may result in some symptoms, indicative of liver inflammation and damage. Appearance of antibodies against the viral surface antigen (HBsAb) in serum is a major indicator of clearance of the virus from the circulation.

As mentioned earlier, 5–10% of adults and most HBV-infected infants are unable to fully clear the virus and develop a chronic HBV infection. Insufficient or exhausted T cell response is believed to be the primary reason for this persistent infection. Low levels of viral DNA and particles are detected beyond 6 months of infection and this is accompanied by occasional flares of viral–immune activity. This continued cycle of inflammation and cell damage, over the years, leads to development of liver cirrhosis and hepatocellular carcinoma in many patients. The status of the disease in a chronic patient can be categorized as immune-tolerant, immune-active, or immune-inactive carrier phase, depending upon the viral activity and host immune response (Trepo et al. 2014).

The course of the disease and its outcome is variable in chronic HBV patients making it tricky to manage. Hence, extensive studies have been done to fully understand the immune pathogenesis of HBV infection and identify possible targets to help manage the infection. The narrow host range of HBV makes it challenging to study. Most of the understanding of the HBV life cycle comes from animal models, i.e., chimpanzees and *Tupaia Belangeri* (Tree shrew) which can be infected by HBV as well as closely related viruses like Duck-HBV and Woodchuck-HBV in their respective host. Many studies are also done on transgenic and humanized mice. Viral biology is also studied in vitro, on either primary human hepatocytes or NTCP expressing transgenic cell lines. Major gaps still exist in the understanding of the pathogenesis and involvement of immune cells in HBV infection (Lamontagne et al. 2016).

14.4 T Cells

T lymphocytes are the major effector cells involved in adaptive immune response to viruses. A robust T cell response clears the infection in acutely infected HBV patients, while weak and inadequate T cell response is believed to result in chronic HBV infection. In a mouse model of HBV infection, the results strongly show that $CD4^+$ T cells (T helper cells) serve as major regulators of the adaptive immune response and the $CD8^+$ T cells (cytotoxic T cells or CTLs) being the main effector cells for HBV clearance along with supporting functions of NK cells and interferons (Yang et al. 2010).

14.4.1 CD4$^+$ Cells or T Helper Cells

CD4$^+$ T helper cells are involved in early priming and development of CD8$^+$ cytotoxic T cell function as well as B cell function of production of antibodies to clear free viral particles (Sant and McMichael 2012). Th1 CD4$^+$ cells have been found to increase during hepatic inflammatory activity and in active phases of chronic HBV infection (Jiang et al. 2002), but not during peak viremia; rather they are required early in the infection for priming of CD8$^+$ T cells for successfully clearing the infection. This was supported by another report (Fisicaro et al. 2009) where they found that CD4$^+$ and CD8$^+$ T cell responses are found to peak, early in the infection, with IFN-γ producing CD8$^+$ cells found at higher frequencies at the height of immune response. Hence, even though CD4$^+$ T cells may not be direct effector cells for viral clearance, they play an important role by early induction of CD8$^+$ T cells. The CD4$^+$ T cells also exert their immunological function by producing cytokines like Interferon (IFN)-γ and Tumor Necrosis Factor (TNF)-α to control the infection and regulate other immune cells (Li et al. 2016a). Chronic HBV infection is characterized by T cell exhaustion and inadequacy of CD4$^+$ T cells function may be another contributing factor to the effector CD8$^+$ T cell exhaustion in chronic viral infections (Ye et al. 2015). Indeed, inhibitory molecule PD-1 (programmed cell death protein 1) has been found to be upregulated in CD4$^+$ T cells in chronic HBV infection (Raziorrouh et al. 2014), hence this could be a contributing factor for the dysfunction of the T cell compartment.

14.4.2 CD8$^+$ T Cells or Cytotoxic T Cells

Early reports show that strong HBV-specific CD8$^+$ T cell response correlates with viral clearance during acute infection. The highest frequency of peripheral HBV-specific CD8$^+$ T cells are found in the acute phase of HBV infection; however, a transient attenuation of the CD8$^+$ proliferation (Maini et al. 1999) and functions, coinciding with increase in the immunosuppressive cytokine, IL-10 during peak viremia is noted (Dunn et al. 2009). An effective CD8$^+$ response is robust, polyclonal and multi-antigen specific, while such a response is undetectable in chronic patients (Rehermann et al. 1995; Stelma et al. 2017). These are the main effector cells that clear the HBV infection by both cytolytic and non-cytolytic mechanisms, as observed, in chimpanzees (Thimme et al. 2003).

The non-cytolytic mechanism of viral inhibition, mediated by IFN-γ and TNF-α, inhibits viral replication and clears viral intermediates from infected cells (Guidotti et al. 1996) without liver damage (Maini et al. 2000) and is the predominant mode of viral clearance in an effective response, while significant cytolytic activity is observed by nonspecific but activated lymphoid cells (Phillips et al. 2010).

The cytotoxic T cell response (CTL) is weaker and less specific in chronic HBV infections with dampening of not only HBV-specific but global T cells functions (Park et al. 2015). The HBV-specific CD8$^+$ T cells show dysfunction correlating with viral levels and they can be detected in patients with lower viremia while rarely found in patients with higher viral load. HBV-specific CTLs, specific for the core antigen (HBcAg), is associated with viral control, while those specific for envelop and polymerase epitopes are more frequent in higher viremia patients, implying that the core specific $CD8^+$ cells are therapeutically more relevant (Webster et al. 2004). Also, HBV mutational escape and evolution is observed during progression of the chronic infection and is believed to be due to the selection pressure from HBV core specific CTLs (Zhang et al. 2018); however, complete understanding of viral escape from $CD8^+$ immune response is still incomplete (Lumley et al. 2018).

The virus-specific $CD8^+$ cells found in circulation of chronic patients show an exhausted phenotype with impaired proliferation, cytokine secretion, expression of inhibitory molecules like PD-1 (Boni et al. 2007), cytotoxic T lymphocyte antigen-4 (CTLA-4) (Schurich et al. 2011), T cell immunoglobulin domain and mucin domain-containing molecule-3 (Tim-3) (Nebbia et al. 2012; Wu et al. 2012) and CD244 (Raziorrouh et al. 2010), which lead to their deletion on interacting with their respective ligands. This exhausted phenotype is a stable form of dysfunction and cannot be reversed by changing the immune environment (Wang et al. 2018). However, their functionality could be improved in vitro, by blocking these inhibitory molecules and receptors highlighting important immunotherapeutic targets in chronic HBV infection. Blocking of PD-1 with antibodies has especially shown promise by increasing IFN-γ producing CTLs in the liver in mice (Maier et al. 2007), and improved viral control in combination with nucleoside analog and DNA vaccine therapy in woodchuks models (Liu et al. 2014b). Therapeutic efficacy in woodchuck model showed no significant liver injury in acute models and good viral control in infected animals expressing PD-1 on $CD8^+$ T cells; however, combination treatment with additional antiviral agents was recommended to achieve a robust and more consistent response for complete immune clearance (Balsitis et al. 2018). Nivolumab, a PD-1 blocking monoclonal antibody has been approved for therapeutic use in many forms of cancer and has recently been approved in the USA for hepatocellular carcinoma caused by various afflictions including HBV infection (Highleyman 2017).

Successful antiviral treatment with pegylated-IFN-α-2b is accompanied with increase in number of circulating HBV-core or envelop-specific $CD8^+$ T cells, along with increased Th1-type cytokines (IL-12, TNF-α, IFN-γ) (Chen et al. 2010). T cells with TCRs (T cell Receptor) which have high avidity for HBV antigens hold lot of potential for adoptive transfer and treatment, and such TCRs have also been cloned and studied in vitro for possible applications in therapy (Wisskirchen et al. 2017). Adoptive transfer of HBV-specific $CD4^+$ and $CD8^+$ T cells along with immunotherapies to prevent T cell exhaustion has been proposed for treatment of chronic viral infection as well as cancer (Kamphorst and Ahmed 2013).

14.4.3 Regulatory T Cells (T regs)

Regulatory T cells or Tregs are a set of naturally occurring and inducible $CD4^+$ $CD25^+$ and FoxP3 expressing T cells with a basic role of controlling hyperactivity of effector T cells. Even though their main role is to prevent autoimmune reactions and reduce collateral tissue damage during infections, an undesirable consequence is immune suppression and pathogen persistence (Belkaid 2007). Probable mechanisms by which they exert the suppressive function is by secretion of immunosuppressive cytokines IL-10, IL-35, TGF-β, regulating maturation of dendritic cells and production of metabolites that inhibit effector T cells with or without direct-cell contact (Li et al. 2016b).

A higher percentage of $CD4^+$ $CD25^+$ Tregs were detected in circulation (Stoop et al. 2005) as well as in the liver of chronic HBV patients, and the increase in proportion was more pronounced in the convalescent phase of infection and correlated with the serum viral load in the patients (Xu et al. 2006). These cells were capable of suppressing activation of $CD8^+$ T cells, while their depletion resulted in increase in HBV-specific $CD8^+$ response, as observed ex vivo (Franzese et al. 2005; Stoop et al. 2005). Tregs in the liver correlate strongly with the viral load and are functionally distinct from peripheral blood Tregs, i.e., these were found to not express CD25, had lower expression of HLA-DR and CTLA-4 (Stoop et al. 2008). HBcAg-specific Tregs are found to decline during acute exacerbations in chronic HBV patients accompanied by increase in HBcAg-specific $CD8^+$ T cells (Feng et al. 2007). It has been suggested that the immunosuppressive function of Tregs is important to minimize extensive histological damage as concluded by Stross et al. (2012) who found that Tregs downregulated the cytokine production and cytotoxicity of the effector T cells, but did not affect the development of HBV-specific effector and memory T cells. Also, Tregs reduced recruitment of macrophages and dendritic cells to the liver, which inadvertently delayed clearance of the virus, in transgenic mouse model (Stross et al. 2012).

In cases of vertical transfer of infection, as seen in the peripheral and cord blood of HBsAg positive newly borns, only the frequency of FoxP3 expressing Tregs was significantly high while the frequency of $CD4^+$ and $CD8^+$ T cells remained unchanged, though the $CD8^+$ T cells were found to be dysfunctional, explaining the development of chronic infections in infected neonates (Shrivastava et al. 2013). Levels of circulating Tregs is also significantly elevated in acute-on-chronic liver failure (ACLF) patients compared to healthy donors and chronic patients, and is proposed to be indicative of the disease severity (Yang et al. 2012). Increased Tregs levels have also been found to increase in frequency as well as impair $CD8^+$ T cell function, promoting disease progression in chronic HBV patients with HCC and proposed to have prognostic value as well as serve as a potential therapeutic target (Fu et al. 2007; Han et al. 2011; Kalathil et al. 2013; Li et al. 2016b). During treatment for HCC with anti-angiogenic agent, Sorafenib, decreased Tregs numbers correlated with overall survival of the HCC patients (Kalathil et al. 2016). Hence, even though Tregs are important for controlling excessive immune-mediated damage, their suppressive function often leads to disease progression in chronic infection setting, making Tregs an important prognostic marker and therapeutic target for immunotherapy in HBV infection (Aalaei-Andabili and Alavian 2012).

14.4.4 T Helper: 17 (Th17) Cells

The T-helper-17 cells are a newly identified lineage of the $CD4^{+}$ T helper cells that predominantly produce IL-17, IL-21, IL-22 and play an important role in protection against extracellular pathogens as well as in inflammation and autoimmunity (Harrington et al. 2005; Park et al. 2005; Bettelli et al. 2007; Ouyang et al. 2008). The Th17 cells in circulation, as well as intrahepatic, increase with disease progression and correlate with viral load, serum ALT, and histopathology index and this may be through activation of mDCs, monocytes, and increased production of pro-inflammatory cytokines in chronic patients (Zhang et al. 2010a). IL-17(a) activates hepatic stellate cells to express fibrogenic factors thereby leading to fibrosis and cirrhosis in an inflammatory environment in the liver (Tan et al. 2013), and this is suggested to be a major mechanism of cirrhosis in severe cases of chronic HBV infection (Sun et al. 2012). TH17 and IL-17 levels are found to be higher in non-survivors and decrease in circulation of recovering patients (Yang et al. 2013).

Th17 and Treg cells play opposite roles of mediating and restraining inflammation and a lower Treg/Th17 ratio indicates extensive liver injury and fibrosis in chronic HBV patients (Li et al. 2012). The serum IL-17 levels and frequency of Th17 cells correlate with ALT levels while nTregs correlate positively with HBV DNA load and inversely with Th17 frequencies (Yang et al. 2016). TGF-β/IL-17 ratio have also been found to markedly increase in decompensated cirrhosis patients and non-survivors, and hence both these ratios can help in monitoring and prognosis of HBV-associated cirrhotic patients (Yu et al. 2014). On administration of NA antiviral therapy, the Treg/Th17 ratio declined in complete responders due to increased Th17 cells and IL-17 and decreased Treg and their cytokines, correlating with decrease in HBV DNA and ALT levels (Yu et al. 2013). Cell-based therapy using autologous bone marrow mesenchymal stem cells, post antiviral therapy in HBV-associated cirrhotic patients, showed improved liver function and an increase in the Treg/Th17 ratio (Xu et al. 2014a).

14.5 B Cells and Bregs

The neutralizing antibodies against HBsAg produced by B cells is essential for clearance of free viral particles and prevention of reinfection (Paul et al. 2017). Chronic HBV patients mostly lack neutralizing antibodies against the viral particle and this explains why the virus persists in these cases (Alberti et al. 1978). B cells have been found to actively present HBsAg to T helper cells more efficiently than classical APCs like macrophages and dendritic cells, in a mice model (Milich et al. 1997). B cells can bind HBcAg and give an IgM response, in a T cell independent fashion; however, these could not class switch to IgG, highlighting the need of T cell's help for this. The same response was not found for HbeAg (Cao et al. 2001). This antigen presentation potential of B cells was also employed to activate HBV-specific cytolytic T lymphocytes especially for chronic conditions, where in B cells from healthy donor was first activated with a soluble CD40 ligand and then pulsed with a peptide from HBcAg and coculture with autologous T cells, showed increase in percentage of induced CTLs (Wu et al. 2010).

The B cells do not seem to be activated in chronic HBV patients as seen by the proportion of $CD86^{+}$ B cells in their blood (Zhao et al. 2015). However, effective treatment can activate the B cell subsets. The B cell subsets remodeling, from naïve and post-germinal to transitional and plasmablasts strongly correlated with decrease in serum HBsAg, as observed in PEG-IFNα treated patients (Aspord et al. 2016). In chronic HBV patients, an exhausted or defective subset of B cells has also been recently characterized. It has been found that HBsAg-specific B cells persist in circulation; however, these show an atypical Memory B cell phenotype (atMBC) which is a functionally defective subset enhanced in conditions of continuous antigen exposure. These cells had enhanced expression of inhibitory molecules like PD-1, were prone to apoptosis, and perhaps diminished ability to differentiate to antibody producing plasma cells and attenuated signaling and cytokine production. These at MBC were found to preferentially accumulate in the infected liver and it is possible that the highly inflammatory microenvironment promotes the development of the defective phenotype. PD-1 blockade was found to partially rescue the premature apoptosis and impaired antibody and cytokine production (Burton et al. 2018). Similar subset of impaired B cells was also defined in another study, where these cells from chronic patients were found to express inhibitory receptors of FcRL family and downregulate genes responsible for antigen presentation. Treatment of B cells from healthy subjects, with HBV core particles, resulted in expression of the inhibitory molecules and this altered phenotype, highlighting another immunosuppressive effect of HBV particles on immune cells (Poonia et al. 2018).

A regulatory subset of B cells which predominantly produces IL-10 was also identified. These were increased in patients, and their frequency and IL-10 levels correlated temporally with liver inflammation. These cells were found to suppress HBV-specific $CD8^{+}$ T cell response in a IL-10 dependent manner and hence play an important role in controlling liver inflammation or in establishment of a tolerogenic environment (Das et al. 2012). This subset was further defined as Bregs and found to be elevated in immune-active chronic HBV patients, which correlated with the serum IL-10, ALT, and AST levels. Also, Bregs and IL-10 levels were higher in inactive carrier group than healthy controls. The expression of TLR2 is elevated and TLR9 is reduced on Bregs of chronic HBV patients compared to healthy controls; however, its significance is yet to be elucidated (Wang et al. 2017).

14.6 Monocytes and Macrophages

Macrophages, Kupffer cells, dendritic cells, and hepatocytes would be few of the first cells to recognize and respond to HBV, through their Toll-like receptors (Ma et al. 2018). Monocytes mediate host antimicrobial defense by migrating to breached tissues in response to chemokines and depending upon interaction with the pathogen, differentiate into macrophages and dendritic cells, along with secretion of immune modulators. Many viruses are known to preferentially infect monocytes, express viral proteins in the cell, and trigger their rapid differentiation to dendritic cells (Hou et al. 2012). Early reports claim that HBV DNA is found in monocytes of chronically infected patients (Yoffe et al. 1986; Guang 1992). As found from in vitro experiments, recombinant HBsAg binds preferentially to B cells and monocytes (Pontisso et al. 1991) especially by interacting and binding to CD14 receptor (Vanlandschoot et al. 2002). Monocytes from chronic HBV patients have been found to be the only antigen presenting cells to contain intracellular HBsAg, which when differentiated ex vivo, was able to cross present and stimulate expansion of HBV-specific T cells (Gehring et al. 2013). However, literature regarding the effect of such interaction of monocytes with different subviral particles of HBV has been ambiguous.

HBcAg induces production of pro-inflammatory cytokines IL-6, IL-12, and TNF-α in monocytic cell line THP-1, by signaling through TLR2 (Cooper et al. 2005). However, there is a reduction of surface expression TLR2 accompanied by reduction of secretion of TNF-α on interaction with HBeAg (Riordan et al. 2006; Visvanathan et al. 2007) and IL-12 on interaction with HBsAg with monocytes (Wang et al. 2013). TLR8 is another important virus-sensing receptor expressed strongly on monocytes but found to be reduced in monocytes from chronic HBV patients with a concomitant reduction in production of antiviral cytokines, also correlating with reduced response to interferon therapy (Deng et al. 2017). Also, while some in vitro studies report an immunosuppressive response of the monocytes on interaction with HbsAg, where in they secrete IL-10 and TNF-α and suppress functions of pDCs (Shi et al. 2012), other similar study observe a robust secretion of inflammatory cytokines IL-6 and TNF by monocytes on treatment with HBsAg, though the serum level of these cytokines in chronic HBV patients did not correlate serum HBV and HBsAg levels (Boltjes et al. 2014). Hence, more studies are required to delineate the exact effect of the monocyte–virus interaction and role of these cells in the disease pathogenesis.

Pro-inflammatory CD16+ subset of monocytes have also been implicated to contribute to liver injury in immune-active patients since these increase in circulation and in liver of these patients, correlating with their serum ALT and extent of liver damage (Zhang et al. 2011). Kupffer cells (KCs), the liver tissue resident macrophages are also expected to have a significant role in the liver microenvironment even though there is sparse information about their role in HBV immunity or pathogenesis. Intrahepatic KCs from chronic HBV patients have been found to be HBsAg positive and more activated than controls (Boltjes et al. 2015). In a mouse model of HBV infection, KCs were found to secrete immunosuppressive IL-10 which could be induced by viral particles to favor viral persistence (Xu et al. 2014a). On the other hand, KCs and lymphoid cells from chronic patients are found to express Fas-L (Tang et al. 2003) while HBsAg sensitizes liver cells to Fas mediated apoptosis. Hence, KCs may play in important role in clearance of infected cells, but at the same time, cause extensive cell death and injury in an infected liver.

14.7 Dendritic Cells

The dendritic cells (DCs) are classical antigen presenting cells and their role is important for antigen presentation and activation of T cells. Two major subtypes of DCs are plasmacytoid DC (pDC) and myeloid DC (mDC) based on their lineage of origin. These differ by expression of certain cell surface markers, with pDCs having a predominant role in antiviral response by producing type I interferons (Kadowaki 2009).

HBV DNA and RNA has been reported to be found in DCs enriched and cultured from chronic HBV patient's blood and these DCs showed impaired functioning like reduced ability to stimulate T cells and reduced secretion of IL-12 (Arima et al. 2003). mDCs show reduction in maturation and secretion of TNF-α while pDCs produce less interferon-α in in vitro experiments (van der Molen et al. 2004). DCs generated from patient monocytes in vitro (moDCs) show impaired T cell stimulatory function and lower IL-12 secretion, which could be restored with antiviral treatment (Beckebaum et al. 2003). Similar functional impairment along with the presence of viral DNA and viral particles within the cells was observed in pDCs from chronic patients (Beckebaum et al. 2002). Even though DCs take up the viral particles, they do not support the viral replication or subsequent steps within the cells (Untergasser et al. 2006). These cells reduce in circulation and enrich in the liver of immune-active chronic HBV patients (Zhang et al. 2007) while antiviral treatment (with adefovir) that reduces the viral load shows some restoration of function as well as number of circulating DCs in the circulation, concomitant with decreased inflammation in the liver (Van der Molen et al. 2006).

The functional impairment of DCs on interaction with viral particles is expected to contribute to the inadequate immune response and persistence of virus in chronic HBV infection. Reviving DC function in chronic infections would be one of the ways to overcome the tolerogenic environment in chronic infections. It has been observed in one study that monocytes retained a depot of HBsAg and on differentiating them into DCs in vitro, they were able to crosspresent the antigen to T cells (Gehring et al. 2013). Hence an idea of using in vitro activated autologous DCs as vaccine in chronic patients has been explored with some success in achieving immune clearance in these patients (Chen et al. 2005; Luo et al. 2010). On treating MoDCs from patient's PBMCs, with HBV sub-viral particles like HBsAg and HBcAg, they were found to have strong capacity to stimulate T cell proliferation and produce IL-12, showing some potential for autologous therapy for chronic HBV patients (Shi et al. 2007). Hence, MoDCs have been considered to help boost HBV-specific T cell response in chronic HBV patients with hepatocellular carcinoma (HCC).

Studies have also been done to understand the interaction of the virus with DCs and mechanism of the dysfunction. Dendritic cells would be one of the first cells to sense the virus through its Toll-like receptor (Ma et al. 2018). It has been found that HBV can downregulate the expression of TL9 and the subsequent IFN-α production from pDCs (Vincent et al. 2011) which is the case in chronic HBV patients (Xu et al. 2012). Targeting activation of Toll-like receptors, specially use of TLR7 ligand

GS-9620, to elicit an antiviral response is an attractive idea and has been tested with some success in Woodchuk HBV model (Menne et al. 2011) and chimpanzee model (Lanford et al. 2013); however, it was not as effective in patients (Janssen et al. 2018). Another approach was the use of nanoparticles containing HBV-CpG and HBsAg which would activate TLR9, and was used with success in mouse in eliciting an enhanced CD4+ T cell response (Lv et al. 2014). TLR 4 and 9 agonists are already used as adjuvants for HBV vaccines while agonists for TLR 3, 7, and 8 show potential and need to be further investigated for use in therapeutics (Ma et al. 2018).

14.8 Natural Killer Cells

Natural killer (NK) cells are from the lymphoid lineage, and are involved mainly in innate immune response against intracellular pathogens and abnormal cells. They exert their antiviral activity by recognition and killing of infected cells and secretion of antiviral cytokines like IFN-γ and TNF-α. They express CD56 and can be categorized as two subsets, i.e., $CD56^{bright}$, which have a predominant function of antiviral cytokine secretion and $CD56^{dim}$, which are more involved in killing target cells (Wu et al. 2015). NK cells also promote release of antiviral cytokines like IFN-α and IL-6 from pDCs. IFN-α in turn upregulates cyotoxic activity of NK cells and IL-6 promotes B cell differentiation (Della Chiesa et al. 2006). NK cells are the most abundant innate immune cells in the liver (Faure-Dupuy et al. 2017) and are one of the first cells to respond to HBV infection in patients which would allow a timely adaptive immune response to be induced (Fisicaro et al. 2009).

Natural killer cells have been found to increase in number post-peak viremia in acute HBV infection, where in a transient reduction of NK cells and T cell responses coincided with a surge in IL-10 immunosuppressive cytokine (Dunn et al. 2009). After peak viremia, an increase in TRAIL (TNF-related apoptosis-inducing ligand)-expressing $CD56^{bright}$ NK cells is observed and is believed to be responsible for clearance of the infected cells (Stelma et al. 2017) while these are also responsible for elimination of HBV-specific $CD8^+$ T cells (Peppa et al. 2013) and $CD4^+$ T cells (Ghosh et al. 2016) which show an exhausted phenotype and express death receptors in chronic infections. Patients who clear an acute infection have fully functional and activated NK cells which correlate with serum ALT levels and inversely with serum viral load, while NK cells from chronic HBV patients produce IL-10 elicited by HBV-induced monocytes (Li et al. 2017), have diminished IFN-γ secretion but enhanced cytolytic activity (Zhao et al. 2012), correlating with liver histopathology and serum ALT levels especially in patients in the immune-active phase (Zhang et al. 2010a, b). This skew in the NK cell function, correlated with serum viral load and antiviral therapy partially restored normal NK cell functioning and interferon production by these cells (Tjwa et al. 2011; Stelma et al. 2015).

Major liver damage is caused by the increased cytolytic activity of NK cells in chronic HBV infection. Hence, it has been proposed to block NKG2D receptor of

NK cells to alleviate the severe liver damage caused by these cells (Huang et al. 2013). PD-1/PD-L1 blockade is another popular potential strategy which would restore IFN-γ production of NK cells, along with alleviating T cell exhaustion. Further study to understand NK cell function in HBV infection will reveal more therapeutic targets for treatment of HBV infection.

14.9 Neutrophils

Neutrophils are the most abundant immune cells which actively eliminate pathogens and target cells. However, over-activation can lead to extensive tissue damage, as happens in hepatitis (Xu et al. 2014b). Reports of direct interaction of the virus and neutrophils is scant, but these cells have a more significant role in the secondary effects of the viral infection. Though one study noted the immunosuppressive effect of recombinant HBeAg by suppression of respiratory burst in neutrophils, as well as their chemotactic functions in vitro (Leu et al. 2014), most studies are focused toward its role in the pathogenesis of injury in hepatitis.

In a mouse model of HBV infection, it was observed that transferring HBV-specific cytotoxic T cells resulted in recruitment of nonspecific inflammatory cells that caused liver pathology. In this setting, depletion of neutrophils reduced the recruitment of antigen nonspecific inflammatory cells, hence ameliorating the severity of liver damage, but did not affect antiviral activity of CTLs. It can be inferred that the neutrophils have a role in the pathogenesis of the disease and may serve as a target for immunotherapeutic approaches (Sitia et al. 2002). Concurrently, blocking a serine protease like neutrophil elastase with an inhibitor, reduced serum ALT levels, and the migration of inflammatory cells in the liver, post injection of antigen-specific CTLs in a HBV mouse model, and hence administration of neutrophil elastase inhibitors may hold therapeutic potential against severe hepatitis (Takai et al. 2005). The recruitment of neutrophils can be due to the expression of CXCL-8 by CTLs which recruits and elicits cytotoxic activity of neutrophils (Gehring et al. 2011) by interacting with the CXCR1 and CXCR2 on neutrophils (Nasser et al. 2009). Hence another approach could be to block CXCR1/CXCR2, which was found to be increased in neutrophils in HBV-related acute-on-chronic liver failure (ACLF) patients. Blocking CXCR1/CXCR2 had significantly reduced cell death in in vitro experiments with hepatic cell lines (Khanam et al. 2017).

The neutrophil-to-lymphocyte ratio (NLR) has been widely accepted as reflection of inflammatory status of the patient before treatment and has been used in the context of hepatitis B infection to predict acute-on-chronic liver failure in patients. Low NLR indicates better prognosis, while higher NLR indicates a high short-term mortality rate (Liu et al. 2014a). The NLR has been found to be a useful predictive marker in patients with HBV-related decompensated cirrhosis (Zhang et al. 2016), ACLF patients treated with artificial liver support system (Fan et al. 2017), in patients receiving PEG interferon treatment (Le et al. 2017), and may be applied to many other similar conditions.

14.10 Conclusions and Future Perspectives

Hepatitis B virus infection is causative of significant number of deaths each year. While a robust immune system manages to clear the infection effectively, a chronic infection is a result of a combination of viral and host factors, of which the immune system plays an integral role. While much of the mechanism of viral clearance or persistence and development of pathology is understood, many aspects still remain obscure. Ambiguity in several aspects exists in the available literature, which may be due to observations from different experimental settings and patient cohorts. The narrow host range and cell specificity of HBV also makes experimental work challenging. However, the undisputed roles of the major immune cells in the pathogenesis, along with the significant observations, have been summarized in this report. The key players involved in the process of immune clearance and immune-pathology would be the logical targets for developing novel therapeutic approaches.

$CD8^+$ T cells are the most important cells involved in clearance of the virus through production of antiviral cytokines like IFN-γ and TNF-α which arrest viral replication, as well as by cytolysis of infected cells and requires help from $CD4^+$ T cells and dendritic cells for its activation. However, in a setting where the viral load far exceeds the $CD8^+$ response, this subset becomes exhausted and express exhaustion markers like PD-1, CTLA-4, Tim-3, and CD244, which in turn marks these cells for removal by natural killer cells. The role of Tregs and Bregs is to prevent over-activation of the immune cells and control extensive histological damage by secretion of immunosuppressive cytokines, during phases of peak T cell activation. However, their sustained activity is believed to contribute to immune tolerance in chronic infections. Th17 is another subset of $CD4^+$ T cells, which secretes pro-inflammatory IL-17 and involved in development of liver fibrosis and Treg/h17 ratio has been found to be of value in prognosis of the disease.

There is scarce literature about role of B cells in chronic infections, while they certainly play a major role in immune clearance in acute, self-resolved infections by secretion of anti-HBs antibodies. The effect of the virus on monocytes and macrophages is also ambiguous since there are reports of these cells interacting with the virus; however, it is not clear if these cells respond with an immunostimulatory or immunosuppressive effect. Dendritic cells have been found to be mostly impaired in chronic infection setting and many research efforts have been to activate them in vitro so that they may activate the T cells on being administered in vivo. Activated pro-inflammatory $CD16^+$ monocytes, tissue resident Kupffer cells, natural killer cells, and neutrophils are believed contribute to liver injury by inducing cell death in the liver, in an attempt to clear infected hepatocytes. The neutrophil-to-lymphocyte ratio is another predictive value used to gauge the inflammatory status of the patient and make a prognosis.

The limitations of the current therapy in treatment of HBV infection, i.e., interferon therapy or nucleos(t)ide analogs, warrants the development of more effective therapies. Advancement in the understanding of the disease pathogenesis re-emphasizes the role of the immune system in clearance of the infection, which when malfunctions, leads to chronic infection, thereby also serving as potential targets for treatment. Most immunotherapeutic efforts is toward enhancing $CD8^+$ T cells function in chronic infections by either reactivating impaired T cells by blocking inhibitory molecules (PD-1, CTLA-4, etc.) or by adoptive transfer of HBV-specific T cells. Antibody response may also be boosted by vaccination and in addition to the prevalent HBsAg recombinant vaccine, the potential of DNA-based vaccines is being evaluated in clinical trials (Fontaine et al. 2015). Some of the significant potential therapeutic strategies targeting the innate immune branch focus on use of TLR agonists and antiviral and immunostimulatory cytokines (Bertoletti and Le Bert 2018). Minimizing liver damage caused due to over-activation of some of the immune cells like NK cells and neutrophils, leading to fibrosis and cirrhosis, may also prove beneficial for recovery of the patient. A single therapeutic strategy may not be sufficient for an effective treatment, depending on the degree of pathology in a patient. Hence a combination of therapies targeting various viral and immune factors promises to be the most effective therapeutic approach for treatment of HBV infection.

Conflict of Interest The authors declare that they have no competing interests.

References

Aalaei-Andabili SH, Alavian SM (2012) Regulatory T cells are the most important determinant factor of hepatitis B infection prognosis: a systematic review and meta-analysis. Vaccine 30(38):5595–5602

Alberti A, Diana S, Sculard GH, Eddleston AL, Williams R (1978) Detection of a new antibody system reacting with Dane particles in hepatitis B virus infection. Br Med J 2(6144):1056–1058

Arima S, Akbar SM, Michitaka K, Horiike N, Nuriya H, Kohara M, Onji M (2003) Impaired function of antigen-presenting dendritic cells in patients with chronic hepatitis B: localization of HBV DNA and HBV RNA in blood DC by in situ hybridization. Int J Mol Med 11(2):169–174

Aspord C, Bruder Costa J, Jacob MC, Dufeu-Duchesne T, Bertucci I, Pouget N, Brevot-Lutton O, Zoulim F, Bourliere M, Plumas J, Leroy V, ANRS HB06 PEGAN study group (2016) Remodelling of B-cell subsets in blood during pegylated IFNα-2a therapy in patients with chronic hepatitis B infection. PLoS One 11(6):e0156200

Balsitis S, Gali V, Mason PJ, Chaniewski S, Levine SM, Wichroski MJ, Feulner M, Song Y, Granaldi K, Loy JK, Thompson CM, Lesniak JA, Brockus C, Kishnani N, Menne S, Cockett MI, Iyer R, Mason SW et al (2018) Safety and efficacy of anti-PD-L1 therapy in the woodchuck model of HBV infection. PLoS One 13(2):e0190058. https://doi.org/10.1371/journal.pone.0190058

Beckebaum S, Cicinnati VR, Dworacki G, Muller-Berghau J, Stolz D, Harnaha J, Whiteside TL, Thomson AW, Lu L, Fung JJ, Bonham AA (2002) Reduction in the circulating pDC1/pDC2 ratio and impaired function of ex vivo-generated DC1 in chronic hepatitis B infection. Clin Immunol 104(2):138–150

Beckebaum S, Cicinnati VR, Zhang X, Ferencik S, Frilling A, Grosse-Wilde H, Broelsch CE, Gerken G (2003) Hepatitis B virus-induced defect of monocyte-derived dendritic cells leads to impaired T helper type 1 response in vitro: mechanisms for viral immune escape. Immunology 109(4):487–495

Belkaid Y (2007) Regulatory T cells and infection: a dangerous necessity. Nat Rev Immunol 7(11):875–888

Bertoletti A, Le Bert N (2018) Immunotherapy for chronic hepatitis B virus infection. Gut Liver 12(5):497–507
Bettelli E, Korn T, Kuchroo VK (2007) Th17: the third member of the effector T cell trilogy. Curr Opin Immunol 19(6):652–657
Boltjes A, Groothuismink ZM, van Oord GW, Janssen HL, Woltman AM, Boonstra A (2014) Monocytes from chronic HBV patients react in vitro to HBsAg and TLR by producing cytokines irrespective of stage of disease. PLoS One 9(5):e97006
Boltjes A, van Montfoort N, Biesta PJ, Op den Brouw ML, Kwekkeboom J, van der Laan LJ, Janssen HL, Boonstra A, Woltman AM (2015) Kupffer cells interact with hepatitis B surface antigen in vivo and in vitro, leading to proinflammatory cytokine production and natural killer cell function. J Infect Dis 211(8):1268–1278
Boni C, Fisicaro P, Valdatta C, Amadei B, Di Vincenzo P, Giuberti T, Laccabue D, Zerbini A, Cavalli A, Missale G, Bertoletti A, Ferrari C (2007) Characterization of hepatitis B virus (HBV)-specific T-cell dysfunction in chronic HBV infection. J Virol 81(8):4215–4225
Burton AR, Pallett LJ, McCoy LE, Suveizdyte K, Amin OE, Swadling L, Alberts E, Davidson BR, Kennedy PT, Gill US, Mauri C, Blair PA, Pelletier N et al (2018) Circulating and intrahepatic antiviral B cells are defective in hepatitis B. J Clin Invest 128(10):4588–4603
Cao T, Lazdina U, Desombere I, Vanlandschoot P, Milich DR, Sällberg M, Leroux-Roels G (2001) Hepatitis B virus core antigen binds and activates naive human B cells in vivo: studies with a human PBL-NOD/SCID mouse model. J Virol 75(14):6359–6366
Chen M, Li YG, Zhang DZ, Wang ZY, Zeng WQ, Shi XF, Guo Y, Guo SH et al (2005) Therapeutic effect of autologous dendritic cell vaccine on patients with chronic hepatitis B: a clinical study. World J Gastroenterol 11(12):1806–1808
Chen J, Wang Y, Wu XJ, Li J, Hou FQ, Wang GQ (2010) Pegylated interferon α-2b up-regulates specific CD8+ T cells in patients with chronic hepatitis B. World J Gastroenterol 16(48):6145–6150
Cooper A, Tal G, Lider O, Shaul Y (2005) Cytokine induction by the hepatitis B virus capsid in macrophages is facilitated by membrane heparan sulfate and involves TLR2. J Immunol 175(5):3165–3176
Das A, Ellis G, Pallant C, Lopes AR, Khanna P, Peppa D, Chen A, Blair P, Dusheiko G, Gill U, Kennedy PT, Brunetto M, Lampertico P, Mauri C et al (2012) IL-10-producing regulatory B cells in the pathogenesis of chronic hepatitis B virus infection. J Immunol 189(8):3925–3935
Della Chiesa M, Romagnani C, Thiel A, Moretta L, Moretta A (2006) Multidirectional interactions are bridging human NK cells with plasmacytoid and monocyte-derived dendritic cells during innate immune responses. Blood 108(12):3851–3858
Deng G, Ge J, Liu C, Pang J, Huang Z, Peng J, Sun J, Hou J, Zhang X (2017) Impaired expression and function of TLR8 in chronic HBV infection and its association with treatment responses during peg-IFN-α-2a antiviral therapy. Clin Res Hepatol Gastroenterol 41(4):386–398
Dunn C, Peppa D, Khanna P, Nebbia G, Jones M, Brendish N, Lascar RM, Brown D, Gilson RJ, Tedder RJ, Dusheiko GM, Jacobs M, Klenerman P, Maini MK (2009) Temporal analysis of early immune responses in patients with acute hepatitis B virus infection. Gastroenterology 137(4):1289–1300
Fan Z, EnQiang C, Yao DL, LiBo Y, Hong L, Lang B, Ping F et al (2017) Neutrophil-lymphocyte ratio predicts short term mortality in patients with hepatitis B virus-related acute-on-chronic liver failure treated with an artificial liver support system. PLoS One 12(4):e0175332
Faure-Dupuy S, Lucifora J, Durantel D (2017) Interplay between the hepatitis B virus and innate immunity: from an understanding to the development of therapeutic concepts. Viruses 9(5):95
Feng IC, Koay LB, Sheu MJ, Kuo HT, Sun CS, Lee C, Chuang WL, Liao SK, Wang SL, Tang LY, Cheng CJ, Tsai SL (2007) HBcAg-specific CD4+CD25+ regulatory T cells modulate immune tolerance and acute exacerbation on the natural history of chronic hepatitis B virus infection. J Biomed Sci 14(1):43–57
Fisicaro P, Valdatta C, Boni C, Massari M, Mori C, Zerbini A, Orlandini A, Sacchelli L, Missale G, Ferrari C (2009) Early kinetics of innate and adaptive immune responses during hepatitis B virus infection. Gut 58(7):974–982

Fontaine H, Kahi S, Chazallon C, Bourgine M, Varaut A, Buffet C, Godon O, Meritet JF, Saïdi Y, Michel ML, Scott-Algara D, Aboulker JP, Pol S, ANRS HB02 Study Group (2015) Anti-HBV DNA vaccination does not prevent relapse after discontinuation of analogues in the treatment of chronic hepatitis B: a randomised trial--ANRS HB02 VAC-ADN. Gut 64(1):139–147

Franzese O, Kennedy PT, Gehring AJ, Gotto J, Williams R, Maini MK, Bertoletti A (2005) Modulation of the CD8+-T-cell response by CD4+ CD25+ regulatory T cells in patients with hepatitis B virus infection. J Virol 79(6):3322–3328

Fu J, Xu D, Liu Z, Shi M, Zhao P, Fu B, Zhang Z, Yang H, Zhang H, Zhou C, Yao J, Jin L, Wang H, Yang Y, Fu YX, Wang FS (2007) Increased regulatory T cells correlate with CD8 T-cell impairment and poor survival in hepatocellular carcinoma patients. Gastroenterology 132(7):2328–2339

Ganem D, Prince AM (2004) Hepatitis B virus infection--natural history and clinical consequences. N Engl J Med 350(11):1118–1129

Gehring AJ, Koh S, Chia A, Paramasivam K, Chew VS, Ho ZZ, Lee KH, Maini MK, Madhavan K, Lim SG et al (2011) Licensing virus-specific T cells to secrete the neutrophil attracting chemokine CXCL-8 during hepatitis B virus infection. PLoS One 6(8):e23330

Gehring AJ, Haniffa M, Kennedy PT, Ho ZZ, Boni C, Shin A, Banu N, Chia A, Lim SG, Ferrari C, Ginhoux F et al (2013) Mobilizing monocytes to cross-present circulating viral antigen in chronic infection. J Clin Invest 123(9):3766–3776

Ghosh S, Nandi M, Pal S, Mukhopadhyay D, Chakraborty BC, Khatun M, Bhowmick D, Mondal RK, Das S, Das K, Ghosh R, Banerjee S, Santra A, Chatterjee M, Chowdhury A, Datta S (2016) Natural killer cells contribute to hepatic injury and help in viral persistence during progression of hepatitis B e-antigen-negative chronic hepatitis B virus infection. Clin Microbiol Infect 22(8):733

Guang LX (1992) Hepatitis B virus infection in monocytes and their functional changes in patients with chronic hepatitis B. Zhonghua Yi Xue Za Zhi 72(11):651–654. 701

Guidotti LG, Ishikawa T, Hobbs MV, Matzke B, Schreiber R, Chisari FV (1996) Intracellular inactivation of the hepatitis B virus by cytotoxic T lymphocytes. Immunity 4(1):25–36

Han YF, Zhao J, Ma LY, Yin JH, Chang WJ, Zhang HW, Cao GW (2011) Factors predicting occurrence and prognosis of hepatitis-B-virus-related hepatocellular carcinoma. World J Gastroenterol 17(38):4258–4270

Harrington LE, Hatton RD, Mangan PR, Turner H, Murphy TL, Murphy KM, Weaver CT (2005) Interleukin 17-producing CD4+ effector T cells develop via a lineage distinct from the T helper type 1 and 2 lineages. Nat Immunol 6(11):1123–1132

Highleyman L (2017, September 27) US FDA approves nivolumab immunotherapy for liver cancer. Retrieved from http://www.infohep.org/page/3175886

Hou W, Gibbs JS, Lu X, Brooke CB, Roy D, Modlin RL, Bennink JR et al (2012) Viral infection triggers rapid differentiation of human blood monocytes into dendritic cells. Blood 119(13):3128–3131

Huang M, Sun R, Wei H, Tian Z (2013) Simultaneous knockdown of multiple ligands of innate receptor NKG2D prevents natural killer cell-mediated fulminant hepatitis in mice. Hepatology 57(1):277–288

Janssen HLA, Brunetto MR, Kim YJ, Ferrari C, Massetto B, Nguyen AH, Joshi A, Woo J, Lau AH, Gaggar A, Subramanian GM, Yoshida EM, Ahn SH, Tsai NCS, Fung S, Gane EJ (2018) Safety, efficacy and pharmacodynamics of vesatolimod (GS-9620) in virally suppressed patients with chronic hepatitis B. J Hepatol 68(3):431–440

Jiang R, Feng X, Guo Y, Lu Q, Hou J, Luo K, Fu N (2002) Early kinetics of innate and adaptive immune responses during hepatitis B virus infection. Chin Med J 115(3):422–424

Kadowaki N (2009) The divergence and interplay between pDC and mDC in humans. Front Biosci (Landmark Ed) 14:808–817

Kalathil S, Lugade AA, Miller A, Iyer R, Thanavala Y (2013) Higher frequencies of GARP(+) CTLA-4(+)Foxp3(+) T regulatory cells and myeloid-derived suppressor cells in hepatocellular carcinoma patients are associated with impaired T-cell functionality. Cancer Res 73(8):2435–2444

Kalathil SG, Lugade AA, Miller A, Iyer R, Thanavala Y (2016) PD-1+ and Foxp3+ T cell reduction correlates with survival of HCC patients after sorafenib therapy. JCI Insight 1(11):e86182

Kamphorst AO, Ahmed R (2013) CD4 T-cell immunotherapy for chronic viral infections and cancer. Immunotherapy 5(9):975–987

Khanam A, Trehanpati N, Riese P, Rastogi A, Guzman CA, Sarin SK (2017) Blockade of neutrophil's chemokine receptors CXCR1/2 abrogate liver damage in acute-on-chronic liver failure. Front Immunol 8:464. https://doi.org/10.3389/fimmu.2017.00464

Koumbi L (2015) Current and future antiviral drug therapies of hepatitis B chronic infection. World J Hepatol 7(8):1030–1040

Lamontagne RJ, Bagga S, Bouchard MJ (2016) Hepatitis B virus molecular biology and pathogenesis. Hepatoma Res 2:163–186

Lanford RE, Guerra B, Chavez D, Giavedoni L, Hodara VL, Brasky KM, Fosdick A, Frey CR, Zheng J, Wolfgang G, Halcomb RL, Tumas DB (2013) GS-9620, an oral agonist of Toll-like receptor-7, induces prolonged suppression of hepatitis B virus in chronically infected chimpanzees. Gastroenterology 144(7):1508–1517

Le PH, Liang KH, Chang ML, Hsu CW, Chen YC, Lin CL, Lin WR, Lai MW et al (2017) Clinical predictors for neutrophil-to-lymphocyte ratio changes in patients with chronic hepatitis B receiving peginterferon treatment. In Vivo 31(4):723–729

Leu C, Lu YC, Peng WL, Chu HT, Hu CP (2014) The hepatitis B virus e antigen suppresses the respiratory burst and mobility of human monocytes and neutrophils. Immunobiology 219(11):880–887

Li J, Qiu SJ, She WM, Wang FP, Gao H, Li L, Tu CT, Wang JY, Shen XZ et al (2012) Significance of the balance between regulatory T (Treg) and T helper 17 (Th17) cells during hepatitis B virus related liver fibrosis. PLoS One 7(6):e39307

Li X, Liu X, Tian L, Chen Y (2016a) Cytokine-mediated immunopathogenesis of hepatitis B virus infections. Clin Rev Allergy Immunol 50(1):41–54

Li W, Han J, Wu H (2016b) Regulatory T-cells promote hepatitis B virus infection and hepatocellular carcinoma progression. Chronic Dis Transl Med 2(2):67–80. https://doi.org/10.1016/j.cdtm.2016.09.001

Li H, Zhai N, Wang Z, Song H, Yang Y, Cui A, Li T, Wang G, Niu J, Crispe IN, Su L et al (2017) Regulatory NK cells mediated between immunosuppressive monocytes and dysfunctional T cells in chronic HBV infection. Gut 67(11):2035–2044

Liang TJ (2009) Hepatitis B: the virus and disease. Hepatology 49(5 Suppl):S13–S21. https://doi.org/10.1002/hep.22881

Liu H, Zhang H, Wan G, Sang Y, Chang Y, Wang X, Zeng H (2014a) Neutrophil-lymphocyte ratio: a novel predictor for short-term prognosis in acute-on-chronic hepatitis B liver failure. J Viral Hepat 21(7):499–507

Liu J, Zhang E, Ma Z, Wu W, Kosinska A, Zhang X, Möller I, Seiz P, Glebe D, Wang B, Yang D, Lu M, Roggendorf M (2014b) Enhancing virus-specific immunity in vivo by combining therapeutic vaccination and PD-L1 blockade in chronic hepadnaviral infection. PLoS Pathog 10(1):e1003856

Lumley SF, McNaughton AL, Klenerman P, Lythgoe KA, Matthews PC (2018) Hepatitis B virus adaptation to the CD8+ T cell response: consequences for host and pathogen. Front Immunol 9:1561

Luo J, Li J, Chen RL, Nie L, Huang J, Liu ZW, Luo L, Yan XJ (2010) Autologus dendritic cell vaccine for chronic hepatitis B carriers: a pilot, open label, clinical trial in human volunteers. Vaccine 28(13):2497–2504

Lv S, Wang J, Dou S, Yang X, Ni X, Sun R, Tian Z, Wei H (2014) Nanoparticles encapsulating hepatitis B virus cytosine-phosphate-guanosine induce therapeutic immunity against HBV infection. Hepatology 59:385–394

Ma Z, Cao Q, Xiong Y, Zhang E, Lu M (2018) Interaction between hepatitis B virus and toll-like receptors: current status and potential therapeutic use for chronic hepatitis B. Vaccine 6(1):6

Maier H, Isogawa M, Freeman GJ, Chisari FV (2007) PD-1:PD-L1 interactions contribute to the functional suppression of virus-specific CD8+ T lymphocytes in the liver. J Immunol 178(5):2714–2720

Maini MK, Boni C, Ogg GS, King AS, Reignat S, Lee CK, Larrubia JR, Webster GJ, McMichael AJ, Ferrari C, Williams R, Vergani D, Bertoletti A (1999) Direct ex vivo analysis of hepatitis B virus-specific CD8(+) T cells associated with the control of infection. Gastroenterology 117(6):1386–1396

Maini MK, Boni C, Lee CK, Larrubia JR, Reignat S, Ogg GS, King AS, Herberg J, Gilson R, Alisa A, Williams R, Vergani D, Naoumov NV, Ferrari C et al (2000) The role of virus-specific CD8(+) cells in liver damage and viral control during persistent hepatitis B virus infection. J Exp Med 191(8):1269–1280

Menne STB, Liu KH, Ascenzi MA, Baldwin BH, Bellezza CA, Cote PJ, Zheng X, Wolfgang G, Turnas D (2011) Antiviral efficacy and induction of an antibody response against surface antigen with the TLR7 agonist GS-9620 in the woodchuck model of chronic HBV infection. J Hepatol 54(1):S441

Milich DR, Chen M, Schödel F, Peterson DL, Jones JE, Hughes JL (1997) Role of B cells in antigen presentation of the hepatitis B core. Proc Natl Acad Sci U S A 94(26):14648–14653

Mohd-Ismail NK, Lim Z, Gunaratne J, Tan YJ (2019) Mapping the interactions of HBV cccDNA with host factors. Int J Mol Sci 20(17):4276. https://doi.org/10.3390/ijms20174276

Murakami Y, Minami M, Daimon Y, Okanoue T (2004) Hepatitis B virus DNA in liver, serum, and peripheral blood mononuclear cells after the clearance of serum hepatitis B virus surface antigen. J Med Virol 72(2):203–214

Nasser MW, Raghuwanshi SK, Grant DJ, Jala VR, Rajarathnam K, Richardson RM (2009) Differential activation and regulation of CXCR1 and CXCR2 by CXCL8 monomer and dimer. J Immunol 183(5):3425–3432. https://doi.org/10.4049/jimmunol.0900305

Nebbia G, Peppa D, Schurich A, Khanna P, Singh HD, Cheng Y, Rosenberg W, Dusheiko G, Gilson R, ChinAleong J, Kennedy P, Maini MK (2012) Upregulation of the Tim-3/galectin-9 pathway of T cell exhaustion in chronic hepatitis B virus infection. PLoS One 7(10):e47648

Ouyang W, Kolls JK, Zheng Y (2008) The biological functions of T helper 17 cell effector cytokines in inflammation. Immunity 28(4):454–467

Park H, Li Z, Yang XO, Chang SH, Nurieva R, Wang YH, Wang Y, Hood L, Zhu Z, Tian Q et al (2005) A distinct lineage of CD4 T cells regulates tissue inflammation by producing interleukin 17. Nat Immunol 6(11):1133–1141

Park JJ, Wong DK, Wahed AS, Lee WM, Feld JJ, Terrault N, Khalili M, Sterling RK, Kowdley KV, Bzowej N, Lau DT, Kim WR, Smith C, Carithers RL, Torrey KW, Keith JW, Levine DL, Traum D, Ho S, Valiga ME, Johnson GS, Doo E, Lok AS, Chang KM, Hepatitis B Research Network (2015) Hepatitis B virus--specific and global T-cell dysfunction in chronic hepatitis B. Gastroenterology 150(3):684–695.e5

Paul S, Dickstein A, Saxena A, Terrin N, Viveiros K, Balk EM, Wong JB (2017) Role of surface antibody in hepatitis B reactivation in patients with resolved infection and hematologic malignancy: a meta-analysis. Hepatology 66(2):379–388

Peppa D, Gill US, Reynolds G, Easom NJ, Pallett LJ, Schurich A, Micco L, Nebbia G, Singh HD, Adams DH, Kennedy PT et al (2013) Up-regulation of a death receptor renders antiviral T cells susceptible to NK cell-mediated deletion. J Exp Med 210(1):99–114

Phillips S, Chokshi S, Riva A, Evans A, Williams R, Naoumov NV (2010) CD8(+) T cell control of hepatitis B virus replication: direct comparison between cytolytic and noncytolytic functions. J Immunol 184(1):287–295

Pontisso P, Morsica G, Ruvoletto MG, Zambello R, Colletta C, Chemello L, Alberti A (1991) Hepatitis B virus binds to peripheral blood mononuclear cells via the pre S1 protein. J Hepatol 12(2):203–206

Poonia B, Ayithan N, Nandi M, Masur H, Kottilil S (2018) HBV induces inhibitory FcRL receptor on B cells and dysregulates B cell-T follicular helper cell axis. Sci Rep 8(1):15296

Raziorrouh B, Schraut W, Gerlach T, Nowack D, Grüner NH, Ulsenheimer A, Zachoval R, Wächtler M, Spannagl M, Haas J, Diepolder HM, Jung MC (2010) The immunoregulatory role of CD244 in chronic hepatitis B infection and its inhibitory potential on virus-specific CD8+ T-cell function. Hepatology 52(6):1934–1947

Raziorrouh B, Heeg M, Kurktschiev P, Schraut W, Zachoval R, Wendtner C et al (2014) Inhibitory phenotype of HBV-specific CD4+ T-cells is characterized by high PD-1 expression but absent coregulation of multiple inhibitory molecules. PLoS One 9(8):e105703. https://doi.org/10.1371/journal.pone.0105703

Rehermann B, Fowler P, Sidney J, Person J, Redeker A, Brown M, Moss B, Sette A et al (1995) The cytotoxic T lymphocyte response to multiple hepatitis B virus polymerase epitopes during and after acute viral hepatitis. J Exp Med 181(3):1047–1058

Riordan SM, Skinner N, Kurtovic J, Locarnini S, Visvanathan K (2006) Reduced expression of toll-like receptor 2 on peripheral monocytes in patients with chronic hepatitis B. Clin Vaccine Immunol 13(8):972–974

Sant AJ, McMichael A (2012) Revealing the role of CD4(+) T cells in viral immunity. J Exp Med 209(8):1391–1395

Schurich A, Khanna P, Lopes AR, Han KJ, Peppa D, Micco L, Nebbia G, Kennedy PT, Geretti AM, Dusheiko G, Maini MK (2011) Role of the coinhibitory receptor cytotoxic T lymphocyte antigen-4 on apoptosis-Prone CD8 T cells in persistent hepatitis B virus infection. Hepatology 53(5):1494–1503. https://doi.org/10.1002/hep.24249

Shi M, Qian S, Chen WW, Zhang H, Zhang B, Tang ZR, Zhang Z et al (2007) Hepatitis B virus (HBV) antigen-pulsed monocyte-derived dendritic cells from HBV-associated hepatocellular carcinoma patients significantly enhance specific T cell responses in vitro. Clin Exp Immunol 147(2):277–286

Shi B, Ren G, Hu Y, Wang S, Zhang Z, Yuan Z (2012) HBsAg inhibits IFN-α production in plasmacytoid dendritic cells through TNF-α and IL-10 induction in monocytes. PLoS One 7(9):e44900

Shrivastava S, TrehanPati N, Patra S, Kottilil S, Pande C, Trivedi SS, Sarin SK (2013) Increased regulatory T cells and impaired functions of circulating CD8 T lymphocytes is associated with viral persistence in Hepatitis B virus-positive newborns. J Viral Hepat 20(8):582–591

Sitia G, Isogawa M, Kakimi K, Wieland SF, Chisari FV, Guidotti LG (2002) Depletion of neutrophils blocks the recruitment of antigen-nonspecific cells into the liver without affecting the antiviral activity of hepatitis B virus-specific cytotoxic T lymphocytes. Proc Natl Acad Sci U S A 99(21):13717–13722

Stelma F, de Niet A, Tempelmans Plat-Sinnige MJ, Jansen L, Takkenberg RB, Reesink HW, Kootstra NA, van Leeuwen EM (2015) Natural killer cell characteristics in patients with chronic hepatitis B virus (HBV) infection are associated with HBV surface antigen clearance after combination treatment with pegylated interferon alfa-2a and adefovir. J Infect Dis 212(7):1042–1051

Stelma F, Willemse SB, Erken R, de Niet A, Sinnige MJ, van Dort K, Zaaijer HL, van Leeuwen E, Kootstra NA, Reesink HW (2017) Dynamics of the immune response in acute hepatitis B infection. Open Forum Infect Dis 4(4):ofx231. https://doi.org/10.1093/ofid/ofx231

Stoop JN, van der Molen RG, Baan CC, van der Laan LJ, Kuipers EJ, Kusters JG, Janssen HL (2005) Regulatory T cells contribute to the impaired immune response in patients with chronic hepatitis B virus infection. Hepatology 41(4):771–778

Stoop JN, Claassen MA, Woltman AM, Binda RS, Kuipers EJ, Janssen HL, van der Molen RG, Boonstra A (2008) Intrahepatic regulatory T cells are phenotypically distinct from their peripheral counterparts in chronic HBV patients. Clin Immunol Soc 129(3):419–427

Stross L, Günther J, Gasteiger G, Asen T, Graf S, Aichler M, Esposito I, Busch DH, Knolle P, Sparwasser T, Protzer U (2012) Foxp3+ regulatory T cells protect the liver from immune damage and compromise virus control during acute experimental hepatitis B virus infection in mice. Hepatology 56(3):873–883

Sun HQ, Zhang JY, Zhang H, Zou ZS, Wang FS, Jia JH (2012) Increased Th17 cells contribute to disease progression in patients with HBV-associated liver cirrhosis. J Viral Hepat 19(6):396–403

Takai S, Kimura K, Nagaki M, Satake S, Kakimi K, Moriwaki H (2005) Blockade of neutrophil elastase attenuates severe liver injury in hepatitis B transgenic mice. J Virol 79(24):15142–15150

Tan Z, Qian X, Jiang R, Liu Q, Wang Y, Chen C, Wang X, Ryffel B, Sun B (2013) IL-17A plays a critical role in the pathogenesis of liver fibrosis through hepatic stellate cell activation. J Immunol 191(4):1835–1844

Tang TJ, Kwekkeboom J, Laman JD, Niesters HG, Zondervan PE, de Man RA, Schalm SW, Janssen HL (2003) Regulatory T cells in HBV and HCV liver diseases: implication of regulatory T lymphocytes in the control of immune response. J Viral Hepat 10(3):159–167

Thimme R, Wieland S, Steiger C, Ghrayeb J, Reimann KA, Purcell RH, Chisari FV (2003) CD8(+) T cells mediate viral clearance and disease pathogenesis during acute hepatitis B virus infection. J Virol 77(1):68–76

Tjwa ET, van Oord GW, Hegmans JP, Janssen HL, Woltman AM (2011) Viral load reduction improves activation and function of natural killer cells in patients with chronic hepatitis B. J Hepatol 54(2):209–218

Trepo C, Chan HLY, Lok A (2014) Hepatitis B virus infection. Lancet 384:2053–2063

Untergasser A, Zedler U, Langenkamp A, Hösel M, Quasdorff M, Esser K, Dienes HP, Tappertzhofen B, Kolanus W, Protzer U (2006) Dendritic cells take up viral antigens but do not support the early steps of hepatitis B virus infection. Hepatology 43(3):539–547

van der Molen RG, Sprengers D, Binda RS, de Jong EC, Niesters HG, Kusters JG, Kwekkeboom J, Janssen HL (2004) Functional impairment of myeloid and plasmacytoid dendritic cells of patients with chronic hepatitis B. Hepatology 40(3):738–746

Van der Molen RG, Sprengers D, Biesta PJ, Kusters JG, Janssen HL (2006) Favorable effect of adefovir on the number and functionality of myeloid dendritic cells of patients with chronic HBV. Hepatology 44(4):907–914

Vanlandschoot P, Roobrouck A, Van Houtte F, Leroux-Roels G (2002) Recombinant HBsAg, an apoptotic-like lipoprotein, interferes with the LPS-induced activation of ERK-1/2 and JNK-1/2 in monocytes. Biochem Biophys Res Commun 297(3):486–491

Vincent IE, Zannetti C, Lucifora J, Norder H, Protzer U, Hainaut P, Zoulim F, Tommasino M, Trépo C, Hasan U et al (2011) Hepatitis B virus impairs TLR9 expression and function in plasmacytoid dendritic cells. PLoS One 6(10):e26315

Visvanathan K, Skinner NA, Thompson AJ, Riordan SM, Sozzi V, Edwards R, Rodgers S, Kurtovic J, Chang J, Lewin S, Desmond P, Locarnini S (2007) Regulation of toll-like receptor-2 expression in chronic hepatitis B by the precore protein. Hepatology 45(1):102–110

Wang S, Chen Z, Hu C, Qian F, Cheng Y, Wu M, Shi B, Chen J, Hu Y, Yuan Z (2013) Hepatitis B virus surface antigen selectively inhibits TLR2 ligand-induced IL-12 production in monocytes/macrophages by interfering with JNK activation. J Immunol 190(10):5142–5151

Wang G, Liu Y, Huang R, Jia B, Su R, Sun Z, Tian C, Xiong Y, Xia J, Yan X, Zhang Z, Wu C (2017) Characteristics of regulatory B cells in patients with chronic hepatitis B virus infection in different immune phases. Discov Med 23(128):295–304

Wang Q, Pan W, Liu Y, Luo J, Zhu D, Lu Y, Feng X, Yang X, Dittmer U, Lu M, Yang D et al (2018) Hepatitis B virus-specific CD8+ T cells maintain functional exhaustion after antigen reexposure in an acute activation immune environment. Front Immunol 9:219. https://doi.org/10.3389/fimmu.2018.00219

Webster GJ, Reignat S, Brown D, Ogg GS, Jones L, Seneviratne SL, Williams R, Dusheiko G, Bertoletti A (2004) Longitudinal analysis of CD8+ T cells specific for structural and nonstructural hepatitis B virus proteins in patients with chronic hepatitis B: implications for immunotherapy. J Virol 78(11):5707–5719

WHO (2017, April) Global hepatitis report. Retrieved from http://www.who.int/hepatitis/publications/global-hepatitis-report2017/en/

Wisskirchen K, Metzger K, Schreiber S, Asen T, Weigand L, Dargel C, Witter K, Kieback E, Sprinzl MF, Uckert W, Schiemann M, Busch DH, Krackhardt AM et al (2017) Isolation and functional characterization of hepatitis B virus-specific T-cell receptors as new tools for experimental and clinical use. PLoS One 12(8):e0182936. https://doi.org/10.1371/journal.pone.0182936

Wu C, Liu Y, Zhao Q, Chen G, Chen J, Yan X, Zhou YH et al (2010) Soluble CD40 ligand-activated human peripheral B cells as surrogated antigen presenting cells: a preliminary approach for anti-HBV immunotherapy. Virol J 7:370. https://doi.org/10.1186/1743-422X-7-370

Wu W, Shi Y, Li S, Zhang Y, Liu Y, Wu Y, Chen Z (2012) Blockade of Tim-3 signaling restores the virus-specific CD8+ T-cell response in patients with chronic hepatitis B. Eur J Immunol 42(5):1180–1191

Wu S, Wang W, Gao Y (2015) Natural killer cells in hepatitis B virus infection. Braz J Infect Dis 19(4):417–425

Xu D, Fu J, Jin L, Zhang H, Zhou C, Zou Z, Zhao JM, Zhang B, Shi M, Ding X, Tang Z, Fu YX, Wang FS (2006) Circulating and liver resident CD4+CD25+ regulatory T cells actively influence the antiviral immune response and disease progression in patients with hepatitis B. J Immunol 177(1):739–747

Xu N, Yao HP, Lv GC, Chen Z (2012) Downregulation of TLR7/9 leads to deficient production of IFN-α from plasmacytoid dendritic cells in chronic hepatitis B. Inflamm Res 61(9):997–1004

Xu L, Gong Y, Wang B, Shi K, Hou Y, Wang L, Lin Z, Han Y, Lu L, Chen D, Lin X, Zeng Q, Feng W, Chen Y (2014a) Randomized trial of autologous bone marrow mesenchymal stem cells transplantation for hepatitis B virus cirrhosis: regulation of Treg/Th17 cells. J Gastroenterol Hepatol 29(8):1620–1628

Xu R, Huang H, Zhang Z, Wang FS (2014b) The role of neutrophils in the development of liver diseases. Cell Mol Immunol 11(3):224–231

Yan H, Zhong G, Xu G, He W, Jing Z, Gao Z, Huang Y, Qi Y, Peng B, Wang H, Fu L, Song M, Chen P, Gao W, Ren B, Sun Y, Cai T, Feng X, Sui J, Li W (2012) Sodium taurocholate cotransporting polypeptide is a functional receptor for human hepatitis B and D virus. elife 1:e00049

Yang PL, Althage A, Chung J et al (2010) Immune effectors required for hepatitis B virus clearance. Proc Natl Acad Sci U S A 107(2):798–802. https://doi.org/10.1073/pnas.0913498107

Yang J, Yi P, Wei L, Xu Z, Chen Y, Tang L, Li L (2012) Phenotypes and clinical significance of circulating CD4(+)CD25(+) regulatory T cells (Tregs) in patients with acute-on-chronic liver failure (ACLF). J Transl Med 10:193. https://doi.org/10.1186/1479-5876-10-193

Yang B, Wang Y, Zhao C, Yan W, Che H, Shen C, Zhao M (2013) Increased Th17 cells and interleukin-17 contribute to immune activation and disease aggravation in patients with chronic hepatitis B virus infection. Immunol Lett 149(1–2):41–49

Yang C, Cui F, Chen LM, Gong XY, Qin B (2016) Correlation between Th17 and nTreg cell frequencies and the stages of progression in chronic hepatitis B. Mol Med Rep 13(1):853–859

Ye B, Liu X, Li X, Kong H, Tian L, Chen Y (2015) T-cell exhaustion in chronic hepatitis B infection: current knowledge and clinical significance. Cell Death Dis 6(3):e1694. https://doi.org/10.1038/cddis.2015.42

Yoffe B, Noonan CA, Melnick JL, Hollinger FB (1986) Hepatitis B virus DNA in mononuclear cells and analysis of cell subsets for the presence of replicative intermediates of viral DNA. J Infect Dis 153(3):471–477

Yu XP, Guo RY, Su ML, Ming DS, Lin CZ, Deng Y, Lin ZZ et al (2013) Dynamic changes of treg and Th17 cells and related cytokines closely correlate with the virological and biochemical response in chronic hepatitis b patients undergoing nucleos(t)ide analogues treatment. Hepat Mon 13(12):e15332. https://doi.org/10.5812/hepatmon.15332

Yu X, Guo R, Ming D, Su M, Lin C, Deng Y, Lin Z, Su Z (2014) Ratios of regulatory T cells/T-helper 17 cells and transforming growth factor-β1/interleukin-17 to be associated with the development of hepatitis B virus-associated liver cirrhosis. J Gastroenterol Hepatol 29(5):1065–1072

Zhang Z, Chen D, Yao J, Zhang H, Jin L, Shi M et al (2007) Increased infiltration of intrahepatic DC subsets closely correlate with viral control and liver injury in immune active pediatric patients with chronic hepatitis B. Clin Immunol 122:173–180

Zhang JY, Zhang Z, Lin F, Zou ZS, Xu RN, Jin L, Fu JL, Shi F, Shi M, Wang HF, Wang FS (2010a) Interleukin-17-producing CD4(+) T cells increase with severity of liver damage in patients with chronic hepatitis B. Hepatology 51(1):81–91

Zhang Z, Zhang S, Zou Z, Shi J, Zhao J, Fan R, Qin E, Li B, Li Z, Xu X, Fu J, Zhang J, Gao B, Tian Z et al (2010b) Hypercytolytic activity of hepatic natural killer cells correlates with liver injury in chronic hepatitis B patients. Hepatology 53(1):73–85

Zhang JY, Zou ZS, Huang A, Zhang Z, Fu JL, Xu XS, Chen LM, Li BS, Wang FS (2011) Hyper-activated pro-inflammatory CD16 monocytes correlate with the severity of liver injury and fibrosis in patients with chronic hepatitis B. PLoS One 6(3):e17484

Zhang H, Sun Q, Mao W, Fan J, Ye B (2016) Neutrophil-to-lymphocyte ratio predicts early mortality in patients with HBV-related decompensated cirrhosis. Gastroenterol Res Pract 2016:4394650

Zhang Y, Wu Y, Deng M, Xu D, Li X, Xu Z, Hu J, Zhang H, Liu K, Zhao Y, Gao F, Bi S, Gao GF, Zhao J, Liu WJ, Meng S (2018) $CD8^+$ T-cell response-associated evolution of hepatitis B virus core protein and disease progress. J Virol 92(17):e02120-17

Zhao J, Li Y, Jin L, Zhang S, Fa R, Sun Y, Zhou C, Shang Q, Li W, Zhang Z, Wang FS (2012) Natural killer cells are characterized by the concomitantly increased interferon-γ and cytotoxicity in acute resolved hepatitis B patients. PLoS One 7(11):e49135

Zhao PW, Ma L, Ji HF, Yu L, Feng JY, Wang J, Liu MY et al (2015) The expression of TLR-9, CD86, and CD95 phenotypes in circulating B cells of patients with chronic viral hepatitis B or C before and after antiviral therapy. Mediat Inflamm 2015:762709

Significance of RNA Sensors in Activating Immune System in Emerging Viral Diseases

15

Preethika Nair and Siddhesh U. Sapre

Abstract

Viruses are known for their utilization of the host machinery to aid their replication within the infected cells. The viral nucleic acid, their intermittent forms contain pathogen-associated molecular patterns (PAMP) which enable the pattern recognition receptors (PRR) to distinguish them from self and mount a response against them. Highly robust targets for the PRRs are the cell entry points and phases of viral replication. RNA-sensing PRR can be classified as endosomal PRR and cytosolic PRR. Both enveloped and non-enveloped viruses utilize endosomal compartments to undergo proteolytic cleavage or pH-dependent conformational changes to ensure membrane fusion. Thus, monitoring the endosomal compartments in order to restrict the virus from penetrating into the cytoplasm is a key antiviral strategy. Endosomal compartments are guarded by Toll-like receptors (TLRs) which are the earliest discovered PRRs. TLR3, 7, and 8 specifically cater to viral RNA sensing. They can detect dsRNA and stable stem structures of ssRNA. Cytosolic PRRs include RIG-I like receptors (RLR) and nucleotide-binding oligomerization domain-containing (NOD)—like receptors (NLR). RLRs are cytosolic helicases which detect viral RNA, ds RNA, short 5′ppp RNA, and RNase L cleaved RNA. Retinoic acid-inducible gene I product (RIG-I), melanoma differentiation-associated antigen 5 (MDA5), and laboratory of genetics and physiology 2 (LGP2) are aspartate-glutamate-any amino acid-aspartate/histidine (DExD/H)-box helicases belonging to the SF2 superfamily. Other RNA helicases like SNRNP200 which belongs to the Ski-2 superfamily, or DDX60 belonging to the Ski-2-like helicase family are also involved in viral RNA sensing. NLRs like NLRP3 and NOD2 are cytosolic RNA-sensing PRRs. In addition to these, RNA-binding proteins like RNase L, protein kinase R (PKR), and interferon-inducible

P. Nair (✉)
NMIMS School of Science, Mumbai, India

S. U. Sapre
National Institute of Virology, Pune, India

P. V. Bramhachari (ed.), *Dynamics of immune activation in viral diseases*,
https://doi.org/10.1007/978-981-15-1045-8_15

transmembrane protein (IFIT) are also present which play pertinent roles to enhance the antiviral immunity during a viral infection. Recognition of PAMPs or DAMPs by PRRs activate various signaling cascades which ultimately triggers the transcription of type I/III IFN, production of various pro-inflammatory cytokines, and various other genes that can ensure an intracellular antiviral state which will aid in containing the viral infection. Furthermore, these interferons can lead to the expression of multiple interferon-stimulated genes (ISGs) which induce antiviral activities controlling the life cycle of the virus, restricting its replication, and transmission to the surrounding cells. This chapter will elaborate on these RNA-sensing PRRs, their structures, agonists, mechanism of activation, and response mounted against viral infection.

Keywords
Host machinery RNA sensors · Viral diseases · Immune activation

15.1 Introduction

Viruses are known for their utilization of the host machinery to aid their replication within the infected cells. The viral nucleic acid, its intermittent forms contain pathogen-associated molecular patterns (PAMP) which enable the pattern recognition receptors (PRR) to distinguish them from self and mount a response against them. Highly robust targets for the PRRs are the cell entry points and phases of viral replication. RNA-sensing PRR can be classified as endosomal PRR and cytosolic PRR. Both enveloped and non-enveloped viruses utilize endosomal compartments to undergo proteolytic cleavage or pH-dependent conformational changes to ensure membrane fusion. Thus, monitoring the endosomal compartments in order to restrict the virus from penetrating into the cytoplasm is a key antiviral strategy. Endosomal compartments are guarded by Toll-like receptors (TLRs) which are the earliest discovered PRRs. TLR3, 7, and 8 specifically cater to viral RNA sensing. They can detect dsRNA and stable stem structures of ssRNA.

Cytosolic PRRs include RIG-I like receptors (RLR) and nucleotide-binding oligomerization domain-containing (NOD)—like receptors (NLR). RLRs are cytosolic helicases which detect viral RNA, ds RNA, short 5′ppp RNA and RNase L cleaved RNA. Retinoic acid-inducible gene I product (RIG-I), melanoma differentiation-associated antigen 5 (MDA5), and laboratory of genetics and physiology 2 (LGP2) are aspartate-glutamate-any amino acid-aspartate/histidine (DExD/H)-box helicases belonging to the SF2 superfamily. Other RNA helicases like SNRNP200 which belongs to the Ski-2 superfamily, or DDX60 belonging to the Ski-2-like helicase family are also involved in viral RNA sensing. NLRs like NLRP3 and NOD2 are cytosolic RNA-sensing PRRs. In addition to these, RNA-binding proteins like RNase L, protein kinase R (PKR), and interferon-inducible transmembrane protein (IFIT) are also present which play pertinent roles to enhance the antiviral immunity during a viral infection.

Recognition of PAMPs or DAMPs by PRRs activate various signaling cascades which ultimately triggers the transcription of type I/III IFN, production of various pro-inflammatory cytokines, and various other genes that can ensure an intracellular antiviral state which will aid in containing the viral infection. Furthermore, these interferons can lead to the expression of multiple interferon-stimulated genes (ISGs) which induce antiviral activities controlling the life cycle of the virus, restricting its replication and transmission to the surrounding cells. This chapter will elaborate on these RNA-sensing PRRs, their structures, agonists, mechanism of activation, and response mounted against viral infection.

15.2 Innate Immunity: Soldiers Against Viruses

Innate immunity provides the first line of defence against the invading pathogens. It exerts its role by distinguishing between self and non-self, subsequently initiating a host response against the invading pathogen. Innate immunity is triggered by the identification of PAMPs (pathogen-associated molecular patterns) via the PRRs (pattern recognition receptors). PRRs are germline-encoded proteins expressed by a variety of cells which can carry out surveillance to detect any pathogenic invasions. PRRs have been classified into various families based on protein domain homology; Toll-like receptors (TLRs), RIG-I like receptors (RLRs), NOD-like receptors (NLRs), C-type lectin receptors (CLRs), AIM2-like receptors (ALRs), and cytosol DNA sensing PRR cyclic GMP-AMP synthase (cGAS). PRRs not only survey the PAMPS but also the DAMPS (danger-associated molecular patterns), which are typically the cellular products produced as a response to cell stress (Anafu et al. 2013; Chen et al. 2017; Chow et al. 2018; Said et al. 2018).

Recognition of PAMPs or DAMPs by PRRs leads to transcription of genes involved in pro-inflammatory responses. The responses mounted against a viral infection, upregulates type I and III interferons, pro-inflammatory chemokines, and various other genes that can ensure an intracellular antiviral state in order to contain the infection. Furthermore, these interferons can lead to the expression of multiple interferon-stimulated genes (ISGs) which induce antiviral activities controlling the life cycle of the virus, restricting its replication and transmission to the surrounding cells.

Viruses are known for their utilization of the host machinery to aid their replication within the infected cells. The viral nucleic acid, their intermittent forms contain PAMPs which enable the PRRs to distinguish them from self and mount a response against them. Highly robust targets for the PRRs are the cell entry points and phases of viral replication. Both enveloped and non-enveloped viruses utilize endosomal compartments to undergo proteolytic cleavage or pH-dependent conformational changes to ensure membrane fusion (Said et al. 2018). Thus, monitoring the endosomal compartments in order to restrict the virus from penetrating into the cytoplasm is a key antiviral strategy. Endosomal compartments are guarded by Toll-like receptors (TLRs) which are the earliest discovered PRRs. TLR3, 7, and 8 specifically cater to viral RNA sensing. Cytosolic PRRs include RIG-I, MDA5, LGP2, NLRP3, and NOD2. RNA sensing is also carried out by other RNA helicases like

SNRNP200 and DDX60 which will be discussed in details. In addition to these, RNA-binding proteins like RNase L, PKR, and IFITs are also present which play pertinent roles to enhance the antiviral immunity during a viral infection.

15.3 RNA-Sensing Endosomal PRRs

15.3.1 Toll-Like Receptors (TLRs)

TLRs are a very important and earliest detected type of PRRs. They comprise of 11 genes on the human genome. Majority of the TLRs are associated with the plasma membrane. However, the TLRs associated with RNA sensing namely, TLR3, 7, and 8 are present in the endosomal compartment. TLRs are type I transmembrane proteins which consists of N-terminal ectodomain or extracellular leucine-rich domain (ECD), middle transmembrane domain (TM), and C-terminal cytoplasmic Toll/IL-1 receptor (TIR) domain. Leucine-rich regions (LRRs) in the ECD are attached into a horseshoe-shaped solenoid structure. α-helices of each LRR forms into the convex structure of the solenoid and the β sheets assemble into the concave surfaces of the solenoid structure. It is a unique property of the TLRs to bind their agonists to the lateral convex surface instead of concave surface.

The formation of M-shaped dimer or multimer is needed for all TLR activation, so that the C-terminal regions of the two TLR ECDs are brought into proximity. It in turn causes the multimerization of cytoplasmic TIR domains, which will recruit downstream adaptors TRIF or MyD88 through homotypic interaction, further forming signaling complex called signalosome and activating downstream transcription factors like NF-κB which induces pro-inflammatory cytokines and interferon regulatory factor (IRF) which in turn induces type I interferon.

15.3.1.1 TLR3: Expression, Structure, and Signaling Pathway

Immune cells like monocytes, macrophages, conventional dendritic cells, NK cells, T and B lymphocytes, mast cells, eosinophils, and basophiles demonstrate the expression of TLR3 in their endosomal compartments (Chen et al. 2017; Said et al. 2018; Gantier and Williams 2011; Hewson et al. 2005; Thomas et al. 2007). They are also expressed in nonimmune cells like keratinocytes, fibroblasts, hepatocytes, astrocytes, and microglia. TLR3 senses dsRNA; however, stable stem structures of ssRNA are also recognized by TLR3. Its structure consists of N-terminal ectodomain or extracellular 23 leucine-rich domain (ECD), middle transmembrane domain (TM), and C-terminal cytoplasmic Toll/IL-1 receptor (TIR) domain. The 23 LRR ECD is responsible for the viral dsRNA binding. Dimerization of the ECD initiates the signaling cascade wherein TIR domain-containing adaptor protein-inducing IFN-β (TRIF) is then recruited and undergoes slight conformational changes to form a signaling complex together with TNF receptor-associated factor 6 (TRAF6), TRAF3, TBK1, IKKε, and IKK. This leads to the activation of IRF3/IRF7 and NF-κB, which results in the production of type 1 IFNs and inflammatory cytokines, respectively.

Further, there are positive regulators to this pathway like S100A9 which acts during the early stages of TLR3 activation by easing the maturation of TLR3-containing early endosomes into late endosomes, etc. Similarly, there are negative regulators such as Rho proteins that decrease the production of pro-inflammatory cytokines upon TLR3 triggering. The pathway is also regulated via post-translational modifications.TLR3 has been implicated in pathogenesis of HCV, infection, HSV-1, HBV, etc. This makes TLR3 an interesting candidate to prod into antiviral therapies.

15.3.1.2 TLR7 and 8: Expression, Structure, and Signaling Pathway

TLR7/8/9 subfamily members are localized at endosomes of the cells. They are present on the endosomes of monocytes, macrophages, plasmocystoid dendritic cells, NK cells, and T and B cells. TLR8 is also present on mast cells and Tregs. Both have very similar ligand recognition and intracellular signaling. TLR7 and 8 are activated by small molecular agonists and GU or U rich single-stranded RNA (ssRNA) (Chen et al. 2017; Gantier and Williams 2011; Crozat and Beutler 2004; Jensen and Thomsen 2012; Jurk et al. 2002; Patel et al. 2014). In viral genomes, untranslated terminal regions (UTR) are GU or U rich. They are highly conserved sequences as they are involved in viral protein translation and RNA replication. Further, they can also recognize phagocytosed vRNA, by the adenosine-to-inosine (A-to-I) editing, which is an important arm of the antiviral response. But 2′-*O*-methylation within an RNA sequence leads to the triggering of TLR8 but not TLR7.

The ECDs of both TLR7 and 8 have 26 LRR motifs in their extracellular domain, which contain multiple insertions such as the Z-loop or undefined region situated between LRRs 14 and 15. Cathepsins and arginine endopeptidase cleave the TLR7 and 8 in the endosomes, at this z-loop. Once cleaved, the fragments are still associated to each other by the intermolecular interactions as they both are crucial for the receptor activation. Recent crystal structures have demonstrated the presence of two binding sites: one for a small chemical stimuli or degradation product of ssRNA and the other which recognizes the viral ssRNA. At steady state they both exist as dimers. However, on association with the agonists, the conformation changes causing multimerization of the cytoplasmic TIR domains. A downstream adaptor MyD88 is recruited through homotypic interaction; a signaling complex called myddosome is formed involving IRAK4, IRAK1, TRAF6, and TRAF3 and downstream transcription factors NF-κB and IRF7 are activated to induce pro-inflammatory cytokines and IFNs, respectively (Fig. 15.1). Furthermore, TLR7 recognizes Streptococcus Group B (SGB) RNA and TLR8 recognize the RNAs from *Escherichia coli, Mycobacteria bovis, Helicobacter pylori*, and *Borrelia burgdorferi*.

15.4 RNA-Sensing Cytosolic PRRs

The first line of cytosolic RNA sensors include the PRR families RLRs and NLRs. RLRs consists of cytosolic helicases: RIG-I, MDA5, and LGP2; other DEXD/H-box helicases like DDX3, DDX60, and SNRP20 (Chow et al. 2018; Dang et al. 2018).

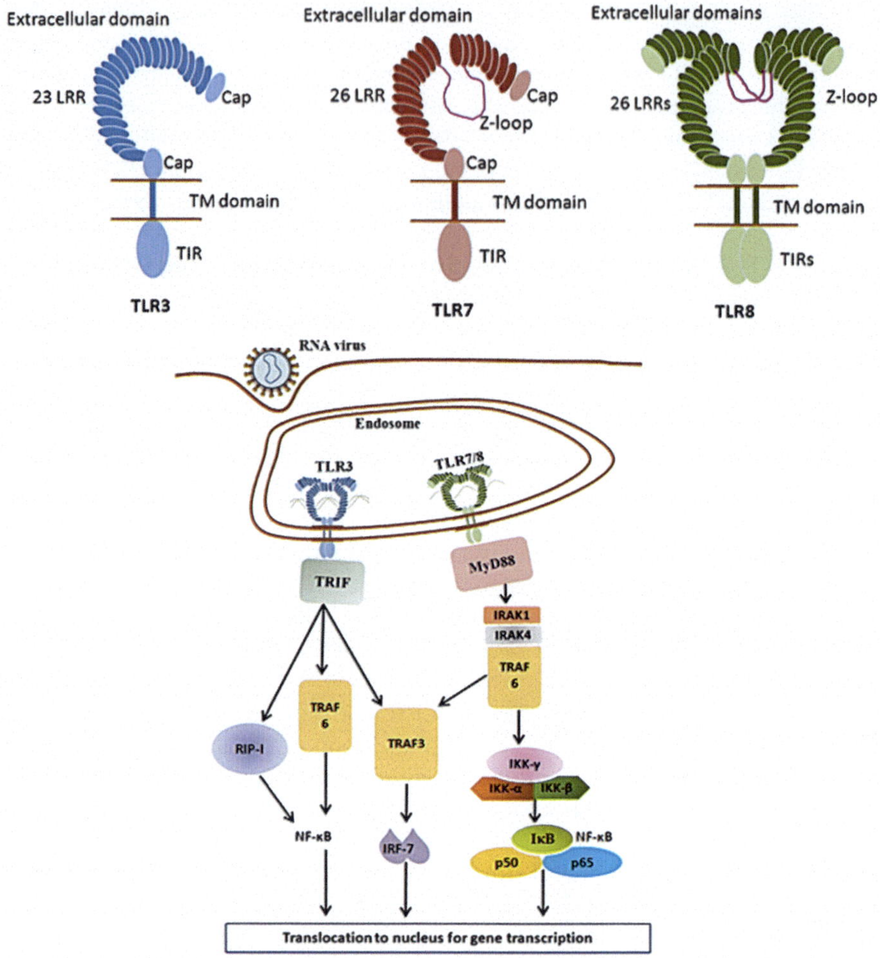

Fig. 15.1 Structure and mechanism of action of endosomal PRRs—TLR3, 7/8

15.4.1 RIG-I Like Receptors (RLRs)

Cytosolic helicases are ubiquitously expressed in low levels in most cell types. They belong to a family of aspartate-glutamate-any amino acid-aspartate/histidine (DExD/H)-box helicases (SF2 superfamily). RLRs are also known as double-stranded RNA (dsRNA)-dependent ATPases. Their function includes sensing the viral RNA in the cytosol, binding to the non-self RNA stably, and imposing the innate immune response against the target RNA. Three members widely studied are the retinoic acid-inducible gene I product (RIG-I), melanoma differentiation-associated antigen 5 (MDA5), and laboratory of genetics and physiology 2 (LGP2).

Overall structure of the RLRs are as follows:

1. Zinc-binding C-terminal domain (CTD), otherwise known as the repressor domain (RD)
2. Attached to the CTD is a central RNA helicase core consisting of two RecA-like helicase domains promotes dsRNA recognition.
3. Further, RIG-I and MDA5 contain at the N-terminus tandem caspase activation and recruitment domains (CARDs) that enables signaling capabilities. LGP2 uniquely lacks this domain.

15.4.1.1 RIG-I

When disengaged, the RIG-I CARDs and RD are bound to the helicase region and this leads to autoinhibition (Chow et al. 2018; Dang et al. 2018; Pichlmair et al. 2006). During viral replication the CTD/RD detects and associates with the viral RNA along with the helicase, exposing the CARDS. These CARDs are no longer autorepressed; the RLRs oligomerize and further associate with the mitochondrial antiviral signaling adaptor called MAVS (also referred to as IPS-1/VISA/cardif). MAVS are located at the cytosolic face of the outer mitochondrial membrane, and this mitochondrial association is necessary for initiating further signaling events. On association the MAVS undergo a prion-like aggregation. This leads to the recruitment of E3 ubiquitin ligases and the downstream effector proteins TNF receptor-associated factor 2 (TRAF2), TRAF3, and TRAF6 which assembles into an active "signalosome." The resulting cascade leads to the phosphorylation and nuclear translocation of key innate immune transcription factors IRF3 and IRF7. Further, it leads to the activation of NF-κB to drive the expression of type I and III interferons, innate immune genes, pro-inflammatory cytokines, and chemokines that will aid in containing the virus infection.

The RD domain is not only critical for RNA sensing but also confers it with the displayed selectivity. RIG-I can recognize 5′ppp dsRNA, 5′pp dsRNA, pU/UC genomic RNA, AU-rich 3 UTR, RNase L cleavage products and circular viral RNA (Chow et al. 2018; Said et al. 2018; Dang et al. 2018; Liu and Gale Jr 2011; Yoneyama and Fujita 2007; Zou et al. 2009). The distinguishing between self and non-self RNA is based on the 5′-7-methylguanosine cap, which is subject to 2′-*O*-methylation.

15.4.1.2 MDA5

RIG-I and MDA5 being homologues showcase the same overall domain structures. The RD domain however recognizes long dsRNA which under normal circumstances is absent in an uninfected cell. The presence of this PAMP signifies vial invasion which activates MDA5 and stimulates a cytokine response similar to the RIG-I. MDA5 binds to the long dsRNA replicative intermediates generated by picornavirus, consisting of its positive-sense genome annealed to the negative-sense antigenome, AU-rich motifs and RNase L cleavage products (Said et al. 2018; Dang et al. 2018; Bruns et al. 2014; Rodriguez et al. 2014).

On binding with the target RNA it undergoes a conformational change and the 2 CARD domains are exposed. The 2 CARDS assemble into a stable tetramer due to its binding with the dsRNA and they form a filament structure. They therefore are independent on the polyubiquitin chain binding; the tetramer itself nucleates the MAVS, which in turn activates TRAF3/TBK1/IKKe/IRF3 and TRAF6/IKK/NF-jB, which drive IFN and pro-inflammatory cytokine expression, respectively.

15.4.1.3 LGP2

Although LGP2 is homologues to RIG-I and MDA5, it lacks the CARD domain which restricts it from binding to the MAVS. However, LGP2 is shown to bind to the termini of blunt-ended dsRNA of different lengths with high affinity, forming complexes with 2:1 stoichiometry. LGP2 has been both implicated as a positive and negative effector of RIG-I and MDA5 response (Said et al. 2018; Jensen and Thomsen 2012; Bruns et al. 2014; Rodriguez et al. 2014). Studies have shown that CTDs of LGP2 and RIG-I are analogous because of which LGP2 CTD interacts with RIG-I to abolish its ability to initiate antiviral signaling. RIG-I also undergoes attenuation of viral sensing. LGP2 has also been believed to utilize its ATP dependent/RNA helicase activity to increase the interaction of nucleic acid PAMPS for improved antiviral RLR signaling. LGP2 has also been implicated for its activity against DICER protein to maintain the cytosolic PAMPs in intact condition for detection by the cytosolic RNA sensors. Therefore, LGP2 overall is a modulator of the innate immune response to a viral infection and not a sensor of PAMPs. LGP2 does not initiate antiviral gene expression.

15.4.1.4 Other DEXD/H-Box Helicases

Certain helicases which are a member of the DEXD/H-box helicases family have been anticipated to function in the RLR-based RNA sensing. DDX3 is one such helicases which associates itself with MAVS, and when it encounters transfected dsRNA, it induces interferon production. DDX3 has also been suggested to aid as a scaffold for the assembly of an RLR signaling complex, promoting innate immunity signaling (Chow et al. 2018; Said et al. 2018; Jensen and Thomsen 2012). DDX3 is being studied further to understand the function and mechanism of action in RNA sensing.

Another DEXD/H-box Helicase DDX60 (Ski-2-like helicase) is implicated in the cytosolic antiviral response. DDX60 maintains the quality of host RNA by acting as an exosome complex to degrade other foreign RNAs. Exosome-mediated degradation also enables the degraded vRNA to be recognized by RIG-I/MDA5. DDX60 also functions as an ISG to suppress viral infection in an infected cell. vRNA and RIG-I/MDA5 interactions are brought together by DDX60 in order to enhance the type 1 interferon production. Thus, DDX60 functions on two levels augmenting the antiviral response by detecting the vRNA and subject them to the RIG-I/MDA5 based RLR pathway; further, they aid in revealing the molecular signature (PAMP) of the vRNA via the exosome-mediated degradation, making DDX60 a immune-stimulatory molecule.

SNRNP200 is another member of the Ski-2 RNA helicase family. It has been recently discovered for promoting vRNA sensing and IRAF3 activation by interacting with TBK-1, found in the RIG-I/MAVS signaling pathway. SNRP200 is a member of the spliceosome complex for removal of introns. During viral infection, the amino terminal of SNRP200 binds to the vRNA, perinuclear relocation occurs and it aids as an adaptor o trigger IRF3 signaling.

15.4.2 NLRs

Nucleotide-binding oligomerization domain-containing (NOD)-like receptors (NLR) are divided into five subfamilies (based on the N terminal effector domains): NLRA (acid activation domain), NLRB (baculovirus inhibitor of apoptosis repeats), NLRC (caspase activation and recruitment domain CARD), NLRP (pyrin domain PYD), and NLRX (unknown domain). NLR is another family of cytosolic RNA-sensing PRRs. All NLRs upon activation lead to inflammasome formation except NOD1 and NOD2. General structure of NLRs are as follows:

1. N-terminal effector domain
2. Followed by nucleotide-binding and oligomerization domain (NOD)
3. C-terminal leucine-rich repeats (LRRs).

NLRC2 (NOD2) recognizes ssRNA viruses. Recognition leads to association with MAVS. This interaction is LRR-dependent and nucleotide binding domains. This triggers the MAVS-dependent pathway which leads to type I IFN and pro-inflammatory cytokine release.

NLRC5 is another NLR which has a distinguished longer LRR domain which thereby forms a helical structure instead of the horseshoe shape. Upon interaction with the viral DNA it induces IFN-mediated JAK-STAT (Janus kinase-signal transducer and activator of transcription) pathway. NLRC5 is an amplifier of the antiviral responses.

NLRP3 on the other hand has a pyrin domain in its N-terminal which enables it to interact with the PYD in the N terminal of apoptosis-associated speck-like protein containing a CARD (ASC). The CARD domain in the C-terminal of ASC further activates caspase-1, leading to the formation of fully functional IL-1 and IL-18. NLPR3 can be activated by lysosomal disintegration, membrane disruption, or generation of ROS caused due to viral PAMPS and DAMPS; therefore, it serves as an indirect sensor of viral invasion (Fig. 15.2).

15.4.3 RNA-Binding Proteins

Multiple RNA-binding proteins are also involved directly or indirectly in the process of vRNA sensing.

15.4.3.1 RNase L/Oligoadenylate Synthase (OAS)

RNase L is an antiviral endoribonuclease. Oligoadenylate synthetase (OAS) catalyzes the conversion of ATP into 2-5-linked oligoadenylates (2-5A) that in turn become the second messengers that bind to and activate RNase L (Chakrabarti et al. 2015; Malathi et al. 2007; Silverman 2007). vRNAs, cellular self-RNA are

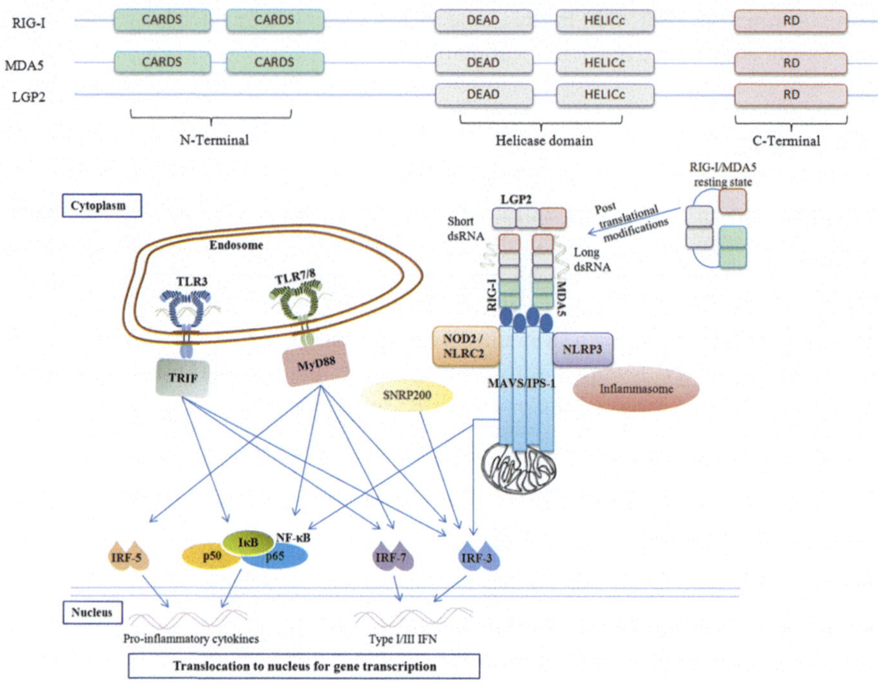

Fig. 15.2 Structure, activation, and mechanism of action of cytosolic PRRs—RIG-I, MDA5, and LGP2. Overview of endosomal and cytosolic RNA sensors

substrates to RNase L. The cleavage products are utilized for increasing the activation of RIG-I/MDA5. These are being studied in order to use them as targets for antiviral therapies (Sarkar 2014).

15.4.3.2 Protein Kinase R (PKR)

PKR functions by inhibiting translation of the viral infected cell (Dauber and Wolff 2009; Zhu et al. 2008). Being a serine/threonine kinase it gets activated on binding to a viral dsRNA/short dsRNA with 5′ppp ends and limited secondary structures (Dauber and Wolff 2009; Gil and Esteban 2000), ultimately inhibiting the eukaryotic translation initiation factor 2A (eIF2A). Antiviral response by PKR is also mediated by stabilizing MDA5.

15.4.3.3 Interferon-Inducible Transmembrane Protein (IFIT)

Multiple viruses induce the interferon response via both innate and adaptive immune responses. This leads to the stimulation of ISGs like IFIT which triggers antiviral cell-intrinsic restriction factors. Though the exact molecular mechanism is yet to be elucidated, they have showcased inhibition of viral translation or the sequestration of viral RNA to inhibit either virus replication or its packaging into new virions (Anafu et al. 2013; Diamond 2014; Vladimer et al. 2014). Furthermore, IFIT acts on the late endosomes blocking the endosomal viral entry. IFIT1 and IFIT5 recognize 5′ppp ssRNA. IFIT2 recognizes U-rich RNAs in vitro, independent of a 5′ppp.

	Endosomal RNA sensors			Cytosolic RNA sensors				RNA-binding proteins		
	TLRs			Rlrs-SF2 family cytosolic helicases		NLRs				
	TLR3	TLR7	TLR8	RIG-I	MDA5	NLRC2 (NOD2)	NLRP3	RNase L	PKR	IFIT
Cellular localization, distribution	Endosomes of immune cells except, non-immune cells like keratinocytes, fibroblasts, hepatocytes, astrocytes and microglia	Endosomes of monocytes, macrophages, plasmocystoid dendritic cells, NK cells, T and B cells	Endosomes of monocytes, macrophages, dendritic cells, NK cells, T and B cells also in mast cells and Tregs	Cytoplasm of all cell types	Cytoplasm of all cell types	Cytoplasm. Macrophages, monocytes, DCs	Cytoplasm Ubiquitously expressed	Nucleoplasm	Activated PKR is fund in both cytoplasm and nucleus	Mitochondria and cytoplasm
Agonists	dsRNA, stable stem structures of ssRNA	GU or U rich single-stranded RNA (ssRNA)	GU or U rich single-stranded RNA (ssRNA)	5′ppp-dsRNA, short dsRNA	Long dsRNA	Bacterial and virus RNA	Pathogen ssRNA/ dsRNA and other distinct set of ligands	vRNAs, cellular self-RNA	Viral dsRNA, short dsRNA with 5′ppp ends and limited secondary structures	5′ppp ssRNA, U-rich RNAs in vitro
Mechanism of activation	dsRNA induced dimerization	Z-loop proteolytic cleavage, and receptor dimer conformational change	Z-loop proteolytic cleavage, and receptor dimer conformational change	Polyubiquitin chain binding mediated receptor tetramerization	Filament formation mediated receptor tetramerization	Tetramerisation	Potassium efflux-NEK7 involved NLRP3 inflammas-ome complex	Binding to the agonist, apoptosis ad interferon dependent pathways	Dimersiation on encountering the agonists leads to auto-phosphorylation	Interferon dependent pathways
Cell signaling pathways	TRIF-TRAF3-TBK1/ IKKe-IRF3 TRIF-TRAF6-IKKs-NF-κB	MyD88-IRAK4/ IRAK1-IRF7. MyD88-IRAF6-IKKs-NF-κB	MyD88-IRAK4/ IRAK1-IRF7. MyD88-IRAF6-IKKs-NF-κB	MAVS-TRAF3-TBK1/ IKKe-IRF3. MAVS-FADD/ TRAF6-IKKs-NF-κB	MAVS-TRAF3-TBK1/ IKKe-IRF3. MAVS-FADD/ TRAF6-IKKs-NF-κB	RIP2-IKKs-NF-κB	ASC-inflammasome caspase-1-IL-1/IL-18			

(continued)

	Endosomal RNA sensors			Cytosolic RNA sensors				RNA-binding proteins		
	TLRs			Rlrs-SF2 family cytosolic helicases		NLRs				
	TLR3	TLR7	TLR8	RIG-I	MDA5	NLRC2 (NOD2)	NLRP3	RNase L	PKR	IFIT
Recognized pathogens	MCMV, HSV-1, EMCV, WNV and enteroviruses	SeV, flu virus, coxsackie virus, vaccinia virus, MV, RSV, retrovirus, SGB	SeV, flu virus, coxsackie virus, vaccinia virus, MV, RSV, retrovirus, *E. coli*, *M. bovis*, *H. pylori*, *B. burgdorferi*	BOV, MV, SeV, NDV, RSV, flu virus, hantavirus, VSV, RV, HCV, JEV, adenovirus, vaccinia virus, HSV, *L. monocytogenes*, *H. pylori*, *S. flexneri*. Rotavirus, dengue virus, WNV, murine hepatitis virus	EMCV, poliovirus and coxasackie virus. Rotavirus, dengue virus, WNV, murine hepatitis virus	RSV, IAV, and HCMV	Flu virus, SeV and bacteria			Influenza A, H5N1, Rotavirus
Crosstalk	Positive regulation by TLR8	Positive regulation by NOD2	Positive regulation by TLR3, NOD2 Negative regulation by TLR7	Negative regulation by TLR3, NOD2	Negative regulation by TLR3	Negative regulation by RIG-I	Positive regulation by MDA5	Positive regulation of RIG-I/ MDA5		Positive regulation by interferon inducing mechanisms

Acknowledgments Authors are grateful to NIV Pune and NMIMS Pune for the support extended.

Conflict of Interest The authors declare that they have no competing interests.

References

Anafu AA, Bowen CH, Chin CR, Brass AL, Holm GH (2013) Interferon inducible transmembrane protein 3 (IFITM3) restricts reovirus cell entry. J Biol Chem 288(24):17261–17271

Bruns AM, Leser GP, Lamb RA, Horvath CM (2014) The innate immune sensor LGP2 activates antiviral signaling by regulating MDA5-RNA interaction and filament assembly. Mol Cell 55(5):771–781

Chakrabarti A, Banerjee S, Franchi L, Loo Y-M, Gale M Jr, Núñez G et al (2015) RNase L activates the NLRP3 inflammasome during viral infections. Cell Host Microbe 17(4):466–477

Chen N, Xia P, Li S, Zhang T, Wang TT, Zhu J (2017) RNA sensors of the innate immune system and their detection of pathogens. IUBMB Life 69(5):297–304

Chow KT, Gale M Jr, Loo Y-M (2018) RIG-I and other RNA sensors in antiviral immunity. Annu Rev Immunol 36:667–694

Crozat K, Beutler B (2004) TLR7: a new sensor of viral infection. Proc Natl Acad Sci 101(18): 6835–6836

Dang W, Xu L, Yin Y, Chen S, Wang W, Hakim MS et al (2018) IRF-1, RIG-I and MDA5 display potent antiviral activities against norovirus coordinately induced by different types of interferons. Antivir Res 155:48–59

Dauber B, Wolff T (2009) Activation of the antiviral kinase PKR and viral countermeasures. Viruses 1(3):523–544

Diamond MS (2014) IFIT1: a dual sensor and effector molecule that detects non-2′-O methylated viral RNA and inhibits its translation. Cytokine Growth Factor Rev 25(5):543–550

Gantier MP, Williams BR (2011) Making sense of viral RNA sensing. Mol Ther 19(9):1578–1581

Gil J, Esteban M (2000) Induction of apoptosis by the dsRNA-dependent protein kinase (PKR): mechanism of action. Apoptosis 5(2):107–114

Hewson CA, Jardine A, Edwards MR, Laza-Stanca V, Johnston SL (2005) Toll-like receptor 3 is induced by and mediates antiviral activity against rhinovirus infection of human bronchial epithelial cells. J Virol 79(19):12273–12279

Jensen S, Thomsen AR (2012) Sensing of RNA viruses–a review on innate immune receptors involved in recognizing RNA virus invasion. J Virol 86(6):2900–2910

Jurk M, Heil F, Vollmer J, Schetter C, Krieg AM, Wagner H et al (2002) Human TLR7 or TLR8 independently confer responsiveness to the antiviral compound R-848. Nat Immunol 3(6):499

Liu HM, Gale M Jr (2011) ZAPS electrifies RIG-I signaling. Nat Immunol 12(1):11

Malathi K, Dong B, Gale M Jr, Silverman RH (2007) Small self-RNA generated by RNase L amplifies antiviral innate immunity. Nature 448(7155):816

Patel MC, Shirey KA, Pletneva LM, Boukhvalova MS, Garzino-Demo A, Vogel SN et al (2014) Novel drugs targeting Toll-like receptors for antiviral therapy. Futur Virol 9(9):811–829

Pichlmair A, Schulz O, Tan CP, Näslund TI, Liljeström P, Weber F et al (2006) RIG-I-mediated antiviral responses to single-stranded RNA bearing 5′-phosphates. Science 314(5801):997–1001

Rodriguez KR, Bruns AM, Horvath CM (2014) MDA5 and LGP2: accomplices and antagonists of antiviral signal transduction. J Virol 88(15):8194–8200

Said EA, Tremblay N, Al-Balushi MS, Al-Jabri AA, Lamarre D (2018) Viruses seen by our cells: the role of viral RNA sensors. J Immunol Res 2018:9480497

Sarkar SN (2014) Could boosting the oligoadenylate synthetase-like pathway bring a new era of antiviral therapy? Futur Virol 9(12):1011–1014

Silverman RH (2007) Viral encounters with 2′,5′-oligoadenylate synthetase and RNase L during the interferon antiviral response. J Virol 81(23):12720–12729
Thomas A, Laxton C, Rodman J, Myangar N, Horscroft N, Parkinson T (2007) Investigating Toll-like receptor agonists for potential to treat hepatitis C virus infection. Antimicrob Agents Chemother 51(8):2969–2978
Vladimer GI, Górna MW, Superti-Furga G (2014) IFITs: emerging roles as key anti-viral proteins. Front Immunol 5:94
Yoneyama M, Fujita T (2007) Function of RIG-I-like receptors in antiviral innate immunity. J Biol Chem 282(21):15315–15318
Zhu R, Zhang Y-B, Zhang Q-Y, Gui J-F (2008) Functional domains and the antiviral effect of the double-stranded RNA-dependent protein kinase PKR from Paralichthys olivaceus. J Virol 82(14):6889–6901
Zou J, Chang M, Nie P, Secombes CJ (2009) Origin and evolution of the RIG-I like RNA helicase gene family. BMC Evol Biol 9(1):85(Continued)

Potentiality of DNA Sensors in Activating Immune System in Emerging Viral Infectious Diseases

16

Siddhesh U. Sapre and Preethika Nair

Abstract

Viruses are obligatory intracellular parasites and hijack the host cell machinery to make more identical copies of it and continue self-propagation. They attach and replicate in the susceptible and permissive hosts and host derived cell lines. They enter the cells either through direct attachment, receptor-mediated endocytosis, or phagocytosis. Hence, to thwart the invasion by viruses, hosts have developed immunity in ascending stages—intrinsic, innate and adaptive immunity. A robust intrinsic and innate immune response governs an effective adaptive immune response, should that be needed. Both enveloped as well as non-enveloped viruses are subject to distinct types of DNA sensors, subject to their site of replication. DNA sensors of viral PAMPs can be classified into three types, based on the location of their PAMPs in the host cellular compartment viz. cell surface, cytoplasmic and nuclear. The host cell membrane both, surface as well as intra cellular, is continuously monitored for the non-host, pathogenic components or PAMPs. Among the intracellular sensors of the viral genome, there are two types—essentially due to the two types of major viral genomes i.e. RNA and DNA sensors. The cytosolic DNA sensors include AIM2, IFI16, cGAS, RNA Pol III, DNA-PK, DDX9, DHX36, DDX41, DDX60, DAI, LRRFIP1, HMGB, ABCF1 and MRE11. PYHIN family of sensors include AIM2, IFI16, IFIX and MNDA. Another recently discovered family of sensor called stimulator of interferon (IFN) genes (STING), specifically houses on the endoplasmic reticulum (ER) and functions in association with its upstream sensor, cGAS. Some DNA sensors shuttle between the cytosol and nucleus pre- and post-extraneous DNA binding. These include IFI16, IFIX, RNA Pol III, etc. There is no exclusive nuclear DNA sensor. Many enzymes known to be present in the cells for their

S. U. Sapre
National Institute of Virology, Pune, India

P. Nair (✉)
NMIMS School of Science, Mumbai, India

P. V. Bramhachari (ed.), *Dynamics of immune activation in viral diseases*,
https://doi.org/10.1007/978-981-15-1045-8_16

obvious primary functions also additionally function as DNA sensors. The DNAse family of sensors include DNAse II and TREX1, which are ubiquitously present in the cell for their housekeeping functions. The RNAse family of sensor includes one member—RNA Pol III. Additionally, DNA-PK also functions to cater to viral DNA sensing. The endosomal DNA sensors include TLR7 and TLR9, which belong to the Toll-like receptor (TLR) family. The DExD/H-box helicase family include the putative DNA sensors recently discovered including DDX9, DHX36, DDX41 and DDX60. Several other sensors remain to be characterised or are less classified viz. DAI, LRRFIP1, HMGB, ABCF1, MRE11. In general, response to a viral RNA or DNA produces three types of responses, namely, production of antiviral cytokines including Types I and III IFNs, release of pro-inflammatory and inflammatory cytokines and chemotactic factors. This chapter discusses the structure, function and mechanism of action of the viral DNA sensors explored till date.

Keywords
Viral DNA sensors · Immune responses · Immune activation

16.1 Introduction

Defence mechanisms employed by the innate branch of immunity include myriad types of cells and soluble molecules in tissues and circulatory system. Such preexisting deployments constantly thwart the attacks on the organism by the pathogenic microorganisms ubiquitously present throughout the biosphere, thus averting the invasion, establishment and multiplication of such agents. In fact, if microbes do institute a niche, the innate immune responses equip early defence, before the involvement of adaptive branch of the immune system.

The peculiarities of the recognition by players of innate immune system have emerged to contest microbes which have outstanding common features. The innate immunity has evolved to perceive the molecular structures which are present on the surface or inside the microorganism in question, as well as the induced pathogenic microbial products. Such biomolecules which spur the innate immune system are often shared by classes of microbes are called pathogen-associated molecular patterns (PAMPS), or even more appropriately, microbial-associated molecular patterns (MAMPs). Such moieties include nucleic acids (single-stranded RNA/DNA), characteristic feature(s) on protein, carbohydrate and lipid products restricted to microorganisms, or even a combination of the biochemical elements. Such building blocks are often indispensable for the survival of microorganisms. Not only is it capable of recognising the foreign biomolecular structures, but also some, which are derived from self: in aberrant conditions like damaged or dying cells. These substances are known as death- or damage-associated molecular patterns (DAMPs).

The cell-associated recognition counterpart that assists in perceiving the molecular patterns are called pattern recognition receptors (PRRs). PRRs constitute various types of cell receptors, present in distinct premises of the cell (surface as well as intracellular), as soluble factors in the circulation and other body secretions.

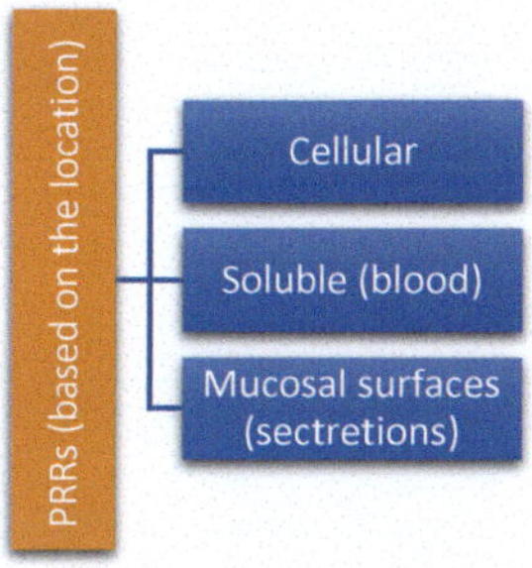

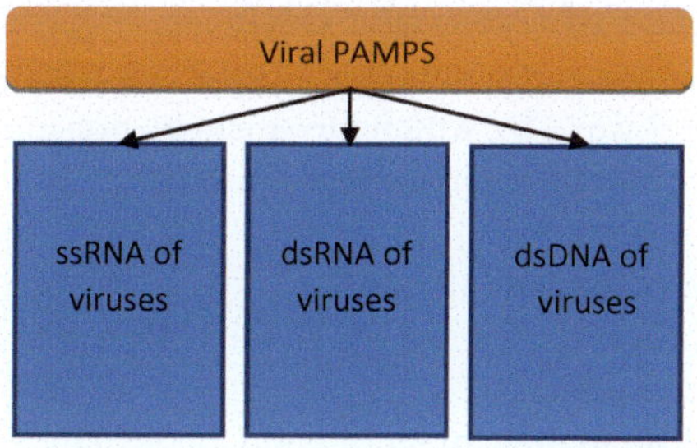

Cell-associated molecules are mostly restricted to the cells of the immune system like phagocytes—Macrophages and Neutrophils; antigen presenting cells (APCs)—Dendritic cells; cells that form the obstacle between the internal milieu and the external environment—epithelial cells; as well as other cells like mast cells and tissue resident cells.

Pathogens like viruses, bacteria, fungi and protozoa can establish themselves within any compartment of the cell viz. cytosol, nucleus, endosomes or on the surface of the cell, tissue, etc. The innate immune system efficiently tackles the microbial invasion in all these compartments by activating the signal transduction pathways downstream of the recognition molecules which ultimately promote the pro-inflammatory and anti-microbe activity.

Broadly though, the pattern recognition receptors can be either cell-associated or soluble.

The cell-associated receptors include:

1. Toll-like receptors (TLRs)
2. NOD-like receptors (NLRs)
3. RIG-I-like receptors (RLRs)
4. Cytosolic DNA sensors (CDSs)
5. C-type lectin-like receptors (CLRs)
6. Scavenger receptors
7. *N*-formyl met-leu-phe receptors

The soluble receptors include:

1. Pentraxins
2. Collectins
3. Ficolins
4. Complement

The scope of discussion on each of the above PRRs is wide. However, since this chapter specifically deals with the DNA-sensing molecules, further discussion would be restricted to the viral DNA sensors.

16.2 Sources of Cytosolic DNA

There are several ways by which DNA can be present in the cell cytoplasm.

These routes include (but are not limited to):

1. Intracellular pathogen infection
2. Impaired ability of clearing exogenous DNA innately metabolised in the endo-lysosomal compartment
3. An asymmetric management of endogenous DNA products and turnover.

Modes of entry of pathogen DNA in the cytoplasm are depicted below.

The following table summarises the currently known DNA sensors in brief—proposed DNA sensor families and the examples:

PRR	Cell types	Site of DNA sensing	Response	References
Toll-like receptor (TLR) family				
TLR9	pDCs	Endosomes	Type I IFN	(Hemmi et al. 2000; Latz et al. 2004, 2007)
PYHIN family				
AIM2	Macrophages, DCs	Cytoplasm	IL-1β, IL-18	(Hornung et al. 2009; Fernandes-Alnemri et al. 2009; Burckstummer et al. 2009; Roberts et al. 2009)
IFI16	Macrophages, endothelial cells	Cytoplasm, nucleus	IFN-β, CXCL10, IL-6, IL-1β	(Unterholzner et al. 2010; Horan et al. 2013; Kerur et al. 2011)
IFIX	Macrophages	Nucleus	IL-1β, IL-18	(Diner et al. 2015a)
MNDA	Less explored			
STING activator family				
cGAS	L929, THP-1, HEK293	Cytoplasm	IFN-β	(Sun et al. 2013)
DNAse family				
DNAse II	Ubiquitous	Lysosomes	DNA degradation	(Okabe et al. 2005)
TREX1	Ubiquitous	Cytoplasm-ER associated	Degradation of DNA elements derived from endogenous retroviruses	(Stetson et al. 2008; Yang et al. 2007)
RNAse family				

PRR	Cell types	Site of DNA sensing	Response	References
RNA Pol III	EBV^+ B cell, macrophage, cell line	Cytoplasm, nucleus	IFN-β	(Ablasser et al. 2009; Chiu et al. 2009)
Protein kinase (PK) family				
DNA-PK	293T, MEFs	Cytoplasm	IFN-λ1, IFN-β, IL-6	(Zhang et al. 2011a; Ferguson et al. 2012)
DExD/H-box helicase family				
DDX9	pDCs	Cytoplasm	TNF-α	(Kim et al. 2010)
DHX36	pDCs	Cytoplasm	IFN-α	(Kim et al. 2010)
DDX41	DCs	Cytoplasm	IFN-α, β	(Zhang et al. 2011b)
DDX60	HeLa cells	Cytoplasm	IFN-β, CXCL10	(Miyashita et al. 2011)
Others/less characterised family				
DAI	Fibroblasts	Cytoplasm	IFN-β, necrosis	(Takaoka et al. 2007; Upton et al. 2012)
LRRFIP1	Less studied	Cytoplasm	Type I IFN induction	(Sabbah et al. 2009; Yanai et al. 2009; Yang et al. 2010)
HMGB		Cytoplasm		
ABCF1		Cytoplasm	Less known	
MRE11	MEFs, DCs	Cytoplasm	IFN-β, CXCL10, IL-6	(Kondo et al. 2013)

16.3 TLR Family

16.3.1 Structure

TLRs belong to the type I integral membrane glycoproteins embedded in the surface membrane or in the endosomal membrane. They possess an extracellular (or intra-endosomal) region—characteristic leucine-rich repeats (LRRs) which are surrounded by cysteine-rich motifs which essentially bind ligand(s), a transmembrane region, and an intracellular (or cytosolic) region—known as the Toll/IL-1 receptor (TIR) involved as a part of the cytoplasmic tail. There are about 18–26 copies of LRRs which vary in different TLRs. Each LRR of TLR protein is composed of around 20–25 amino acids and multiple such LRRs make up a typical question mark hook shaped protein scaffold which adapts for ligand binding. Both, the concave as well as convex surfaces, are involved in ligand binding. The N-terminus of a TLR is at its LRR end, whereas the C-terminus is at its TIR end.

16.3.2 Function

There are nine different functional TLRs in humans (TLR1-9). The function of TLR-10 is unknown. TLRs always act as dimers. Some of them form heterodimers like TLR-1 and TLR-2, whereas others form homodimers like TLR-3, TLR-4,

TLR-5, TLR-5, TLR-6, TLR-7, TLR-8, TLR-9 and TLR-10. TLR-1, 2, 4, 5 and 6 are expressed on the surface of the membrane, whereas TLR-3, 7, 8 and 9 are expressed inside the endosomal membrane. TLR-4 utilises accessory proteins MD2, LPS-binding protein (LBP) and CD14.

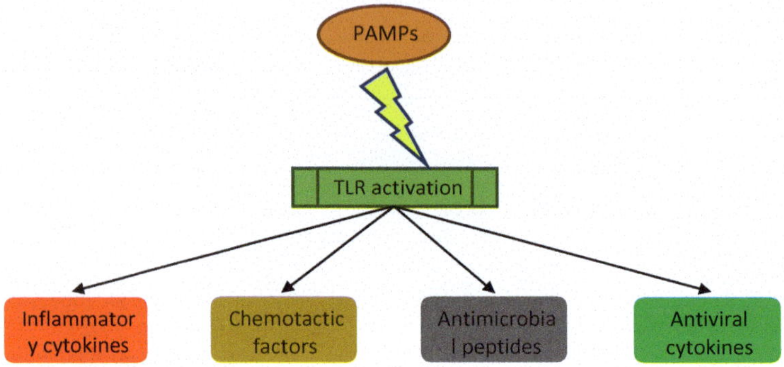

Amongst all, TLR-9 recognises unmethylated CpG dinucleotides. In the human genome, the DNA methyltransferases heavily methylate cytosine residues. However, in the genome of bacteria and many viruses, CpG dinucleotides remain unmethylated and thus serve as PAMP. Delivery of TLR-9 (and TLR-3, 7, 8) from the endoplasmic reticulum to the endosome relies on UNC-93B, a protein composed of 12 transmembrane domains. Deficiency of UNC-93B1 due to rare human mutations increases the susceptibility to herpes simplex encephalitis.

16.3.3 Mechanism of Action

Ligand binding to the TLR brings about dimerization of the respective TLRs involved. This brings about the cytoplasmic tails to come in proximity. TIR domain containing adaptor proteins are now recruited, which bind the TIR domains in the cytoplasmic tails. Further, this brings about recruitment and activation of distinct protein kinases, which activate the transcription factors. Nuclear factor κ-B (NF-κB), activation protein-1 (AP-1), interferon response factor 3 (IRF3) and IRF7 are the major transcription factors which are activated by TLR signalling pathways. NF-κB and AP-1 are responsible for the expression of genes encoding inflammatory response molecules, including (but not limited to) inflammatory cytokines, endothelial adhesion molecules and chemokines. IRF3 and IRF7 stimulate the production type I interferons (IFNs), which are central to antiviral innate immune responses. Different TLRs use different combination of adaptors and signalling intermediates and thus mediate unique downstream effects. TLR9 uses the MyD88-dependent, TRIF-independent pathway and activates both NF-κB and IRFs. Hence, they induce both inflammatory and antiviral responses.

Two protein domains of MyD88 allow it to function as an adaptor protein: a TIR domain at its carboxy terminus that associates with the TIR domains of the TLR cytoplasmic tails and a death domain at its amino terminus, which associates with the death domains present in other intracellular signalling proteins. It is worthwhile to note that both the domains of MyD88 are essential for signalling. The MyD88 death domain recruits and activates IL-1 receptor associated kinase (IRAK4) and IRAK1, which are both serine-threonine protein kinases. The complex involving IRAK, TIR of cytoplasmic tails and MyD88 executes two functions—executing the enzymes that produce a signalling scaffold, employing the scaffold to recruit other molecules which are then phosphorylated by the IRAKs. The formation of signalling scaffold is a multi-step process: The IRAK complex brings in the enzyme tumour necrosis factor receptor-associated factor 6 (TRAF6), which functions as an E3 ubiquitin ligase in association with TRIKA1 (composed of UBC13, which is an E2 ubiquitin ligase with a cofactor for Uve1A). TRAF6 and UBC13 together have the function of polyubiquitination. This polyubiquitin has linkages between lysine63 of pervious subunit and C terminus of the next, leading to K63 linkages. The same process can be initiated on vivid proteins, including TRAF6 or the multi-ubiquitin chains can exist independently as free linear chains which can be extended to form polyubiquitin chains, which can bind to other signalling proteins. Further, the scaffold brings in signalling complex that is made up of TAB1, TAB2-ubiquitin binding adaptor molecules, and TAK1-a serine-threonine kinase. IRAK complex phosphorylates TAK1, which propagates signalling by activation of certain MAPKs like c-Jun terminal kinase (JNK) and MAPK14 (p38 MAPK). This brings about downstream activation of AP-1 family of transcription factors which transcribe cytokine genes.

TAK1 additionally phosphorylates and activates the IκB kinase (IKK) complex. IKKα, IKKβ and IKKγ constitute the IKK complex (the IKKγ is also known as NF-Kb essential modifier or NEMO). NEMO binds to polyubiquitin chains, which leads to IKK complex close to TAK1. TAK1 then activates IKKβ by its phosphorylation. IKKβ then phosphorylates inhibitor of κB (IκB), which is a cytoplasmic protein that natively binds to the transcription factor NF-κB. IκB is made up of two subunits—p50 and p65. The binding of IκB prevents the movement of NF-κB to the nucleus from cytoplasm. Post phosphorylation of IKK, the IκB is released from the functional subunit of NF-κB, which leads to translocation of NF-κB to nucleus, where it leads to transcription of genes for pro-inflammatory cytokines like TNF-α, IL-1β and IL-6. It is noteworthy that the effect of TLR activation varies depending on the cell type in which it occurs.

The nucleic acid sensing TLRs, including TLR9 activate IRF family of proteins. Natively present in the cytoplasm, they only get activated upon phosphorylation of serine and threonine residues in their C terminal. Upon activation, they move to the nucleus to act as transcription factors. Among all the IRFs, IRF3 and IRF7 are particularly important for TLR signalling and expression of antiviral type I IFNs. For TLR9 signalling in plasmacytoid dendritic cells, MyD88 exclusively is used as an adaptor protein. The TIR domain of MyD88 employs IRAK1/IRAK4 complex as described earlier. However, the IRAK complex carries out a different function beyond recruiting TRAFs which generates a signalling scaffold. In these cells, IRAK1 can also interact with IRF7, which is highly expressed by plasmacytoid

Table 16.1 Containing location of different TLRs, their ligand (Janeway)

Toll-like receptor	Functional association	Cellular or subcellular location	Ligand	Adaptor proteins
TLR1	TLR1:TLR2 heterodimer	Monocytes, dendritic cells, mast cells, eosinophils, basophils	Lipomannans (mycobacteria), Lipoproteins (diacyl lipopeptides; triacyl lipopeptides); Lipoteichoic acids (gram-positive bacteria); cell-wall β-glucans (bacteria and fungi)	MyD88/ MAL
TLR2	TLR2-TLR6 heterodimer			
TLR3	TLR3-TLR3 homodimer	Macrophages, dendritic cells, intestinal epithelium	Double-stranded RNA (viruses), poly I:C	TRIF
TLR4	TLR4-TLR4 homodimer	Macrophages, dendritic cells, mast cells, eosinophils	LPS (gram-negative bacteria)	MyD88/ MAL; TRIF/ TRAM
TLR5	TLR5-TLR5 homodimer	Intestinal epithelium, macrophages, dendritic cells	Flagellin (bacteria)	MyD88
TLR7	TLR7-TLR7 homodimer	Plasmacytoic dendritic cells, macrophages, eosinophils, B cells	Single-stranded RNA (viruses)	MyD88
TLR8	TLR8-TLR8 homodimer	Macrophages, neutrophils	Single-stranded RNA (viruses)	MyD88
TLR9	TLR9-TLR9 homodimer	Plasmacytoid dendritic cells, eosinophils, B cells, basophils	DNA with unmethylated CpG (bacteria and herpesviruses)	MyD88
TLR10	TLR3-TLR3 homodimer	Plasmacytoid dendritic cells, eosinophils, B cells, basophils	Unknown	Unknown

dendritic cells. This enables IRAK1 to phosphorylate IRF7 which leads to induction of type I IFNs (Table 16.1).

16.4 PYHIN Family

There are 4 PYHIN proteins in humans, and 13 in mice. Two human proteins in this class, namely AIM2 and IFN-γ inducible (IFI16) have been predicted based on the studies so far, to be necessary for different DNA-modulated immune responses and perhaps, also function as DNA sensors (Hornung et al. 2009; Roberts et al. 2009;

Unterholzner et al. 2010). PYHIN family of proteins can also activate the formation of inflammasome.

Inflammasome Inflammasomes is a complex structure which consists of many proteins which form in the cytosol upon stimulation with cytosolic PAMPs and DAMPs, the outcome of which is production of active forms of IL-1β and IL-18. IL1-β and IL-18 are actually produced as inactive precursors and their activation is dependent on their proteolytic cleavage by the enzyme caspase-1. These cytokines are then released from the cell, which then promote inflammatory responses. Inflammasomes are made up of oligomers of a sensor, caspase-1, and an adaptor which links the interaction between the rest of the two components. The oligomeric complexes only form upon stimulation with DNA detected by sensors.

16.4.1 Structure of PYHIN

The PYHIN family proteins contain an N-terminal pyrin (PY) domain and an H inversion (HIN) domain

16.4.1.1 AIM2

Absent in melanoma (AIM2) is one such example of a PYHIN family of proteins. The HIN region of AIM2 recognises dsDNA genome and triggers caspase 1 activation. This occurs through the simultaneous interaction of pyrin domain with ASC. Primarily AIM2 is involved in the inflammasome formation (Schattgen and Fitzgerald 2011). Microbial DNA is one such PAMP. It is usually located in the cytosol and is key to responses in vitro to vaccinia virus (Janeway). However, it ubiquitously oligomerises upon stimulation. The downstream aggregates so formed recruit and activate a protease-caspase-1 that leads to the maturation of pro-inflammatory cytokines IL-1β and IL-18 which ultimately culminates into a programmed cell death, termed 'pyroptosis' (Bergsbaken et al. 2009; Miao et al. 2011).

16.4.1.2 IFI16

Interferon-inducible protein 16 (IFI16) was recently characterised as the first viral DNA sensor to function within the nucleus. It is a member of PYHIN protein family (Diner et al. 2015a; Schattgen and Fitzgerald 2011). IFI16 has two HIN domains. At its C terminus, it has two HIN200 domains which bind to DNA. At its N-terminal, is a pyrin (PY) domain which mediates homotypic interactions within the molecules as well as cooperative assembly of small subunits of IFI16 (Li et al. 2013; Morrone et al. 2014). This binding is sequence-independent manner (Jin et al. 2012). It primarily functions in the nucleus, where it is located. It recognises viral dsDNA.

The function of IFI16 depends on the type of cell in which it functions. In the immune cells, IFI16 binds to cytosolic viral DNA, engaging sting and induces IFN production (Unterholzner et al. 2010; Horan et al. 2013; Jakobsen et al. 2013). However, in non-immune cells, IFI16 majorly functions in the nucleus by localising there (Diner et al. 2015a; Li et al. 2013, 2012; Orzalli et al. 2012). A plausible explanation for this is that the there exists a multi-partite nuclear localisation signal on IFI16 and is necessary for its transit between nucleus and cytoplasm (Li et al. 2012). Especially in case of herpesvirus infections, both IFI16 and STING are required for inducing the expression of IFN and IFN-stimulated genes (Orzalli et al. 2012). The differences in IFI16 DNA-sensing are possibly due to the cell type-dependent process (Diner et al. 2015b). Prompt responses are required in case of immune cells, and thus, DNA-sensing components localised in the cytoplasm makes more sense, whereas, in non-immune cells, the IFI16 might play some housekeeping functions but additionally may respond to the successful viral infections in the nucleus. Additionally, IFI16 also mounts inflammatory and apoptotic responses to foreign DNA through inflammasome—a multiprotein assembly (Kerur et al. 2011; Johnson et al. 2013; Ansari et al. 2013; Singh et al. 2013; Monroe et al. 2014). In reality, the evidence that IFI16 elicits both, type I IFN response as well as inflammation is contradictory, since type I IFNs are known to exhibit anti-inflammatory effect (Theofilopoulos et al. 2005; Billiau 2006; Guarda et al. 2011). Recently, IFI16 was shown to play a direct role in inhibiting the formation of both, AIM2 and NLR family inflammasomes (Veeranki et al. 2011). Hence, it is quite possible that the role of IFI16 in inflammasome responses is cell type dependent and other yet unknown factors do play a role too. Recent evidences suggest that IFI16 and cGAS may function in harmony to execute immune signalling to nuclear foreign DNA (Orzalli et al. 2015). This also suggests that IFI16 is a dominant nuclear DNA sensor, whereas cGAS has auxiliary functions like stabilising the IFI16 to enable or prolong signal efficiency.

16.4.1.3 IFIX

Some of the PYHIN family members, namely, IFIX and MNDA also majorly localise in the nucleus. Yet, their functions were not known in the early phase and were suspected to have some immunological function. IFIX associates with antiviral factors and its expression is inversely associated with the capacity of herpesvirus replication (Diner et al. 2015a) as it binds to the DNA of the virus and remains localised in the nucleus. It binds DNA substrates in a sequence-dependent fashion and leads to type I IFN response (Diner et al. 2015a). IFIX expression, like IFI16, is dependent on the type of cell and tissue, thus making it likely that its function varies from one cell type to other (Ding et al. 2004; Haque et al. 2014).

16.5 STING Activator Family

16.5.1 STING (*a.k.a.* MITA, ERIS and TMEM173)

STING is an ER localised transmembrane adaptor protein that is anchored by an amino-terminal tetraspan transmembrane domain. Its carboxy-terminal domain extends into the cytoplasm and interacts to form an inactive form of STING—the STING homodimer. Type I IFN response is crucial for successful defence against viral pathogens. Stimulator of IFN genes (STING) is one such pathway vital to the mechanism of dsDNA-induced stimulation of type I IFN responses. If a viral dsDNA happens to exist in the cytosol, post entry and uncoating, it activates the enzyme cGAS (cyclic GMP-AMP synthase) that generates cyclic guanosine monophosphate-adenosine monophosphate (cGAMP), a signalling molecule. cGAS contains a protein motif as a part of the nucelotidyltransferase (NTase) family of enzymes, including adenylate cyclase and distinct DNA polymerases (Janeway). cGAS has an affinity for DNA and readily attaches to the cytosolic DNA. This interaction stimulates the cGAS enzymatic activity that leads to generation of cGAMP from GTP and ATP in the cytoplasm. cGAMP binds to both the subunits of STING dimer and activates STING signalling.

It activates TANK binding kinase 1 (TBK1) through interaction with cGAMP. TBK1 phosphorylates and activates a transcription factor IRF3 which induces expression of type I IFN genes. Besides these, the STING is also known to respond to other cytosolic DNA sensors like DAI and IFI16. Additionally, STING induces autophagy. In the innate immune system, autophagy is a potential mechanism of delivering cytosolic microbes to lysosome where they are acted upon by the proteolytic enzymes.

It is quite interesting to know that STING, MAVS and TRIF, all have similar amino acid sequence motif at their carboxy termini

16.6 DNAse Family

16.6.1 DNAse II

Cells possess DNAses, which degrade the unwanted DNA in the compartment it is not expected to be present. DNAse II is restricted to lysosomes which digests pathogenic and by-products of dead cells which enter this compartment (Okabe et al. 2005). Originally, this mechanism has evolved for the degradation of the cells undergoing programmed cell death, i.e. apoptosis, which are usually taken up by macrophages. In cells which lack DNAse II activity, DNA may stimulate aberrant responses by entering cytoplasm, consequently stimulating the cytosolic DNA sensors (Okabe et al. 2005).

16.6.2 TREX1

Yet another cellular DNAse is three primer repair endonuclease 1(TREX1) that is present in association with the endoplasmic reticulum (ER) in the cytoplasm. Under housekeeping conditions, basal amount of DNA tends to accumulate in the cytosol. TREX1 regularly degrades such DNA in the cytosol.

16.7 RNAse Family

Another putative cytosolic DNA sensor discovered is RNA Polymerase III (PolIII). It uses AT-rich and herpesvirus DNA as a template to produce 5′triphosphate RNAs. RNAs so produced get detected by the cytosolic RNA sensors like RIG-I (Ablasser et al. 2009; Chiu et al. 2009). However, the broad mechanism being the Pol III as a potential DNA senor remains to be completely defined.

16.8 Protein Kinase (PK) Family

Some proteins which are usually involved in DNA damage repair also serve as DNA sensors. DNA-PK (DNA-dependent protein kinase which is composed of Ku70, Ku80 and DNA-PKc).

16.9 DExD/H-Box Helicase Family

Both, RNA, as well as DNA helicases constitute the DExD/H-box (DDX) protein family which has it characteristic DExD/H-box domain. DDXs are multi-functional and control the gene induction at multiple points that includes signal transduction pathways, gene promoters, mRNA splicing, and translation regulation. Several DNA sensors have been identified in the DDX family (Kim et al. 2010; Zhang et al. 2011b; Yoneyama et al. 2004; Schroder et al. 2008).

16.9.1 DDX9 and DHX36

DHX9 and DHX36 are the other two DExD/H-box helicases which bind to CpG DNA and interact with MyD88 (Kim et al. 2010). The inflammatory response and type I interferon response depend partly on DHX9 and DHX36, respectively.

16.9.2 DDX41

DEAD box polypeptide 1 (DDX41) is closely related to RIG-I. It appears to signal through the STING. It has been reported that DDX41 interacts with dsDNA, both, in vitro and in vivo. In some cells, it has also been shown to be crucial to the DNA-dependent activation of type I IFN production involving STING and TBK1 (Zhang et al. 2011b). In cells where the IFI16 expression is less or restricted, the DDX41 acts as the primary DNA sensor of cytoplasmic DNA which further induced the IFN induction and IFI16 expression, the latter of which, amplifies the innate immune responses (Zhang et al. 2011b). The pattern of DNA sensor expression across different types of cells is vital to define the type of sensor which mediates the DNA primed innate immune responses. Also, recently, DDX41 has also been shown to directly bind the cyclic di-nucleotides (CDNs) and thus, indirectly, the IFN response induced was DDX41 dependent (Parvatiyar et al. 2012)

16.9.3 DDX60

It is a novice antiviral factor among the pre-existing list of DExD/H box helicases and functions in association with RIG-I, MDA5 and LGP2 to induce the type I IFN response (Miyashita et al. 2011).

16.10 Other/Less Characterised Family

There are many unknown and known but less characterised candidate DNA sensors. They might play some vital role in other cellular processes, but their role as a DNA sensor is yet to be confirmed. Very less is known regarding their mechanism of recognition and signalling, or their in vivo function.

16.10.1 DAI

Another protein is DAI (*a.k.a.* ZBP1) (Takaoka et al. 2007). The peculiarity of this protein is that the response of DAI as a DNA sensor is decided by the cell type (Unterholzner et al. 2010; Upton et al. 2012, 2010; DeFilippis et al. 2010; Ishii et al. 2008). It interacts with the dsDNA and drives type I IFN response (DeFilippis et al. 2010).

16.10.2 LRRFIPI

It senses the DNA present in the cytoplasm and phosphorylates β-catenin. B-catenin translocates to the nucleus to induce IFN-β production.

16.10.3 HMGB

Three subtypes of HMGB, viz. HMGB1, HMGB2 and HMGB3 are known to respond to cytosolic DNA. Primarily, ABCF1 binds to the cytosolic DNA. This complex then binds to HMGB2 and IFI16 to stimulate further innate immune response (Yanai et al. 2009; Yang et al. 2010; Lee et al. 2013; Goubau et al. 2013)

16.10.4 ABCF1

They function in harmony with the HMGB group of sensors.

16.10.5 MRE11

Meiotic recombination 11 homolog a (MRE11A) has the capability to sense dsDNA in the cytosol and activates STING pathway.

16.11 Sensing of Viral Components by DNA Sensors

There are numerous mechanisms through which the DNA sensors in the host cells can detect the viral genomic components and trigger a cascade of downstream reactions. Some of them have been discussed below:

16.11.1 Herpesviridae

Till date, the role of DNA sensors has been best studied extensively using herpesviruses, HSV, in particular (Paludan et al. 2011). Quite evidently, currently known-well, established, as well as prospective DNA sensors have been shown to counteract herpesvirus infections. The very primary DNA sensor was TLR9, followed by the contemporary DNA sensors. To state a few, the HSV-1 infection in human origin primary MDMs (monocyte-derived macrophages) induces the production of pro-inflammatory cytokines like IL-1β, which imitates the response post HSV-1 infection along IFI16 axis (Horan et al. 2013). However, this response is altered when the cell type differs. HSV-1 infection of primary murine dendritic cells (DCs) invokes a type I IFN response with the protein DDX41 involved (Zhang et al. 2011b). IRF3 is known to drive the type I IFN response as well. DAI drives IRF3 activation as well. However, contrastingly, DAI (or ZBP1) is shown to play a central role in inducing necroptosis, post infection with MCMV (Upton et al. 2012).

As discussed earlier, STING is central to all the DNA induced antiviral responses. This coincides with the study which demonstrated that STING deficiency in mice causes increased susceptibility to HSV-1 infection (Ishikawa et al. 2009). While it is not known if the cell-specific response applies to all the cell-lines, the above documented work irrefutably states that the response to herpesvirus infections in primary cells is cell type specific.

16.11.2 Retroviridae

HIV is the most widely studied retrovirus for known reasons on its complex pathogenesis, drug resistance and co-evolution with the host. As is known, HIV replication proceeds through and RNA–DNA double-stranded intermediate in the process of its RNA being converted to a dsDNA form. Hence, both, RNA as well as DNA sensors, play a crucial role in the detection of retroviral genome upon target cell entry (Solis et al. 2011; Berg et al. 2012; Doitsh et al. 2010; Yan et al. 2010). A recent study, intended to understand the type I IFN response against HIV proved that ssDNA, rather than dsDNA form of HIV, is more potent at inducing the type I IFN response. TREX1 is a 3′5′-exonuclease that degrades the unintegrated cytosolic cDNA, making it unavailable for sensing to the DNA sensors. Eliminating the expression of TREX1 leads to upregulation in type I IFN response and was more so in case of ssRNA intermediate of HIV, rather than dsDNA form. Undoubtedly, thus, cell does employ the intracellular sensors of DNA that culminates in type I IFN response. Although the definite sensors involved remain to be defined to date, they sure function through a STING-TBK1-IRF3 axis.

Other virus families and the respective DNA sensors have been summarised in Table 16.2

16.11.3 Mechanisms Employed By Viruses to Evade DNA Sensors

Viral genomes are subject to recognition by DNA sensors only if they happen to expose their characteristic features, leading to the consequent recognition as non-self genomic entity by the host DNA sensing mechanisms. However, myriad viruses evade and subvert the host immune responses so as to render these mechanisms of detecting the viral components ineffective.

One such strategy is to simply obscure their genome and replication intermediates involving potential ligand to the DNA sensors:

Viruses belonging to the *Herpesviridae* family inhibit DNA sensors like DAI, DHX9 and IFI16 responses.

1. HSV-1 encodes a protein ICP0, an E3 ubiquitin ligase, engages proteasomal degradation of IFI16 and also prevents its nuclear relocalisation. These further limits the activation of IRF3 (Orzalli et al. 2012).
2. HCMV possesses pUL83/pp65, a matrix protein. It inhibits ISG induction. In addition to this, it is known to interact with IFI16 (Cristea et al. 2010).

Table 16.2 Human nuclear replicating DNA viruses and implicated DNA sensors

Virus	Implicated DNA sensors
Adenovirus C (types 1, 2, 5 and 6)	cGAS, TLR9
Hepatitis B virus	AIM2, cGAS
Cytomegalovirus	AIM2, DAI/ZBP1, IFI16, TLR7, TLR9
Epstein–Barr virus	RNA Pol III, IFI16, TLR9
Herpes simplex virus type I	cGAS, DAI/ZBP1, DDX41, DDX60, DHX9, DHX36, DNA-PK, RNA Pol III, IFI16, TLR9, IFIX
Herpes simplex virus type II	DNA-PK, TLR9
Herpesvirus 8 (Kaposi sarcoma-associated virus)	IFI16
Varicella zoster virus	NLRP3, TLR9
Papillomavirus (>170 types)	TLR9

3. Proteases of viral origin like the NS2B3 protease, coronavirus papain-like proteases subvert STING in a direct manner (Aguirre et al. 2012; Sun et al. 2012; Yu et al. 2012).

Acknowledgements Authors are grateful to NIV Pune and NMIMS Pune for the support extended.

Conflict of Interest The authors declare that they have no competing interests.

References

Ablasser A et al (2009) RIG-I-dependent sensing of poly(dA:dT) through the induction of an RNA polymerase III-transcribed RNA intermediate. Nat Immunol 10(10):1065–1072
Aguirre S et al (2012) DENV inhibits type I IFN production in infected cells by cleaving human STING. PLoS Pathog 8(10):e1002934
Ansari MA et al (2013) Constitutive interferon-inducible protein 16-inflammasome activation during Epstein-Barr virus latency I, II, and III in B and epithelial cells. J Virol 87(15):8606–8623
Berg RK et al (2012) Genomic HIV RNA induces innate immune responses through RIG-I-dependent sensing of secondary-structured RNA. PLoS One 7(1):e29291
Bergsbaken T, Fink SL, Cookson BT (2009) Pyroptosis: host cell death and inflammation. Nat Rev Microbiol 7(2):99–109
Billiau A (2006) Anti-inflammatory properties of Type I interferons. Antivir Res 71(2-3):108–116
Burckstummer T et al (2009) An orthogonal proteomic-genomic screen identifies AIM2 as a cytoplasmic DNA sensor for the inflammasome. Nat Immunol 10(3):266–272
Chiu YH, Macmillan JB, Chen ZJ (2009) RNA polymerase III detects cytosolic DNA and induces type I interferons through the RIG-I pathway. Cell 138(3):576–591
Cristea IM et al (2010) Human cytomegalovirus pUL83 stimulates activity of the viral immediate-early promoter through its interaction with the cellular IFI16 protein. J Virol 84(15):7803–7814
DeFilippis VR et al (2010) Human cytomegalovirus induces the interferon response via the DNA sensor ZBP1. J Virol 84(1):585–598
Diner BA et al (2015a) The functional interactome of PYHIN immune regulators reveals IFIX is a sensor of viral DNA. Mol Syst Biol 11(1):787
Diner BA, Lum KK, Cristea IM (2015b) The emerging role of nuclear viral DNA sensors. J Biol Chem 290(44):26412–26421
Ding Y et al (2004) Antitumor activity of IFIX, a novel interferon-inducible HIN-200 gene, in breast cancer. Oncogene 23(26):4556–4566
Doitsh G et al (2010) Abortive HIV infection mediates CD4 T cell depletion and inflammation in human lymphoid tissue. Cell 143(5):789–801
Ferguson BJ et al (2012) DNA-PK is a DNA sensor for IRF-3-dependent innate immunity. elife 1:e00047
Fernandes-Alnemri T et al (2009) AIM2 activates the inflammasome and cell death in response to cytoplasmic DNA. Nature 458(7237):509–513
Goubau D, Deddouche S, Reis e Sousa C (2013) Cytosolic sensing of viruses. Immunity 38(5):855–869
Guarda G et al (2011) Type I interferon inhibits interleukin-1 production and inflammasome activation. Immunity 34(2):213–223

Haque A et al (2014) Mouse pyrin and HIN domain family member 1 (pyhin1) protein positively regulates LPS-induced IFN-beta and NO production in macrophages. Innate Immun 20(1):40–48
Hemmi H et al (2000) A toll-like receptor recognizes bacterial DNA. Nature 408(6813):740–745
Horan KA et al (2013) Proteasomal degradation of herpes simplex virus capsids in macrophages releases DNA to the cytosol for recognition by DNA sensors. J Immunol 190(5):2311–2319
Hornung V et al (2009) AIM2 recognizes cytosolic dsDNA and forms a caspase-1-activating inflammasome with ASC. Nature 458(7237):514–518
Ishii KJ et al (2008) TANK-binding kinase-1 delineates innate and adaptive immune responses to DNA vaccines. Nature 451(7179):725–729
Ishikawa H, Ma Z, Barber GN (2009) STING regulates intracellular DNA-mediated, type I interferon-dependent innate immunity. Nature 461(7265):788–792
Jakobsen MR et al (2013) IFI16 senses DNA forms of the lentiviral replication cycle and controls HIV-1 replication. Proc Natl Acad Sci U S A 110(48):E4571–E4580
Jin T et al (2012) Structures of the HIN domain: DNA complexes reveal ligand binding and activation mechanisms of the AIM2 inflammasome and IFI16 receptor. Immunity 36(4):561–571
Johnson KE, Chikoti L, Chandran B (2013) Herpes simplex virus 1 infection induces activation and subsequent inhibition of the IFI16 and NLRP3 inflammasomes. J Virol 87(9):5005–5018
Kerur N et al (2011) IFI16 acts as a nuclear pathogen sensor to induce the inflammasome in response to Kaposi Sarcoma-associated herpesvirus infection. Cell Host Microbe 9(5):363–375
Kim T et al (2010) Aspartate-glutamate-alanine-histidine box motif (DEAH)/RNA helicase A helicases sense microbial DNA in human plasmacytoid dendritic cells. Proc Natl Acad Sci U S A 107(34):15181–15186
Kondo T et al (2013) DNA damage sensor MRE11 recognizes cytosolic double-stranded DNA and induces type I interferon by regulating STING trafficking. Proc Natl Acad Sci U S A 110(8):2969–2974
Latz E et al (2004) TLR9 signals after translocating from the ER to CpG DNA in the lysosome. Nat Immunol 5(2):190–198
Latz E et al (2007) Ligand-induced conformational changes allosterically activate Toll-like receptor 9. Nat Immunol 8(7):772–779
Lee MN et al (2013) Identification of regulators of the innate immune response to cytosolic DNA and retroviral infection by an integrative approach. Nat Immunol 14(2):179–185
Li T et al (2012) Acetylation modulates cellular distribution and DNA sensing ability of interferon-inducible protein IFI16. Proc Natl Acad Sci U S A 109(26):10558–10563
Li T, Chen J, Cristea IM (2013) Human cytomegalovirus tegument protein pUL83 inhibits IFI16-mediated DNA sensing for immune evasion. Cell Host Microbe 14(5):591–599
Miao EA, Rajan JV, Aderem A (2011) Caspase-1-induced pyroptotic cell death. Immunol Rev 243(1):206–214
Miyashita M et al (2011) DDX60, a DEXD/H box helicase, is a novel antiviral factor promoting RIG-I-like receptor-mediated signaling. Mol Cell Biol 31(18):3802–3819
Monroe KM et al (2014) IFI16 DNA sensor is required for death of lymphoid CD4 T cells abortively infected with HIV. Science 343(6169):428–432
Morrone SR et al (2014) Cooperative assembly of IFI16 filaments on dsDNA provides insights into host defense strategy. Proc Natl Acad Sci U S A 111(1):E62–E71
Okabe Y et al (2005) Toll-like receptor-independent gene induction program activated by mammalian DNA escaped from apoptotic DNA degradation. J Exp Med 202(10):1333–1339
Orzalli MH, DeLuca NA, Knipe DM (2012) Nuclear IFI16 induction of IRF-3 signaling during herpesviral infection and degradation of IFI16 by the viral ICP0 protein. Proc Natl Acad Sci U S A 109(44):E3008–E3017
Orzalli MH et al (2015) cGAS-mediated stabilization of IFI16 promotes innate signaling during herpes simplex virus infection. Proc Natl Acad Sci U S A 112(14):E1773–E1781
Paludan SR et al (2011) Recognition of herpesviruses by the innate immune system. Nat Rev Immunol 11(2):143–154

Parvatiyar K et al (2012) The helicase DDX41 recognizes the bacterial secondary messengers cyclic di-GMP and cyclic di-AMP to activate a type I interferon immune response. Nat Immunol 13(12):1155–1161
Roberts TL et al (2009) HIN-200 proteins regulate caspase activation in response to foreign cytoplasmic DNA. Science 323(5917):1057–1060
Sabbah A et al (2009) Activation of innate immune antiviral responses by Nod2. Nat Immunol 10(10):1073–1080
Schattgen SA, Fitzgerald KA (2011) The PYHIN protein family as mediators of host defenses. Immunol Rev 243(1):109–118
Schroder M, Baran M, Bowie AG (2008) Viral targeting of DEAD box protein 3 reveals its role in TBK1/IKKepsilon-mediated IRF activation. EMBO J 27(15):2147–2157
Singh VV et al (2013) Kaposi's sarcoma-associated herpesvirus latency in endothelial and B cells activates gamma interferon-inducible protein 16-mediated inflammasomes. J Virol 87(8): 4417–4431
Solis M et al (2011) RIG-I-mediated antiviral signaling is inhibited in HIV-1 infection by a protease-mediated sequestration of RIG-I. J Virol 85(3):1224–1236
Stetson DB et al (2008) Trex1 prevents cell-intrinsic initiation of autoimmunity. Cell 134(4):587–598
Sun L et al (2012) Coronavirus papain-like proteases negatively regulate antiviral innate immune response through disruption of STING-mediated signaling. PLoS One 7(2):e30802
Sun L et al (2013) Cyclic GMP-AMP synthase is a cytosolic DNA sensor that activates the type I interferon pathway. Science 339(6121):786–791
Takaoka A et al (2007) DAI (DLM-1/ZBP1) is a cytosolic DNA sensor and an activator of innate immune response. Nature 448(7152):501–505
Theofilopoulos AN et al (2005) Type I interferons (alpha/beta) in immunity and autoimmunity. Annu Rev Immunol 23:307–336
Unterholzner L et al (2010) IFI16 is an innate immune sensor for intracellular DNA. Nat Immunol 11(11):997–1004
Upton JW, Kaiser WJ, Mocarski ES (2010) Virus inhibition of RIP3-dependent necrosis. Cell Host Microbe 7(4):302–313
Upton JW, Kaiser WJ, Mocarski ES (2012) DAI/ZBP1/DLM-1 complexes with RIP3 to mediate virus-induced programmed necrosis that is targeted by murine cytomegalovirus vIRA. Cell Host Microbe 11(3):290–297
Veeranki S et al (2011) IFI16 protein mediates the anti-inflammatory actions of the type-I interferons through suppression of activation of caspase-1 by inflammasomes. PLoS One 6(10):e27040
Yan N et al (2010) The cytosolic exonuclease TREX1 inhibits the innate immune response to human immunodeficiency virus type 1. Nat Immunol 11(11):1005–1013
Yanai H et al (2009) HMGB proteins function as universal sentinels for nucleic-acid-mediated innate immune responses. Nature 462(7269):99–103
Yang YG, Lindahl T, Barnes DE (2007) Trex1 exonuclease degrades ssDNA to prevent chronic checkpoint activation and autoimmune disease. Cell 131(5):873–886
Yang P et al (2010) The cytosolic nucleic acid sensor LRRFIP1 mediates the production of type I interferon via a beta-catenin-dependent pathway. Nat Immunol 11(6):487–494
Yoneyama M et al (2004) The RNA helicase RIG-I has an essential function in double-stranded RNA-induced innate antiviral responses. Nat Immunol 5(7):730–737
Yu CY et al (2012) Dengue virus targets the adaptor protein MITA to subvert host innate immunity. PLoS Pathog 8(6):e1002780
Zhang X et al (2011a) Cutting edge: Ku70 is a novel cytosolic DNA sensor that induces type III rather than type I IFN. J Immunol 186(8):4541–4545
Zhang Z et al (2011b) The helicase DDX41 senses intracellular DNA mediated by the adaptor STING in dendritic cells. Nat Immunol 12(10):959–965

Advanced Immunotechnological Methods for Detection and Diagnosis of Viral Infections: Current Applications and Future Challenges

17

Pallaval Veera Bramhachari, Ganugula Mohana Sheela, A. M. V. N. Prathyusha, M. Madhavi, K. Satish Kumar, Neelapu Nageswara Rao Reddy, and Chanda Parulekar Berde

Abstract

Diagnosis and identification of viruses is an important component of diagnostic virology laboratory. Although various modes of diagnostic methods are now available at disposal, a vast majority of the diseases across the globe remain undiagnosed. This is largely due to the overlapping undifferentiated set of symptoms across myriad set of RNA and DNA viral diseases. As such, it becomes critical to take into consideration several factors for viral diagnosis ranging from the type and quality of specimen collected, time of specimen collection, mode of transport, accuracy, specificity, sensitivity, and the type of diagnostic method used. This chapter broadly emphasizes various methods on diagnostic virology ranging from the classical methods of diagnosis to the most recently developed molecular methods of detection of virus.

P. V. Bramhachari · G. Mohana Sheela · A. M. V. N. Prathyusha (✉)
Department of Biotechnology, Krishna University, Machilipatnam, India

M. Madhavi
Department of Zoology, Nizam College, Osmania University, Hyderabad, Telangana, India

K. Satish Kumar
Department of Zoology, Adikavi Nannaya University, Rajamundry, Andhra Pradesh, India

N. N. R. Reddy
Department of Biochemistry and Bioinformatics, GITAM Institute of Science, Gandhi Institute of Technology and Management (GITAM) (Deemed-to-be-University), Visakhapatnam, Andhra Pradesh, India

C. P. Berde
Gogate Jogalekar College, Ratnagiri, Maharashtra, India

P. V. Bramhachari (ed.), *Dynamics of immune activation in viral diseases*,
https://doi.org/10.1007/978-981-15-1045-8_17

Keywords
RNA and DNA viral diseases · Diagnostics · Accuracy · Specificity · Sensitivity

17.1 Introduction

Viruses are obligate intracellular parasites and their detection and identification is an imperative component clinically. Viral infectious signify a major portion in public health perspectives with thousands of deaths annually. Notwithstanding from highly contagious infections and serious pandemics to common influenza episodes, clinical diagnosis of viral infection often relies on early detection. Therefore, effectual detection of viruses is indispensable aid to avert transmission, initiate befitting therapy, and scrutinize response to treatment which leads to effective disease control and management (Souf 2016).

Viral diagnosis is a dynamic process. Although prophylaxis is better than cure, an accurate diagnosis of any virus fundamental to the infection is equally vital. Generally, diagnostic tests are categorized into three groups: direct detection, isolation of virus, and serology tests. Direct examination methods are merely faster and examined directly for presence of virus particles, virus antigen, or viral nucleic acids. Immunofluorescence assay is extensively used for rapid identification of virus infections through detection of virus antigen or virus-specific antibodies in clinical specimens. Viral diagnosis is a crucial revolution and renders more accessible and makes it potential to standardize the recently developed diagnostic methods. Serology essentially comprises bulk of work of any virology laboratory. A serological diagnosis can be identified by increasing titers of antibody among acute and convalescent stages of infection, or by the detection of Immunoglobulin M (IgM).

The diagnosis of viral infections were enhanced noticeably all through 1990s, with the onset of highly sensitive methods, viz. enzyme-linked immunosorbent assay (ELISA) and PCR are superseded for this reason. Nevertheless, during the last 20 years this technique is being utilized, due to its unique nature, making diagnosis probable through visualization of virus. Then after, introduction of ELISA then revolutionized viral diagnosis by simplifying detection and shortening the time (Torrance and Jones 1981). Molecular diagnostic tools for viral diagnosis has trialed speedy advancements in last few years (Hayden and Persing 2001), and revolutionized diagnosis of infectious diseases, particularly viral diseases. The use of amplification techniques, viz. PCR, RT-PCR, or NASBA (nucleic acid sequence-based amplification) (Van Belkum and Niesters 1995) for detection of virus, genotyping, and quantification have several advantages such as high reproducibility and sensitivity, including broad dynamic range (Ebner et al. 2005, 2006). Several molecular diagnostic techniques were recently swapped by fully automated devices that use less time, maneuver smaller volumes of liquids, and provide quantified results with improved accuracy. The current review emphasizes the advanced immunotechniques, how the specific characteristics of diagnostic methods revolutionized field of viral diagnosis clinically from a decade.

17.2 Recent Advances in Viral Disease Diagnosis

17.2.1 Advanced Methods in Diagnostic Virology

17.2.1.1 Microscopy

Transmission electron microscopy (TEM) is a merely imaging technique that permits direct image of viruses attributable to its nanoscale resolution. This is a classical technique owing to its advantage of direct visualization of virus. Amid the 1960s and 1990s, TEM contributed a breakthrough and doled out as a diagnostic tool for recognizing numerous virus unswervingly in any biological samples. For that reason, in modern years, role of TEM in diagnosis of anonymous infectious agents in specific viral outbreaks or transmission clusters shifted from regular use to an initial screening test. TEM is also considered an important method for assessment of viral safety in biopharmaceutical products. Application of TEM understands the unknown viruses for which there is an imminent risk of contamination, and TEM will undeniably be functional for documenting the subsistence of viruses or virus-like particles in these cells and the products derivative from them (Roingeard et al. 2019).

17.2.2 Advanced Serological Immunoassays

The immunoassays are primarily antibody/antigen dependent assays. This principle works on immunization with the antigens and activating, or infection. Therefore, measuring of IgG immunoglobulin and quantifying the antibodies are preferentially used as diagnostic markers. The antigens or antibodies are coated with conjugated labels like metals, radioactive isotopes, fluorescent tags, respectively. As a part of modern research on immunotechniques, a diagnostic approach for chronic hepatitis C infection (CHC), detects specific antibody to HCV (anti-HCV) (indirect tests) and assays that can detect, quantify, or characterize components of HCV viral particles, viz. HCV RNA and core antigen (direct tests). Quantification of HCV core antigen (cAg) as a one-step procedure has been indispensable. Hence, in order to evaluate the performance of cAg quantification in diagnosing CHC and how it is predisposed by concomitant HIV or HBV infections, cross-sectional validation assays, i.e., HCV Ag quantification as a one-step procedure in diagnosing CHC in Cameroon was designed to abridge the diagnostic process (Duchesne et al. 2017). It is pertinent to note that a multiplex microsphere immunoassay (MIA) recently developed possess diagnostic power to incarcerate viral envelope protein, which evokes to be strong cross-reactive antibodies to other *flaviviruses* and differential power of viral nonstructural proteins NS1 and NS5 was. Interestingly this serologic assay needs to be employed in rapid clinical diagnosis of ZIKV and/or dengue virus infections for screening immune responses in vaccine trials (Wong et al. 2017). It is noteworthy that oral fluid is a noninvasive biospecimen that can harbor pathogen-specific antibodies and reach potential to replace blood-based testing protocols. Therefore, a saliva-based oral fluid immunoassay was developed to assess past and recent hepatitis E virus (HEV) infections from noninvasive sampling methods. The sensitivity and specificity of this assay was comparable to serum-based ELISAs. This salivary assay could improve our understanding of the ecology and natural history of HEV (Pisanic et al. 2017).

Diagnosing ZIKV remains a great challenge, as detection of viral RNA is only possible merely few days after onset of symptoms. Conversely, novel high-throughput image-based fluorescent neutralization method for identification of ZIKV was thoroughly evaluated and developed which reported higher sensitivity than Plaque reduction neutralization test (PRNT) and MAC-ELISA, respectively. This test might employ for clinical diagnosis, clinical trials, and confirmation and seroprevalence studies of ZIKV infection (Koishi et al. 2018). In one of the recent studies, detection of serum HEV antigen (Ag) is deemed to be sensitive and promising biomarker for HEV antigen diagnosis with HEV RNA in both acute and chronic genotypes. Strikingly an antigen assay was recently evaluated for diagnosing HEV genotypes with higher sensitivity than commercial anti-HEV IgM and HEV RNA ELISA tests (Zhang et al. 2019). Nonetheless, recent studies on respiratory syncytial virus (RSV) developed Luciferase Immunoprecipitation Systems (LIPS) assay to detect IgG Antibodies against Human RSV G-Glycoprotein. Moreover, Human RSV G-Glycoprotein also acts as biomarker for natural exposure or immunization. RSV genes encoding native and mutated G (mG) proteins from subgroups A and B strains were cloned, expressed as luciferase-tagged proteins, and experimented separately to spot anti-RSV-G specific IgG antibodies employing a high-throughput luciferase immunoprecipitation system (LIPS-G). It was pertinent to note that RSV monoclonal antibodies and polyclonal antisera explicitly bound in LIPS-G_A and/or -G_B assays (Crim et al. 2019).

The diagnosis of (ZIKV) and dengue virus (DENV) infections against viral envelope protein and nonstructural proteins (NS) was developed using flavivirus multiplex microsphere immunoassay (MIA). However MIA could not differentiate more recent from past infections, which represents a key diagnostic challenge; therefore, in a most recent report an immunoglobulin G (IgG)-based avidity assay was developed for its diagnostic performance to accurately differentiate between recent ZIKA and past dengue virus infections. This assay was found useful in patients with high risk of ZIKA complications, viz. pregnant women and monitoring immune responses in vaccine trials (Furuya et al. 2019). Consecutively to develop serological diagnosis of ZIKV-IgA and ZIKV-IgG, avidity assays were evaluated to characterize ZIKA infections in want of viremia. These assay facilitated construed low avidity of IgG and IgA results, enhanced the serological diagnosis of ZIKV (Amaro et al. 2019). In another study, homologous proteins of diverse flaviviruses exhibited high degrees of sequence uniqueness, mainly within subgroups. This led to prevalent immunological cross-reactivity. Therefore, a proportional deconvolution of complex B cell responses against ZIKV and other flavivirus were deliberated by screening with a microarray chip-based high-resolution serological analysis primed from overlapping peptides covering the whole amino acid sequence of ZIKV genomic polyprotein was developed. Additionally with advent of this assay several infections, viz. dengue, yellow fever, tick-borne encephalitis, and West Nile viruses shall be diagnosed (Hansen et al. 2019).

17.2.2.1 ELISA-Based Immunodetection

Enzymes are extensive tool for diagnosing virus which have various applications like enzyme immune assay, ELISA. Enzyme immune assay has different applications like fluorescence polarization immune assay (FPIA), micro-particle immune assay (MEIA), chemiluminescent (CLIA). Enzyme immune assays work with antigen–antibody interaction with the conjugated tags like fluorescent tags, chemiluminescent tags which are complemented with substrates like polarized light and fluorescent substrates. As a part of most advanced immunotechniques, an ultrasensitive colorimetric assay called magnetic nano(e)zyme-linked immunosorbent assay (MagLISA) was developed, wherein silica-shelled magnetic nanobeads (MagNBs) and gold nanoparticles were pooled to monitor influenza A virus up to femtogram per milliliter concentration (Oh et al. 2018). Sensitive and specific detection of Crimean–Congo hemorrhagic fever virus (CCHFV) was developed employing specific IgM and IgG antibodies in human sera using recombinant CCHFV nucleoprotein as antigen in μ-capture and IgG immune complex (IC) ELISA tests (Emmerich et al. 2018). Recently, truncated forms of HeV and NiV (Hendra and Nipah virus) G proteins as well as full-length NiV nucleocapsid (N) protein were used for detection of Hendra and Nipah virus specific antibodies in pigs. These recombinant proteins were expressed through diverse expression systems and an indirect ELISA was developed for detection (Fischer et al. 2018). A rapid diagnostic platform for colorimetric differential detection of DENV and CHIKV viral infections was recently developed with a possibility to alter clinical diagnosis of acute febrile illnesses in resource-limited settings. This platform principally facilitates consistent and accurate multiplexed detection of chikungunya and dengue IgM/IgG antibodies in human clinical samples within short stint (Wang et al. 2019).

17.2.2.2 Immunofluorescence-Based Immunodetection

Immunofluorescence (IF) is extensively used for speedy detection of viral infections clinically started in early 1970s. IF is used for diagnosis of virus antigen and virus-specific IgG/IgA/IgM antibody in clinical specimens. In this technique, fluorescein-labeled antibody to stain specimens containing specific virus antigens, were used for UV illumination. As a part of modern research on immunotechniques, new recombinant rabies virus expressing green fluorescent protein (rRV-GFP) is more rapid, simpler, and less expensive detection and for quantification of virus neutralizing antibodies in blood sera. This technique simplified multistep Rapid Fluorescent Focus Inhibition Test (RFFIT) procedure by purging immunostaining step (Qin et al. 2019a, b).

17.2.2.3 PCR/RT-PCR-Based Immunodetection

Nucleic acid (DNA and RNA) amplification assays are conventionally known as polymerase chain reactions (PCRs). Distinct PCRs exists based on type of nucleic acid and information known a propos the sequences of genomes for immunodetection. Rous Sarcoma Virus (RSV) is one of the most significant causative agents of respiratory tract infection in children and related with high morbidity and mortality. However, very little is known with reference to effects of respiratory viral infections. A reverse transcription recombinase polymerase amplification assay (RT-RPA) is nucleic acid probe based on novel isothermal amplification technique which has been widely employed to detect human RSV. The results exemplified that concurrence rates between RT-RPA assay and qRT-PCR assay for clinical samples was 96%, demonstrating that RT-RPA assay holds better diagnostic presentation on clinical samples in remote rural areas in developing countries (Xi et al. 2019). Using quantitative PCR (qPCR) for common respiratory viruses and for two genes (*CCL8*/*CXCL11*) is recognized to be extremely upregulated in viral infections. Notably, RNA-seq virus detection achieved 86% sensitivity compared to qPCR-based screening in asthmatic children which consequently drives immune cell airway infiltration, cellular remodeling, and alteration of asthmogenic gene expression (Wesolowska-Andersen et al. 2017). Nevertheless as a part of latest advancements on immunotechniques, isothermal reverse transcription and recombinase polymerase amplification (RT-RPA) of synthetic RNA (Ebola virus) employing paper microfluidics devices was developed initially. Later on based on RNA detection and multiplexed analysis for Ebola virus diagnostics were optimized and demonstrated with high sensitivity. Additionally, nine-spot multilayered device achieving parallel detection of three distinct RNA sequences opens a route in the direction for detection of multiple viral pathogens (Magro et al. 2017).

Rabies virus (RABV) is one of the most significant global zoonotic pathogens. Two sensitive real-time quantitative RT-PCR assays were developed and validated for large-spectrum detection of RABV, with a focus on African isolates. The primer and probe sets were targeted for highly conserved regions of nucleoprotein (N) and polymerase (L) genes. Effective detection and high sensitivity of these assays can be effectively functional in general research and used in diagnostic process and epizootic surveillance (Faye et al. 2017). However, in recent times a fluorescent reverse transcription loop-mediated isothermal amplification (RT-LAMP) employing quenching probes for detection of Middle East respiratory syndrome coronavirus was developed. Additionally, detection efficacy of QProbe RT-LAMP was comparable to that of RT-PCR assay (Azhar 2018). Furthermore this assay can as well be used as authoritative diagnostic tool for rapid detection and surveillance of MERS-CoV infections (Shirato et al. 2018). Consecutively to assist detection of ZIKV infections, and distinguish these infections from DENV and CHIKV, Trioplex real-time RT-PCR assay was recently developed. However the performance of Trioplex real-time RT-PCR assay was particularly employed for the detection of ZIKV, DENV and CHIKV viruses. Simultaneous testing of more than one specimen type from each patient affords a superior diagnostic sensitivity of this technique (Santiago et al. 2018).

An in situ hybridization (RNA-ISH) assay was recently developed to detect viral hemorrhagic septicemia virus (VHSV), an OIE listed piscine rhabdovirus, in infected fish cells with fathead minnow (FHM) as model cell line. Two antisense RNA probes targeting fragments of N and G genes were amplified by RT-PCR employing VHSV-specific primers trailed by transcription reaction in presence of digoxigenin dUTP has competently localized VHSV mRNAs in infected cells. The diagnostic sensitivity of RNA-ISH assay was better than immunocytochemistry, qRT-PCR and $TCID_{50}$ (Qadiri et al. 2019). Development and evaluation of a new one-step, real-time RT-PCR assay was developed for detecting latest H9N2 influenza viruses competent of causing human infection. The sensitivity of one-step, real-time RT-PCR assay was generally determined to be used in vitro transcribed RNA, devoid of any cross-reactivity against RNA from H1–15 subtypes of influenza viruses and other viral respiratory pathogens with no nonspecific reactions (Saito et al. 2019).

17.2.2.4 Spectroscopy-Based Immunodetection

In the current trends mass spectrometry (MS) is a benchmark for qualitative and quantitative diagnosis of viruses clinically. In clinical laboratories, MALDI (matrix-assisted laser desorption ionization) and ES (electrospray) ionization methods are most frequently used as they allow ionization of analyte in considerable amounts. The combination (RT-PCR/ESI-MS) was able to detect viral pathogens (acute viral respiratory infections and influenza A viruses) usually for those viruses which are undetected by regular testing methods as well as provides rapid and detailed data (about types and subtypes of virus) in short period (Deyde et al. 2010; Chen et al. 2011). Yet another study reported that near-infrared spectroscopy (NIRS) is a rapid, reagent-free, and cost-effective tool used to detect ZIKV noninvasively in heads and thoraces of intact *Aedes aegypti* mosquitoes with high prediction accuracies relative to quantitative RT-qPCR reaction. Perhaps this technique could be extended upon for identifying probable arbovirus hotspots to guide spatial prioritization of vector control (Fernandes et al. 2018). Recently developed surface plasma resonance (SPR) spectroscopy was developed to be a valuable optical biosensor and potential method for diagnosis of dengue virus E-protein and also for identification of antibodies to DENV antigen. The diagnosis limit, sensitivity, and selectivity of SPR sensing in DENV antigen was amazingly high. This technique was introduced as a novel 3D-PAMAM-SAM-Au multilayer thin film for future research on SPR sensing applications (Omar and Fen 2018).

17.2.2.5 miRNA-Based Immunodetection

miRNA are conserved small noncoding RNA with 19–24 nucleotides which regulates post-transcriptional modification. miRNAs are transcribed by RNA polymerase II into pre-miRNA which is processed by Dorsa/DGCR-8 and Dicer and transported to cytoplasm. RISC in cytoplasm processes pre-miRNA into mature miRNA and regulates post-transcriptional activities. As a part of the most advanced immunotechniques, exosomal microRNAs were recently studied as potential diagnostic markers for various malignancies, including hepatocellular carcinoma (HCC). Serum exosomal microRNAs combined with alpha-fetoprotein as diagnostic markers of hepatocellular carcinoma (Wang et al. 2018). In another study, synthetic miRNA-based approach was developed to express neutralizing antibodies directly in lung via aerosol, to prevent from human RSV and influenza infections. Engineered mRNA-expressed antibodies prevented RSV infection. It is noteworthy that an expressing membrane-anchored broadly neutralizing antibody in lungs could potentially be promising pulmonary prophylaxis approach (Tiwari et al. 2018). In a most recent

Table 17.1 Advanced immunotechnological techniques for detection and diagnosis of viral infections

S. No.	Techniques	Application to diagnosis of emerging viral infections
1.	Serological tests	Diagnostic approach for chronic hepatitis C infection, ZIKV, RSV, DENV, tick-borne encephalitis, and West Nile viruses
2.	Enzyme assays (fluorescence polarization immune assay (FPIA), microparticle immune assay (MEIA), chemiluminescent (CLIA)	Identification of structural proteins of Hendra and Nipah virus, Crimean–Congo hemorrhagic fever virus (CCHFV), chikungunya
3.	PCR-based immunodetection (RT-RPA, RT-LAMP, RNA-ISH, RT-PCR, Trioplex real-time PCR)	Detection of Ebola virus, Rabies virus, Middle East respiratory syndrome coronavirus, hemorrhagic septicemia virus, influenza viruses
4.	Mass spectrometry (MS/MS)	Prognosis of DENV antigen, ZIKV
5.	miRNA-based immunodetection	Circulating microRNA biomarkers in detection of arboviruses; hepatitis infection
6.	Next-generation sequencing (NGS)	Provides more exact and precise time estimates of infection, crucial for HIV-1 surveillance
7.	Metagenomics (panpathogen metagenomic sequencing assay)	Diagnostic standard techniques used in diagnosis and genetic analysis of influenza and other clinical respiratory viruses
8.	Immunosensors (sandwich-type electrochemiluminescence (ECL) immunosensor)	Diagnostic test for HBV, swine flu (H1N1) infection, H7N9 virus with high sensitivity
9.	Microfluidic technology	Used to test the convenience of neutralizing antibodies to explore impact on virus–cell interactions
10.	CRISPR/Cas system	Prognosis of coxsackievirus, hepatitis B virus, Zika virus
11.	Nanoparticles-based immunodetection	Nanoparticle-based lateral flow immunoassay as point-of-care diagnostic tool for infectious agents and diseases

study, endogenous microRNAs (miRNA) are evolutionarily conserved and their presence in biological fluids signifies regulatory role of circulating miRNAs in pathogenesis, immune responses, and viral infections. On the other hand, noninvasive diagnostic approach, using biomarkers, currently plays a central role in early diagnosis of viral infections. Given the fact, a recent report depicted numerous circulating microRNA biomarkers viz. miR155 and miR1260 in influenza; miR12, miRVP3p, and miR184 in arboviruses; and miR29b and miR125 in hepatitis infection for diagnostic function, respectively (Ojha et al. 2019) (Table 17.1).

17.2.2.6 Next-Generation Sequencing (NGS)-Based Immunodetection

Next-generation sequencing (NGS) is one of the noteworthy achievements recorded in the current era. Ahead from genome sequencing of known organisms, permitted breakthrough of new viruses dependable for unknown human diseases, for tracking viral outbreaks and pandemics as influenza to comprehend their emergence and transmission profiles. Viral diversity from next-generation sequencing of HIV-1 samples provides more exact and precise estimates of time since infection, consequently, the infection regencies are also crucial for HIV-1 surveillance and understanding of viral pathogenesis. NGS-derived average pair-wise diversity exhibited higher sensitivity and specificity compared to fraction of ambiguous nucleotides (Carlisle et al. 2019).

17.2.2.7 Metagenomics-Based Immunodetection

It is noteworthy that metagenomic next-generation sequencing assay (mNGS) for pan-pathogen detection has been effectively tested in patients with acute illness of several viral etiologies. In this connection a customized bioinformatics pipeline, SURPI+, was recently designed to quickly analyze mNGS data, generate an automated summary of detected pathogens, and provide a graphical user interface for evaluating and interpreting results (Miller et al. 2019). Panpathogen Metagenomic sequencing assay for SLEV (St. Louis encephalitis virus) infection in CSF (cerebrospinal fluid) is an unbiased approach to infectious disease testing, although several challenges still remain relating to test availability, interpretation, and validation that were reported. However, recently metagenomic next-generation sequencing was employed to diagnose fatal case of meningoencephalitis caused by SLEV (Chiu et al. 2017). In another study, metagenomic viral sequencing is the potential diagnostic test for influenza which also provides insights on transmission, drug resistance, evolution, and simultaneously detects other viruses. It is pertinent that Oxford Nanopore Technology was employed to metagenomic sequencing of respiratory samples. This technology operated with very high sensitivity compared to current diagnostic standard techniques and certainly this approach may show a great promise for nanopore platform to be used in diagnosis and genetic analysis of influenza and other clinical respiratory viruses (Lewandowski et al. 2019) (Table 17.1).

17.2.2.8 Monoclonal Antibodies-Based Immunodetection

In the recent past designing diagnostic and therapeutic platforms based on aptamer technology is undoubtedly a potential approach in viral infections. Nevertheless the oligonucleotide aptamers which are potential alternatives for monoclonal antibodies based detection could be aimed against any protein in infected cells and any components of viral particles are deemed as probable novel diagnostic molecules against viral hepatitis. It is noteworthy that these aptamer molecules could be a favorable substitute for monoclonal antibody in near future (Mirian et al. 2017).

17.2.2.9 Immunosensors-Based Immunodetection

A sensitive and selective electrochemical immunosensor for label-free ZIKV protein detection was recently developed, employing functionalized interdigitated microelectrode of gold (IDE-Au) array. This ZIKV immune-sensing chip can be integrated with miniaturized potentiostat (MP)-interfaced with smart phone for rapid ZIKV-infection detection is obligatory for early stage diagnostics at point-of-care application (Kaushik et al. 2018). However, a cost-effective and portable graphene-enabled biosensor to detect ZIKV with a highly specific immobilized monoclonal antibody was recently developed. Field effect biosensing (FEB) with monoclonal antibodies covalently linked to graphene enables the real-time, quantitative detection of native ZIKV antigens. This assay is first-of-its-kind graphene-enabled ZIKA biosensor which makes it an ideal candidate for the development as a medical diagnostic test (Afsahi et al. 2018).

An accurate and timely diagnosis of any new reassortment of avian influenza is very crucial for controlling disease outbreaks. In view of this, a simple strategy for rapid and sensitive detection of H7N9 virus was achieved by employing an intensity-modulated surface plasmon resonance (IM-SPR) biosensor technique integrated with newly generated monoclonal antibody. This novel antibody demonstrates noteworthy specificity to identify H7N9 virus compared to homemade target-captured ELISA, qRT-PCR, and rapid influenza diagnostic test (RIDT) with high sensitivity (Chang et al. 2018).

In a recent study, sandwich-type electrochemiluminescence (ECL) immunosensor was developed for ultrasensitive determination of HBV surface antigen. The primary antibody of HBs (Ab_1) was immobilized on surface of the carboxyl-modified magnetic nanoparticles (MNPs). Then, the PAMAM dendrimer with many amine functional groups was employed as carrier for immobilizing CdTe@CdS quantum dots (QDs) and the secondary antibody (Ab_2) amplified ECL signal of QDs considerably improved sensitivity. Strikingly, this ECL sensor was designed based on signal amplification with dendrimer–quantum dots structures (Babamiri et al. 2018). In another study, hemagglutinin (HA), a glycoprotein present on the surface of influenza A subtype virus H1N1, virus binds to human cells with sialic acid on membrane of upper respiratory tract. For early detection of swine flu (H1N1) infection in human, an impedimetric hemagglutinin gene-based biosensor was developed by immobilizing amino-labeled single-stranded DNA probe onto cysteine modified gold surface of the screen printed electrode for early and rapid detection of H1N1 (swine flu) in human (Mohan et al. 2019).

17.2.2.10 Microfluidic Technology-Based Immunodetection

Microfluidic technique permitted research of viral measurement of fusion kinetics, viral infectivity, and screening of viral responses to neutralizing molecules. Besides that, microfluidic platforms also signify promising and innovative clinical tools with applications in clinical diagnostics including drug screening. Microfluidics specifically emphasizes an array of techniques concerned with the precise control and manipulation of fluids, within microscopic channels (Whitesides 2006). This moderately clear-cut perception underpins a range of biological research techniques, from flow cytometric and DNA analyses to enzyme and immunoassays (Duncombe et al. 2015). A recent study investigating the use of an integrated microfluidic system for diagnostics utilized aptamers against IAV (H1N1) for viral detection (Tsang et al. 2016) was studied. Remarkably, two recent studies into the infectivity of murine norovirus (MNV) also utilized droplet-based microfluidic platforms to test the convenience of neutralizing antibodies and explored their impact on virus–cell interactions (Fischer et al. 2018). Since viral infections are complex and highly dynamic process, appreciably affected by the physical and chemical environment, studies into infection biology should preferably occur in microfluidics system

17.2.2.11 CRISPR/Cas System-Based Immunodetection

The use of CRISPR/Cas system for RNA-based gene therapy is currently raising numerous potential therapeutic applications. CRISPR/cas9 is the latest and unique RNA targeted gene therapy successfully applied in 2007 and observed in *Staphylococcus pneumonia*. The CRISPR is RNA sequence repeats which targets the foreign DNA cleavage by binding to the PAM flanking sequences which mediates the endonuclease called Cas through guide RNA (g-RNA) and responsible for the double strand breaks in the host or foreign DNA and silences the gene expression by nonhomologous end joining (NHEJ). There are different kinds which are mediated by different cas proteins. Notably like class 2, type II has cas9 where g-RNA complements cas protein and other system, eventually class II and type V complements with cas12a and the CRISPR codes crRNA acts as guide RNA by complementing the type V cas 12 protein complex. There are other 29 CRISPER/cas systems that were identified (Makarova et al. 2015). The application of CRISPR/CAS in viral diagnostics makes a breakthrough and it increased specificity. The type V CRISPR-CAS12a is used for the detection of the viral DNA. The CAS12a is designed specific to the viral DNA. The ssDNA is tagged with a Reporter which has fluorophore and quencher on both ends. The cleavage activity of viral DNA by cas12a confers the cleavage of the ssDNA. As a result flourophore reporter, the cleavage of ssDNA emits the fluorescence which is detected by quenchers and further amplified by recombinase polymerase PCR. The technique was characterized by DETECTR. There are other methods like SHERLOCK (Gootenberg et al. 2018) which works consiently based on the Cas13 system. This tool is specially applied in the ZIKA virus diagnosis (Table 17.1).

A novel arrayed CRISPR screen is aptly based on the plasmid library expressing single-guide RNA (sgRNA) and disrupted 1514 genes, encoding kinases, proteins related to endocytosis, and Golgi-localized proteins, individually using 4542 sgRNAs. This CRISPR screen uncovered host factors indispensable for infection by coxsackievirus B3 (CVB3) which comprises human pathogens causing diverse diseases. This technique is more sensitive as compared to arrayed screens based on siRNA-mediated knockdown (Kim et al. 2018).The lenti viral-based CRISPR screen-based sg-RNA plasmid pool can be employed as potential diagnostic tool for HPV and can target other viral diseases. An automated POC system for EBV detection with RNA-guided RNA endonuclease Cas13a, employing its collateral RNA degradation subsequent to its activation was recently developed by Qin and colleagues. Followed by an automated microfluidic mixing and hybridization, nonspecific cleavage products of Cas13a were allowed to quantify by a custom-integrated fluorometer. This CRISPR-Cas13a based diagnostic method is rapid, amplification-free, simple, and sensitive, thus establishing a key technology toward a useful POC diagnostic platform (Qin et al. 2019a, b).

17.2.2.12 Nanoparticles-Based Immunodetection

The prologue of a new class of nanoscale materials with manifold exceptional properties and functions has sparked series of breakthrough applications in biomedical and diagnostic applications. It is pertinent to note that some recent advances in nanoparticle-based lateral flow immunoassay as point-of-care diagnostic tool for infectious agents and diseases has been recently developed for the detection of infectious viral agents. Lateral flow immunoassay (LFIA) technology is a paper-based, point-of-care strip biosensor designed to detect a specific analyte in virus-infected samples (Banerjee and Jaiswal 2018). AuNP-based detection techniques were reported by various groups of clinically relevant viruses with a unique focus on applied types of bio-AuNP hybrid structures, virus detection targets, and assay modalities and formats were recently developed (Draz and Shafiee 2018).

17.3 Future Perspectives and Conclusions

The most recently developed diagnostic viral techniques are redesigning the field of clinical virology, which could contribute to reduction in incidence of serious infectious viral diseases. The foremost advantages of molecular techniques are its higher sensitivity and specificity matched up with other diagnostic methods, viz. serological assays and culture methods, as well as its rapidity and possibility of automation. Nevertheless, the technological capabilities alone are inadequate if not sustained by health promotion policies to boost the consciousness a propos the significance of early detection of infectious viral diseases outbreak and its spread. In conclusion, good quality diagnosis has a cost that only developed countries can afford in regular practice so far, and this is delaying the execution of new-fangled methods in developing world and in disease endemic areas. Conversely, it is anticipated that asserted efforts may persist toward developing new high-quality tests inexpensive in low-income countries, which would considerably reinforce disease control strategies.

Acknowledgments The authors are grateful to their respective academic institutions for the support extended.

Conflict of Interest The authors declare that they have no competing interests.

References

Afsahi S, Lerner MB, Goldstein JM, Lee J, Tang X, Bagarozzi DA Jr, Goldsmith BR (2018) Novel graphene-based biosensor for early detection of Zika virus infection. Biosens Bioelectron 100:85–88

Amaro F, Sánchez-Seco MP, Vázquez A, Alves MJ, Zé-Zé L, Luz MT, De Ory F (2019) The application and interpretation of IgG avidity and IgA ELISA tests to characterize Zika virus infections. Viruses 11(2):179

Azhar EI (2018) Development of fluorescent reverse transcription loop-mediated isothermal amplification (RT-LAMP) using quenching probes for the detection of the Middle East respiratory syndrome coronavirus. J Virol Methods 258:41–48

Babamiri B, Hallaj R, Salimi A (2018) Ultrasensitive electrochemiluminescence immunosensor for determination of hepatitis B virus surface antigen using CdTe@ CdS-PAMAM dendrimer as luminescent labels and Fe3O4 nanoparticles as magnetic beads. Sensors Actuators B Chem 254:551–560

Banerjee R, Jaiswal A (2018) Recent advances in nanoparticle-based lateral flow immunoassay as a point-of-care diagnostic tool for infectious agents and diseases. Analyst 143(9):1970–1996

Carlisle LA, Turk T, Kusejko K, Metzner KJ, Leemann C, Schenkel C, Yerly S (2019) Viral diversity from next-generation sequencing of HIV-1 samples provides precise estimates of infection recency and time since infection. J Infect Dis 220(2):254–265

Chang YF, Wang WH, Hong YW, Yuan RY, Chen KH, Huang YW, Wang SF (2018) Simple strategy for rapid and sensitive detection of Avian influenza A H7N9 virus based on intensity-modulated SPR biosensor and new generated antibody. Anal Chem 90(3):1861–1869

Chen KF, Rothman RE, Ramachandran P, Blyn L, Sampath R, Ecker DJ, Gaydos CA (2011) Rapid identification viruses from nasal pharyngeal aspirates in acute viral respiratory infections by RT-PCR and electrospray ionization mass spectrometry. J Virol Methods 173(1):60–66

Chiu CY, Coffey LL, Murkey J, Symmes K, Sample HA, Wilson MR, Stryke D (2017) Diagnosis of fatal human case of St. Louis encephalitis virus infection by metagenomic sequencing, California, 2016. Emerg Infect Dis 23(10):1964

Crim RL, Kumari S, Jayanti P, Audet S, Kulkarni A, Beeler J (2019) Development of luciferase immunoprecipitation systems (LIPS) assay to detect IgG antibodies against human respiratory syncytial virus G-glycoprotein. Vaccine 7(1):16

Deyde VM, Sampath R, Garten RJ, Blair PJ, Myers CA, Massire C, Klimov AI (2010) Genomic signature-based identification of influenza A viruses using RT-PCR/electro-spray ionization mass spectrometry (ESI-MS) technology. PLoS One 5(10):e13293

Draz MS, Shafiee H (2018) Applications of gold nanoparticles in virus detection. Theranostics 8(7):1985

Duchesne L, Njouom R, Lissock F, Tamko-Mella GF, Rallier S, Poiteau L, Pawlotsky JM (2017) HCV Ag quantification as a one-step procedure in diagnosing chronic hepatitis C infection in Cameroon: the ANRS 12336 study. J Int AIDS Soc 20(1):21446

Duncombe TA, Tentori AM, Herr AE (2015) Microfluidics: reframing biological enquiry. Nat Rev Mol Cell Biol 16(9):554–567

Ebner K, Suda M, Watzinger F, Lion T (2005) Molecular detection and quantitative analysis of the entire spectrum of human adenoviruses by a two-reaction real-time PCR assay. J Clin Microbiol 43:3049–3053

Ebner K, Rauch M, Preuner S, Lion T (2006) Typing of human adenoviruses in specimens from immunosupressed patients by PCR-fragment length analysis and real-time quantitative PCR. J Clin Microbiol 44:2808–2815

Emmerich P, Mika A, von Possel R, Rackow A, Liu Y, Schmitz H, Berisha L (2018) Sensitive and specific detection of Crimean-Congo Hemorrhagic Fever Virus (CCHFV)—Specific IgM and IgG antibodies in human sera using recombinant CCHFV nucleoprotein as antigen in μ-capture and IgG immune complex (IC) ELISA tests. PLoS Negl Trop Dis 12(3):e0006366

Faye M, Dacheux L, Weidmann M, Diop SA, Loucoubar C, Bourhy H, Faye O (2017) Development and validation of sensitive real-time RT-PCR assay for broad detection of rabies virus. J Virol Methods 243:120–130

Fernandes JN, Dos Santos LM, Chouin-Carneiro T, Pavan MG, Garcia GA, David MR, Sikulu-Lord MT (2018) Rapid, noninvasive detection of Zika virus in Aedes aegypti mosquitoes by near-infrared spectroscopy. Sci Adv 4(5):eaat0496

Fischer K, Diederich S, Smith G, Reiche S, dos Reis VP, Stroh E, Balkema-Buschmann A (2018) Indirect ELISA based on Hendra and Nipah virus proteins for the detection of henipavirus specific antibodies in pigs. PLoS One 13(4):e0194385

Furuya AK, Hunt D, George KS, Dupuis AP, Kramer LD, Shi PY, Wong S (2019) Use of the immunoglobulin G avidity assay to differentiate between recent Zika and past dengue virus infections. Clin Sci 133(7):859–867

Gootenberg JS, Abudayyeh OO, Kellner MJ, Joung J, Collins JJ, Zhang F (2018) Multiplexed and portable nucleic acid detection platform with Cas13, Cas12a, and Csm6. Science 360(6387):439–444

Hansen S, Hotop SK, Faye O, Ndiaye O, Böhlken-Fascher S, Pessôa R, Schmidt-Chanasit J (2019) Diagnosing Zika virus infection against a background of other flaviviruses: studies in high resolution serological analysis. Sci Rep 9(1):3648

Hayden R, Persing DH (2001) Diagnostic molecular microbiology review. Curr Clin Top Infect Dis 21:323–348

Kaushik A, Yndart A, Kumar S, Jayant RD, Vashist A, Brown AN, Nair M (2018) A sensitive electrochemical immunosensor for label-free detection of Zika-virus protein. Sci Rep 8(1):9700

Kim HS, Lee K, Kim SJ, Cho S, Shin HJ, Kim C, Kim JS (2018) Arrayed CRISPR screen with image-based assay reliably uncovers host genes required for coxsackievirus infection. Genome Res 28(6):859–868

Koishi AC, Suzukawa AA, Zanluca C, Camacho DE, Comach G, dos Santos CND (2018) Development and evaluation of a novel high-throughput image-based fluorescent neutralization test for detection of Zika virus infection. PLoS Negl Trop Dis 12(3):e0006342

Lewandowski K, Xu Y, Pullan ST, Lumley SF, Foster D, Sanderson N, Vipond R (2019) Metagenomic nanopore sequencing of influenza virus direct from clinical respiratory samples. bioRxiv, 676155

Makarova KS, Wolf YI, Alkhnbashi OS, Costa F, Shah SA, Saunders SJ, Horvath P et al (2015) An updated evolutionary classification of CRISPR–Cas systems. Nat Rev Microbiol 13(11):722

Magro L, Jacquelin B, Escadafal C, Garneret P, Kwasiborski A, Manuguerra JC, Tabeling P (2017) Paper-based RNA detection and multiplexed analysis for Ebola virus diagnostics. Sci Rep 7(1):1347

Miller S, Naccache SN, Samayoa E, Messacar K, Arevalo S, Federman S, Ingebrigtsen D (2019) Laboratory validation of a clinical metagenomic sequencing assay for pathogen detection in cerebrospinal fluid. Genome Res 29(5):831–842

Mirian M, Khanahmad H, Darzi L, Salehi M, Sadeghi-Aliabadi H (2017) Oligonucleotide aptamers: potential novel molecules against viral hepatitis. Res Pharm Sci 12(2):88

Mohan H, Gill PS, Kumar A (2019) Hemagglutinin gene based biosensor for early detection of swine flu (H1N1) infection in human. Int J Biol Macromol 130:720–726

Oh S, Kim J, Tran VT, Lee DK, Ahmed SR, Hong JC, Lee J (2018) Magnetic nanozyme-linked immunosorbent assay for ultrasensitive influenza A virus detection. ACS Appl Mater Interfaces 10(15):12534–12543

Ojha R, Nandani R, Pandey RK, Mishra A, Prajapati VK (2019) Emerging role of circulating microRNA in the diagnosis of human infectious diseases. J Cell Physiol 234(2):1030–1043

Omar NAS, Fen YW (2018) Recent development of SPR spectroscopy as potential method for diagnosis of dengue virus E-protein. Sens Rev 38(1):106–116

Pisanic N, Rahman A, Saha SK, Labrique AB, Nelson KE, Granger DA, Heaney CD (2017) Development of an oral fluid immunoassay to assess past and recent hepatitis E virus (HEV) infection. J Immunol Methods 448:1–8

Qadiri SSN, Kim SJ, Krishnan R, Kim JO, Kim WS, Oh MJ (2019) Development of an in-situ hybridization assay using riboprobes for detection of viral haemorrhagic septicemia virus (VHSV) mRNAs in a cell culture model. J Virol Methods 264:1–10

Qin P, Park M, Alfson K, Tamhankar M, Carrion R, Patterson J, Du K (2019a) Rapid and fully microfluidic Ebola virus detection with CRISPR-Cas13a. ACS Sens 4:1048–1054

Qin S, Volokhov D, Rodionova E, Wirblich C, Schnell MJ, Chizhikov V, Dabrazhynetskaya A (2019b) A new recombinant rabies virus expressing a green fluorescent protein: a novel and fast approach to quantify virus neutralizing antibodies. Biologicals 59:56–61

Roingeard P, Raynal PI, Eymieux S, Blanchard E (2019) Virus detection by transmission electron microscopy: still useful for diagnosis and a plus for biosafety. Rev Med Virol 29(1):e2019

Saito S, Takayama I, Nakauchi M, Nagata S, Oba K, Odagiri T, Kageyama T (2019) Development and evaluation of a new real-time RT-PCR assay for detecting the latest H9N2 influenza viruses capable of causing human infection. Microbiol Immunol 63(1):21–31

Santiago GA, Vázquez J, Courtney S, Matías KY, Andersen LE, Colón C, Muñoz-Jordan JL (2018) Performance of the Trioplex real-time RT-PCR assay for detection of Zika, dengue, and chikungunya viruses. Nat Commun 9(1):1391

Shirato K, Semba S, El-Kafrawy SA, Hassan AM, Tolah AM, Takayama I, Azhar EI (2018) Development of fluorescent reverse transcription loop-mediated isothermal amplification (RT-LAMP) using quenching probes for the detection of the Middle East respiratory syndrome coronavirus. J Virol Methods 258:41–48

Souf S (2016) Recent advances in diagnostic testing for viral infections. Bioscience Horizons: Int J Stud Res 9

Tiwari PM, Vanover D, Lindsay KE, Bawage SS, Kirschman JL, Bhosle S, Santangelo PJ (2018) Engineered mRNA-expressed antibodies prevent respiratory syncytial virus infection. Nat Commun 9(1):3999

Torrance L, Jones RAC (1981) Recent developments in serological methods suited for use in routine testing for plant viruses. Plant Pathol 30(1):1–24

Tsang MK, Ye W, Wang G, Li J, Yang M, Hao J (2016) Ultrasensitive detection of Ebola virus oligonucleotide based on upconversion nanoprobe/nanoporous membrane system. ACS Nano 10(1):598–605

Van Belkum A, Niesters HG (1995) Nucleic acid amplification and related techniques in microbiological diagnosis and epidemiology. Cell Mol Biol 41:615–623

Wang Y, Zhang C, Zhang P, Guo G, Jiang T, Zhao X, Tian Y (2018) Serum exosomal micro RNA s combined with alpha-fetoprotein as diagnostic markers of hepatocellular carcinoma. Cancer Med 7(5):1670–1679

Wang R, Ongagna-Yhombi SY, Lu Z, Centeno-Tablante E, Colt S, Cao XE, Erickson D (2019) A rapid diagnostic platform for colorimetric differential detection of dengue and chikungunya viral infections. Anal Chem 8:5415–5423

Wesolowska-Andersen A, Everman JL, Davidson R, Rios C, Herrin R, Eng C, Fingerlin TE (2017) Dual RNA-seq reveals viral infections in asthmatic children without respiratory illness which are associated with changes in the airway transcriptome. Genome Biol 18(1):12

Whiteside TL (2006, February) Immune suppression in cancer: effects on immune cells, mechanisms and future therapeutic intervention. In Seminars in cancer biology (Vol. 16, No. 1, pp. 3–15). Academic Press, Cambridge

Wong SJ, Furuya A, Zou J, Xie X, Dupuis AP II, Kramer LD, Shi PY (2017) A multiplex microsphere immunoassay for Zika virus diagnosis. EBioMedicine 16:136–140

Xi Y, Xu CZ, Xie ZZ, Zhu DL, Dong JM, Xiao G (2019) Development of a reverse transcription recombinase polymerase amplification assay for rapid detection of human respiratory syncytial virus. Mol Cell Probes 45:8–13

Zhang H, Rao H, Wang Y, Wang J, Kong X, Ji Y, Luo B (2019) Evaluation of an antigen assay for diagnosing acute and chronic hepatitis E genotype 4 infection. J Gastroenterol Hepatol 34(2):458–465

Current Advances in Multi-Epitope Viral Vaccines Development and Research

18

A. M. V. N. Prathyusha and Pallaval Veera Bramhachari

Abstract

Viral diseases are major public health concern and cause significant morbidity and mortality globally. The following dreadful viruses viz. Influenza virus, Ebola virus (EBOV), and Sudan virus are the most recent viruses to cause a global health concern. Despite recent progress in reduced deaths by viral infection, need for new molecular, immunoinformatic tools, together with safe plus effective vaccines is prerequisite due to the pros and cons of prophylactic and therapeutic vaccines in failure to provide optimal protection. The challenges for viral vaccine advancements are not restricted for recognition of suitable antigens or adjuvants and delivery methods, nonetheless cover technical and manufacturing hurdles in transforming a vaccine to clinic. Research and process improvement is technological basis which prompts production of new vaccines, essential for commercialization. In this review, we emphasize the present status and recent advances in designing and developing viral vaccines.

Keywords

Viral vaccines · Multi-epitopes · Inluenza virus · Ebola virus (EBOV) · Hepatitis E virus

18.1 Introduction

During the past three decades several notable zoonotic virus infections, viz. HIV, Nipah (NiV) viruses, Hendra (HeV), Ebola, avian influenza, Marburg filoviruses, Lassa virus (LASV), Crimean–Congo hemorrhagic fever (CCHF) viruses and many more have emerged suddenly from anonymity and became serious health threats

A. M. V. N. Prathyusha · P. V. Bramhachari (✉)
Department of Biotechnology, Krishna University, Machilipatnam, India

P. V. Bramhachari (ed.), *Dynamics of immune activation in viral diseases*,
https://doi.org/10.1007/978-981-15-1045-8_18

globally, provoking concern regarding their sustained epidemic transmission in immunologically naive human population. With each new threat rapid need for development of effective and efficient vaccines have been significant part for public health concern. Indeed, vaccines are considered powerful strategy for prevention of emerging viral infections, since, in most of the cases, further options for treatment or effectiveness of therapeutics are limited or nonexistent. Vaccines for number of viral diseases with major health concern are not yet available. However, struggles are still enduring to develop fully effective vaccine strategies for recurrent and emerging viruses and other current threats need to be addressed. In addition, resurgent interest remains, in development of new vaccines strategies against pandemic viruses (Kanekiyo et al. 2019).

Traditional vaccine development approaches are amenable yet for emerging viruses, while application of molecular techniques in virology has profoundly influenced our understanding of virus biology. New approaches and technologies hold an essential role in modern vaccine development and in silico approaches, multi-epitope (B and T cell epitopes, peptide, subunit) vaccines using immunoinformatics depicts a major role to address challenges in vaccine development. In silico approaches have attained great acceptance with current advancements in genome and protein sequence databases. With the advancements in immunoinformatics, limitations related to prediction of accurate B cell epitopes and human class I and II restricted T cell epitopes are surmounted. This strategy can assemble vaccines with epitopes for optimal presentation by phagocytic processing machinery. Furthermore, several promising approaches are heading from lab to clinic to address unresolved challenges in viral vaccine development. It is notable that multi-epitope (ME) vaccines possesses the ability to stimulate ample repertoires of immune responses, along with dealing to pathogens effectively with genetic variations. Several approaches of vaccination relying to replicating, attenuated, and nonreplicating virus vectors have become useful vaccine delivery platforms. This chapter emphasizes a few recent multi-epitope vaccine approaches for effective vaccines encompassing some current vaccine platform strategies.

18.2 Multi-Epitope Vaccines

Epitope-based vaccines (epitope vaccines) signify novel approach for generation of specific immune response and prevention of unfavorable responses against other epitopes in complete antigen. EVs have several advantages over other forms of vaccines, particularly with regard to safety, ease of production, storage and distribution, without cold chain issues. They also offer opportunity to vaccinate against several pathogens or multiple epitopes from same pathogen. Potential advantages of epitope-based vaccines also include increased safety, opportunity to rationally engineer epitopes for increased potency and breadth, and ability to focus immune responses on conserved epitopes.

18.3 Multi-Epitope Subunit Vaccines

Subunit vaccines comprise merely antigenic part of virus with potential to promote immune response while overcoming problems associated with conventional vaccines. Exploitation of vaccination strategy in combating virus is a major concern since synthesis of conventional vaccines often has several obstacles. Hasan et al. developed novel reverse vaccinology approach that aims to link immunogenetics, immunogenomics using bioinformatics to identify new vaccine targets. Detection of T cell and B cell epitopes and HLA (human leukocyte antigen) using in silico approaches reinforced a possibility in potent vaccine candidate discovery. Vaccine candidates may be identified as foreign molecules after injection to body. Therefore, prediction of antigenicity is a crucial part in synthesis of subunit vaccines. Furthermore, assessment of hydrophilicity is a significant criterion for B cell epitope prediction. Hasan and group reported a design to synthesize nonallergic, immunogenic, and thermostable chimeric ME monovalent subunit vaccine against avian influenza A (H7N9) virus, and Marburg virus using vaccinomics approach. Viral proteome was assessed for epitopes with high antigenicity using TMHMM, VaxiJen v2.0 server (transmembrane topology screening), AllerTOP, AllergenFP, PA^3P, Allermatch servers (allergenicity and toxicity analysis), IEDB's epitope conservancy analysis tool (population coverage assessment), and MGL Tools (molecular docking). Two proteins, envelope glycoprotein (GP) and matrix protein (VP40), generated potent T cell epitopes and are recognized to be most antigenic proteins in Marburg virus. PEP-FOLD de novo approach was used in prediction of 3D-peptide structures from amino acid sequences of top ranked epitopes. Finally, three vaccines were designed most effectively using suitable adjuvants with PADRE sequence, combined in sequential manner with highly immunogenic (CTL, HTL, and BCL) respectively. Such constructed vaccines with predicted epitopes were validated experimentally using animal models for their nonallergic reactions and immunogenic potential (Hasan et al. 2019a, b). Using similar approaches subunit vaccines may be produced against Chikungunya virus (Narula et al. 2018), Kaposi's sarcoma-associated herpesvirus (KSHV) (Chauhan et al. 2019) respectively.

18.4 Multi-Epitope DNA Vaccines

DNA vaccines imply a convenient method to design vaccines with tailored epitopes being administered into plasmid vectors with desirable memory response. Furthermore, to evade unfavorable immunodominant epitopes from pathogens, chimeric multi-epitope-based DNA vaccine was designed against subgroup J avian leukosis virus in chickens (Xu et al. 2016). Bounds and his group using in silico algorithms designed DNA construct entailing HLA class II-restricted T cell epitopes obtained from GP and NP (nucleocapsid protein) of EBOV (Ebola virus), SUDV (Sudan virus), and structural proteins of VEEV (Venezuelan equine encephalitis virus). Identified epitopes with high specificity were examined for binding ability to soluble HLA molecules. Then vaccinated BALB/c mice with ME-DNA vaccine developed and evaluated for immune response using interferon (IFN)-g ELISpot analyses. These studies provide a concept for designing an ME immunogen along with evaluation focusing on immune response concerning preferred T cell epitopes (Bounds et al. 2017).

18.5 Multi-Epitope Peptide Vaccines

Peptide vaccination can induce both humoral- and cell-mediated immune response by stimulating T cell immunity in both humans and animals with minimal side effects. Peptide vaccines are short immunogenic peptide fragments which can elicit targeted immune response avoiding the chance of allergenic responses. Several peptide vaccines were under progress, viz. vaccine for hepatitis C virus (HCV), human immunodeficiency virus (HIV), foot and mouth disease, malaria, influenza, swine fever, human papilloma virus (HPV), and anthrax (Verma et al. 2018). With advancements in computational biology and immunoinformatics approaches, designing effective strategies for prediction of antigenic epitopes became easier. A peptide-based vaccine is majorly presented via class II MHC molecules and gets processed by endocytic pathway. Immune response can also be induced by cytotoxic T cells (CTL) by cross-presentation where exogenous antigens are processed and presented onto class I MHC. To improve antigenic presentation, cell penetrating peptides (CPPs) was one of the promising approach to penetrate peptides efficiently into cells using cationic peptides (TAT). Commonly used methods for CPP designing and tagging to antigen were: (1) chemical linkage through covalent bonds, and (2) recombinant fusion constructs by bacterial expression vectors. Gross et al. tagged a C-terminal viral protein R (Vpr55-91 and Vpr55-82) fragment of human papillomavirus to CPP which paved way for practical ME immunization for neoantigen vaccination in cancer patients and is effective now clinically (Gross et al. 2019).

18.6 Vector-Based ME Vaccines

Site-specific recombination with Cre-recombinase based multi-epitope (ME) vaccine was designed using relatively conserved immunogenic domains of antigenically distinct strains of the H5, H7, and H9 avian influenza viruses. Three domains M2 ectodomain (M2e), hemagglutinin (HA) fusion domain (HFD), T cell epitope of nucleoprotein (TNP) and HA α-helix domain (HαD) were separated by linkers and inserted into Human Adenovirus (Ad) vector. BALB/c mice were vaccinated and evaluated for immune responses and protection efficacy using hemagglutination inhibition, viral neutralization, ELISA, and ELISpot assays. Such ME vaccine approach provided broad protection against avian influenza virus (Hassan et al. 2017). Yasmin and group have designed an RNA-dependent RNA polymerase (L) epitope-based ME vaccine using immunoinformatics. Two conserved envelope glycoproteins GP1 and GP2 of EBOV were recognized as targets for synthesis of epitope vaccine using various softwares. Collected EBOV glycoproteins were sequenced using further examination for proteins with good immunogenicity by in silico approaches. Such predicted B and T cell epitopes can confer long-lasting immunity against EBOV with better ability of protection (Yasmin and Nabi 2016).

18.7 Current Vaccine Platform Strategies

Strategies for growth of triumphant vaccines are due to design of an antigen delivery system that optimizes antigen presentation and stimulates broad protective immune responses. Recent advances in vector delivery technologies, immunology, basic virology, genetics, and molecular cell biology have led to in-depth understanding of cellular mechanisms by which vaccines should stimulate the adaptive immune response, thus presenting novel strategies of vaccination. Classic approach to vaccine development is still acquiescent to emerging viruses, the application of molecular techniques in virology has overwhelmingly predisposed our perception of virus biology, and vaccination schemes based on replication strategies, attenuated and nonreplicating virus vector approaches that have turned out to be valuable vaccine platforms. Several virus-like particle (VLP)-based vaccines are currently commercializes in global market, including vaccines against hepatitis B virus and human papillomavirus (Roldao et al. 2010). Furthermore, through genetic fusion or chemical conjugation, VLPs are attractive carrier proteins of foreign antigens, since they can efficiently display them within a host immune system (Tissot et al. 2010; Plummer and Manchester 2011, Brune et al. 2016). An important advantage of VLP-based vaccine platforms is that VLPs can present antigens in a dense, repetitive manner, thus effectively enabling the cross-linking of B cell receptors (BCRs) (Zabel et al. 2014).

18.8 Innovation Challenges and Opportunities

Uncertainty of the public health priority and demand for some targets may be unclear, which increases uncertainty of potential return on investment (ROI). Furthermore, new vaccine targets are scientifically more complex and challenging. The challenges presented may require substantial investment in new tools, standards, analysis methods, other novel approaches to demonstrate safety and effectiveness of vaccine and also limited knowledge in science to progress optimal vaccines. However, there is a limited understanding of viral pathogenesis and immune responses against some targeted infectious viral agents, lack of optimal immune response to potential vaccine candidates, and a limited understanding for protection mechanism against some vaccine candidates. Conducting clinical trials to evaluate safety and efficacy of certain preventive vaccines may be particularly challenging for several reasons, including: relatively low disease incidence (congenital cytomegalovirus disease; neonatal group B streptococcal disease); limited infrastructure in affected geographic areas (EBV vaccine); ethical considerations (assessing novel approaches to pertussis vaccination of pregnant women to prevent neonatal pertussis which is recommended); or new settings (hospital acquired infections).

18.9 Conclusion and Future Perspectives

Increasing incidence of viral infections development of effective and efficient vaccines has mandated a significant part of public health concern. Strikingly, bioinformatics, immunogenetics, immunoinformatics, reduces time and covers the need, efforts should be expended in designing of new viral vaccines. However, epitope vaccines made progress in development of more effective vaccines. Interpreting vaccine design, despite of using immunoinformatics and in silico approaches remain at development stage yet. Furthermore, moderate successes might lead to significant progress on ME vaccine efficiency against human and animal pathogens, particularly in terms of optimizing T cell epitope polypeptide construct in vaccine assembly. This further improves cellular processing in epitope presentation and protection against highly variable viral pathogens.

Acknowledgment The authors gratefully acknowledge Krishna University, Machilipatnam for the support extended.

Conflict of Interest The authors declare that they have no competing interests.

References

Bounds CE, Terry FE, Moise L, Hannaman D, Martin WD, De Groot AS, Schmaljohn CS (2017) An immunoinformatics-derived DNA vaccine encoding human class II T cell epitopes of Ebola virus, Sudan virus, and Venezuelan equine encephalitis virus is immunogenic in HLA transgenic mice. Hum Vaccin Immunother 13(12):2824–2836

Brune KD, Leneghan DB, Brian IJ, Ishizuka AS, Bachmann MF, Draper SJ, Biswas S, Howarth M (2016) Plug-and-display: decoration of virus-like particles via isopeptide bonds for modular immunization. Sci Rep 6:19234

Chauhan V, Rungta T, Goyal K, Singh MP (2019) Designing a multi-epitope based vaccine to combat Kaposi Sarcoma utilizing immunoinformatics approach. Sci Rep 9(1):2517

Gross DA, Leborgne C, Chappert P, Masurier C, Leboeuf M, Monteilhet V, Kichler A (2019) Induction of tumor-specific CTL responses using the C-terminal fragment of Viral protein R as cell penetrating peptide. Sci Rep 9(1):3937

Hasan M, Azim KF, Begum A, Khan NA, Shammi TS, Imran AS, Urme SRA (2019a) Vaccinomics strategy for developing a unique multi-epitope monovalent vaccine against Marburg marburgvirus. Infect Genet Evol 70:140–157

Hasan M, Ghosh PP, Azim KF, Mukta S, Abir RA, Nahar J, Khan MMH (2019b) Reverse vaccinology approach to design a novel multi-epitope subunit vaccine against avian influenza A (H7N9) virus. Microb Pathog 130:19–37

Hassan AO, Amen O, Sayedahmed EE, Vemula SV, Amoah S, York I, Mittal SK (2017) Adenovirus vector-based multi-epitope vaccine provides partial protection against H5, H7, and H9 avian influenza viruses. PLoS One 12(10):e0186244

Kanekiyo M, Ellis D, King NP (2019) New vaccine design and delivery technologies. J Infect Dis 219(Supplement_1):S88–S96

Narula A, Pandey RK, Khatoon N, Mishra A, Prajapati VK (2018) Excavating chikungunya genome to design B and T cell multi-epitope subunit vaccine using comprehensive immunoinformatics approach to control chikungunya infection. Infect Genet Evol 61:4–15

Plummer EM, Manchester M (2011) Viral nanoparticles and virus-like particles: platforms for contemporary vaccine design. Wiley Interdiscip Rev Nanomed Nanobiotechnol 3:174–196

Roldao A, Mellado MCM, Castilho LR, Carrondo MJ, Alves PM (2010) Virus-like particles in vaccine development. Expert Rev Vaccines 9(10):1149–1176

Tissot AC, Renhofa R, Schmitz N, Cielens I, Meijerink E, Ose V, Jennings GT, Saudan P, Pumpens P, Bachmann MF (2010) Versatile virus-like particle carrier for epitope based vaccines. PLoS One 5:e9809

Verma S, Sugadev R, Kumar A, Chandna S, Ganju L, Bansal A (2018) Multi-epitope DnaK peptide vaccine against S. Typhi: an in silico approach. Vaccine 36(28):4014–4022

Xu Q, Cui N, Ma X, Wang F, Li H, Shen Z, Zhao X (2016) Evaluation of a chimeric multi-epitope-based DNA vaccine against subgroup J avian leukosis virus in chickens. Vaccine 34(33):3751–3756

Yasmin T, Nabi AN (2016) B and T cell epitope-based peptides predicted from evolutionarily conserved and whole protein sequences of Ebola virus as vaccine targets. Scand J Immunol 83(5):321–337

Zabel F, Mohanan D, Bessa J, Link A, Fettelschoss A, Saudan P, Kundig TM, Bachmann MF (2014) Viral particles drive rapid differentiation of memory B cells into secondary plasma cells producing increased levels of antibodies. J Immunol 192:5499–5508

The manufacturer's authorised representative in the EU is Springer Nature Customer Service Centre GmbH, Europaplatz 3, 69115 Heidelberg, Germany. If you have any concerns regarding our products, please contact ProductSafety@springernature.com

Printed and bound by CPI Group (UK) Ltd, Croydon, CR0 4YY
30/09/2026
02225232-0002